I0037302

Thermoset Composites: Preparation, Properties and Applications

Edited by

Anish Khan[1,2], Showkat Ahmad Bhawani[3], Abdullah M. Asiri[1,2], Imran Khan[4]

[1] Department of Chemistry, King Abdulaziz University, Jeddah-21589, Saudi Arabia

[2] Center of Excellence for Advanced Materials Research, King Abdulaziz University, Jeddah 21589, P.O. Box 80203, Saudi Arabia

[3] Department of Chemistry, Faculty of Resource Science and Technology, Universiti Malaysia Sarawak (UNIMAS), 94300 Sarawak, Malaysia

[4] Applied Science and Humanities Section, University Polytechnic, Faculty of Engineering and Technology, Aligarh Muslim University, Aligarh, 202002- India

Copyright © 2018 by the authors

Published by **Materials Research Forum LLC**
Millersville, PA 17551, USA

All rights reserved. No part of the contents of this book may be reproduced or transmitted in any form or by any means without the written permission of the publisher.

Published as part of the book series
Materials Research Foundations
Volume 38 (2018)
ISSN 2471-8890 (Print)
ISSN 2471-8904 (Online)

Print ISBN 978-1-945291-86-9
ePDF ISBN 978-1-945291-87-6

This book contains information obtained from authentic and highly regarded sources. Reasonable efforts have been made to publish reliable data and information, but the author and publisher cannot assume responsibility for the validity of all materials or the consequences of their use. The authors and publishers have attempted to trace the copyright holders of all material reproduced in this publication and apologize to copyright holders if permission to publish in this form has not been obtained. If any copyright material has not been acknowledged please write and let us know so we may rectify this in any future reprints.

Distributed worldwide by

Materials Research Forum LLC
105 Springdale Lane
Millersville, PA 17551
USA
http://www.mrforum.com

Manufactured in the United States of America
10 9 8 7 6 5 4 3 2 1

Table of Contents

Preface

Composites are of high demand and, therefore, a field of high current research interest. The most valuable and applicable composites are thermoset composites because they have boundless adaptability in regard to design, shape and structure. They can be shaped into many-sided segments and can have extensive variety of densities and different compound details in order to meet desired properties.

The current book introduces thermoset composites and its properties and applications. The focus of the book is the preparation and applications of such thermoset composites for industrial applications. Thermoset composites are the material of the current era and are in high demand. Thermoset composites have multidisciplinary use in science and technology and that is why these composites are different from the rest of the materials currently on the market. The special characteristic of the book is that it presents unified knowledge of such thermoset composites on the basis of characterization, design, manufacture, and applications. This book will benefit lecturers, students, researchers and industrialist who are working in the field of thermoset polymers and composites for application in particular and material science in general.

We are highly thankful to contributors of different book chapters who provided us their valuable innovative ideas and knowledge in this edited book. We attempt to gather information related to thermoset composites for different application from diverse fields around the world (Malaysia, India, Korea, USA, Saudi Arabia, and South Africa) and finally complete this venture in a fruitful way. We greatly appreciate the contributor's commitment to their support to compile their ideas.

Anish Khan, Dhowkat Ahmad Bhawani,

Abdullah M. Asiri, Imran Khan

Thermoset Composite
Materials Research Foundations 38 (2018)

Materials Research Forum LLC
doi: http://dx.doi.org/10.21741/9781945291876

Chapter 1

Energy Absorption of Natural Fibre Reinforced Thermoset Polymer Composites Materials for Automotive Crashworthiness: A Review

Mohamed Alkateb[1], S.M. Sapuan[1,2,3,*], Z. Leman[1], M.R. Ishak[3,4] and Mohammad Jawaid[2]

[1]Department of Mechanical and Manufacturing Engineering, Universiti Putra Malaysia 43400 UPM, Serdang Selangor, Malaysia

[2]Laboratory of Bio Composite Technology, Institute of Tropical Forestry and Research Products (INTROP), Universiti Putra Malaysia 43400 UPM, Serdang Selangor, Malaysia

[3]Aerospace Manufacturing Research Centre (AMRC), Universiti Putra Malaysia 43400 UPM, Serdang Selangor, Malaysia

[4]Departments of Aerospace Engineering, Universiti Putra Malaysia, 43400 UPM, Serdang, Selangor, Malaysia

*sapuan@upm.edu.my

Abstract

Energy absorption capacity of composite materials is important in order to develop safety measurements for human beings in a car accident. Energy absorption, fiber type, matrix type, fiber structure, the shape of the pieces, processing conditions, are all important parameters. Changes in these parameters can cause particular subsequent changes in the energy absorption of the composite material of up to two times. Previous studies focused on how to introduce natural fibers into industrial applications and the replacement of synthetic fibres with natural fiber materials. In this paper, a detailed review of the energy absorption properties of the polymer composite material discussed. In order to understand the effects of certain parameters for the energy absorption capacity of a good composite material an attempt is made to classify the work in the field of energy absorption for composite materials that is published in the literature.

Keywords

Energy Absorption, Thermoset Polymer, Natural Fiber, Axial Crush, Composite

Contents

1.1 Introduction

Due to the advantages of polymers over conventional materials in various applications, it has substituted many of traditional materials and especially metal ones. Because they are easy to process, high productivity, low cost and versatile, they are used in many applications. However, for some specific uses, some mechanical properties, such as strength and toughness of polymer materials are inadequate. Various approaches have been developed to improve such properties. In most of these applications (crashworthiness), the properties of the polymer are modified with fillers and fibers to suit the requirements of high strength / high modulus. Fiber-reinforced polymers have better specific properties compared to conventional materials and find applications in various fields [1].

There is an important difference between penetration resistance and impact (crashworthiness) resistance. Either in passenger cars, the ability to survive by absorbing the impact energy of an occupant is called the "crashworthiness" of the structure. Impact resistance is related to the absorption of energy by a controlled fracture mechanism and modes that can maintain gradual decay of the load profile during absorption. However, penetration resistance is associated with total absorption without permeation of the projectile or fragment.

Crashworthiness is today one of the important factors in designing transportation means such as automobiles, rail cars, and aeroplanes. This is because it concerns vehicle structural integrity and its ability to absorb crash as well as providing a protective shell around the occupants. Crashworthy constructions must be designed to absorb the impact energy in a controlled manner and the passenger compartment must allow the passengers to rest without high deceleration which can cause serious injuries, especially brain injury.

In the last century, extensive studies have proven the high ability of composite materials in the field of collapsible energy absorber devices. It is also evident that composite materials meet design requirements by the vehicles manufacturers as well as customers demand for a safe vehicle with low fuel consumption and high pay load. As a consequence, more, metals parts will be replaced by composite ones for weight saving and increased reliability. However, the challenge is to find a suitable polymeric composite material with specific features for a suitable structural application.

According to the current legislation of automobiles requires, design of vehicles must have the ability to absorb an impact in the event of an impact at speeds of up to 15.5 m/sec (35 mph) with a solid, immovable object. The high efficiency of an impact energy absorber device may be defined as its ability to decelerate smoothly the occupant compartment within the allowable limit of 20 G [2]. However, optimum energy absorbed management from practical collapsible energy absorber device is characterized by having a very small elastic energy and the area under its load-displacement curve is represented in a rectangular form with long sides (i.e. a constant force). It is evident for all practical collapsible energy absorber devices that initially their resistance response records very high load till reaching its full capacity after which definitely different degrees of unstable response takes place [3]. Due to the black and white design of energy absorber device, one can define the desirable energy absorber device as the one with suppressed energy absorption during the elastic or pre-initial crush failure stage not to exceed the safe allowable limits. Moreover, its post crush stage should have a very stable response during the post-crush stage. In such design and for gross deformation, the overall stability of the energy absorber device is important as well as its energy absorbing capability and load carrying capacity.

Attempts have been made to use composite materials in the development of energy dissipation devices driven by the need to overcome the negative impact of both size and mass to improve fuel economy. The size and mass of the vehicle provide certain protection but may have a negative inertial effect. In addition to the properties of stiffness-to-weight ratio and strength-to-weight ratio, fatigue resistance and corrosion resistance, the ability to condition the composite makes them very attractive in crashworthiness. The challenge is the use of specific features of geometry and material to achieve greater safety while at the same time realizing greater safety without adversely affecting the overall economics of production. Increasingly metal parts are being replaced by polymer composites to reduce the overall weight of the car and improve the fuel economy of the vehicle.

Unlike metals, especially compression, most composite materials are generally not ductile to loads but are characterized by ductile responses. Metallic structures collapse under buckling collapsing crushing or impacting in an accordion type method with a wide range of plastic deformation, but it is believed that the breakage of fibers, the cracking of the matrix, the decoupling of the fiber matrix cause separation between layers. The actual mechanism and the order of damage are highly dependent on the lamina orientation, type of trigger, geometry of the structure and crush speed, all of them can be designed appropriately to develop high energy absorption mechanisms.

1.2 Materials

Natural fibers are raw materials directly obtained and can be spun into filaments, yarns or ropes. Natural fibers are classified either on the basis of their origin, from animals, plants, or minerals. It's increasingly popular in the automotive industry and several other industries in recent years, due to the high performance of mechanical properties, low cost, low density, and significant processing benefits compared to most synthetic fibers and availability. Table 1 illustrates natural fibers in the world and their world production. By using natural materials and modern construction technology, construction waste is reduced, energy efficiency is improved, and the concept of sustainability is promoted.

Natural fiber polymer composite (NFPC) is a composite of high-strength natural fibers embedded in a polymer matrix like oil palm, jute, kenaf, sisal and flax [5]. Usually, polymers can be divided into two categories, thermoplastic and thermoset since the structure of the thermoplastic matrix material is composed of one or two-dimensional molecules, these polymers tend to be softer in the range of elevated heat and tend to roll back its properties through cooling. On the other hand, thermosetting polymers can be defined as highly crosslinked polymers cured using heat alone or with heat and pressure and/or light irradiation. This structure gives the thermosetting polymer good properties, such as the high flexibility to adjust desired final properties, high strength, and modulus of elasticity [6,7].

The thermoplastic resins widely used for biofibers are polyethylene [8], polypropylene (PP) [9], and polyvinyl chloride (PVC); here's polyester, epoxy resins and phenolic are mostly utilized thermosetting matrices, different factors can affect the characteristics and performance of NFPCs. Hydrophilicity nature and loading of natural fiber [10] also effect on the composite properties [11]. Usually, high fiber loading is required to achieve good properties of NFPC [12]. Generally, increasing the fiber content improves the tensile properties of the composite [13]. Another important factor that considerably affects the properties and surface properties of the composite is the process parameter utilized.

Table 1. Natural fibers in the world and their world production [4].

Fiber source	World production (ton)
Bamboo	30.000
Sugar cane bagasse	75.000
Jute	2300
Kenaf	970
Flax	830
Grass	700
Sisal	375
Hemp	214
Coir	100
Ramie	100
Abaca	70

1.3 Thermoset and thermoplastic composites

As composite materials continue to be adopted in more industries, fiber reinforced plastics can found in products that people interact with every day such as cars and sports equipment, etc. Fiber reinforced plastics consist of reinforcing fibers surrounded by a plastic matrix. Several types of fibers can be used including industrial and natural that give the material its high tensile strength. The matrix provides compressive strength to the composite and, in the case of fiber reinforced plastics, it can be produced using thermosetting or thermoplastic.

Thermoset polymers are polymers that are cured into a solid form and cannot be returned to their original uncured form. Composites made with thermoset matrices are strong and have very good fatigue strength. They are extremely brittle and have low impact-toughness making. They are commonly used for high-heat applications because the thermoset matrix doesn't melt like thermoplastics. Thermoset composites are generally cheaper and easier to produce because the liquid resin is very easy to work with. Thermoset composites are very difficult to recycle because the thermoset cannot be remolded or reshaped; only the reinforcing fiber used can be reclaimed.

Thermoplastic polymers are a polymer that can be molded, melted, and remolded without changing its physical properties thermoplastic polymers are tougher and less brittle than thermosets, with good impact resistance and damage tolerance.

1.4 Matrix

Matrix is essential ingredients to embed fibres and provide a supporting medium for them. It is the ability of the matrix to transfer stresses which determines the degree of realization of mechanical properties of fibres and final performance of the resultant composites. Stress-strain behavior and adhesion properties are important properties, which control the ability of the matrix to transfer stresses. A lot of research is being carried out on the basic understanding of the relationship between properties and production of tough, strong, stiff, and environment resistant composite structures. This has helped in the development of composites having acceptable properties. The epoxy resins are still the work- horse of advanced polymer composites today [14].

Since the matrix can be melted the composite materials are easier to repair and can be remolded and recycled easily. Thermoplastic composites are less dense than thermosets making them a viable alternative for weight critical applications. The thermoplastic composites manufacturing process is more energy intensive due to the high temperatures and pressures needed to melt the plastic and impregnate fibers with the matrix making thermoplastic composites costlier than thermosets. These two similar materials have different properties that both will continue to be used for different applications for very different reasons, future products are likely to be a combination of both.

Previous studies have shown that thermoplastic matrix composites (Carbon/PEEK or Glass/PEEK) exhibited higher compression and compression of impact (CAI) properties with a higher strain to failure compared to thermoset composites [15,16]. One of the recent studies show that carbon fabric laminates with different thermoplastic resins (PEEK and PPS) provided smaller delaminated areas than the laminate with epoxy resin after low-speed impact testing; this result is due to the tougher matrix system of the thermoplastic composite material [17]. However, the high resin viscosity of the thermoplastic resin is a problem for impregnating the reinforcement fiber into a tightly woven or unidirectional composite [18–21]. On the other hand, thermoset composite materials are easy to process because of low viscosity of the resin, and that leads to a lower void content. Material costs and tool costs for processing thermoplastic composites are higher than in thermoset composites [22]. Diverse types of epoxy resins are generally utilized as a result of their adaptability, worthy mechanical properties, high consumption resistance, and their straightforward cure procedure and less influenced by water and warmth. Engineering fields of epoxy resins applications are the airplanes, car commercial ventures, airframe rocket applications and fiber wound structures.

1.5 Test methodologies

The crash test can be performed under two conditions: quasi-static and dynamic conditions.

1.5.1 Quasi-static test

In a quasi-static test, the specimen is crushed at a constant speed. The quasi-static test may not be a true simulation of the actual collision situation because in the actual crash state, the crash speed decreases from the initial crash speed and finally to rest. Many materials used to design crashworthy structures are sensitive to speed. In other words, the energy absorption capacity depends on the speed of the crash. Therefore, determination of materials as good after quasi-static testing them does not ensure satisfactory performance as collision able structure during an actual crash.

1.5.2 Dynamic test

Dynamic test conditions can be simulated using conventional drop tower test rigs. The test specimen is mounted on the impact plate such that the axis of the tube is parallel to the direction of the travel of the dead weight. The drop weight platform is raised to a predetermined height depending on the impact energy and speed required. The dropped platform is released with a grip latch pin mechanism that is extracted with two manually activated electronic solenoids. The signals from the load cells located beneath the impact plate are fed to the analog-digital converter. From the digitized data the required load-displacement response is recorded [23].

The disadvantage of the impact test is that the crushing process takes place in a fraction of a second. Therefore, it is difficult to study crushing unless you have expensive equipment like a high-speed camera.

1.6   Crashworthiness design

Crashworthiness is the capacity of a structure and any of its segments to protect the occupants in survivable accidents. So also, in the automotive industry, crash value hints a measure of the vehicle's basic capacity to plastic-partner misshape but then keep up an adequate survival space for its occupants in accidents including sensible deceleration loads. Limitation frameworks and inhabitant bundling crash value assessment is determined by a blend of tests and explanatory strategies. Crashworthiness parameters for each specimen can be determined from the load-displacement as displayed in Figure 1.

In the past 20 years, research defines the impact resistance as the ability of the vehicle to save the crew from damage in a sudden accident. Therefore, impact resistance will be an important factor in the automatic movement and the vehicle design ethos.

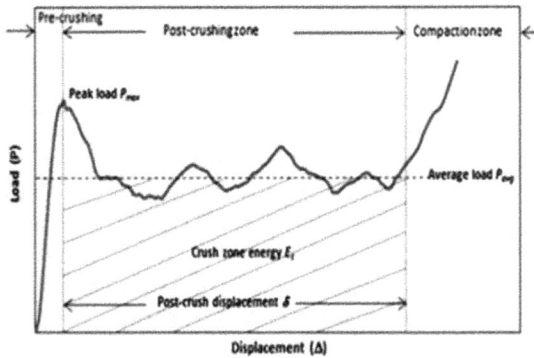

Figure 1 Typical load-displacement response [23].

To understand the absorption of energy, a few parameters like fiber type, matrix type, fiber architecture, fiber content, shape, specimen geometry, processing conditions, fiber volume fraction, and testing speed that affect the crashworthiness of a composite tube (e.g., the peak load, crash force efficiency, average crash load, and specific energy absorption) are selected. Values for some of these parameters that help the tested tubes avoid fast buckling have been determined in previous studies. Figure 2. shows the effect of the fiber content on the tensile strength and modulus. Furthermore, tubes with numerous shapes, such as circular tubes [24-26], square tubes [27], elliptical cones composite tubes [28], radial corrugated composite and composite corrugated tubes [29], hexagonal and hexagonal ring systems [30], and cone–tube–cone composite systems [31] have been designed and used in the experiments. The fiber content is a very important factor that affects a composite structure and its mechanical properties. Jawaid et al. [32] concluded that the tensile properties increased with the amount of jute fiber in hybrid composites. Davoodi et al. investigated mechanical properties of hybrid kenaf/glass reinforced epoxy composites for passenger car bumper beam. They found advantages in terms of mechanical properties compared to other common bumper beam materials [33]. El Shekeil et al. [34] investigated the effect of the fiber content (at 20, 30, 40, and 50% by mass) on the tensile, flexural, and impact properties of kenaf bast fiber-reinforced epoxy composites. The composite specimens were prepared using the melt-mixing and compression molding methods. The sample with a fiber content of 30% had the highest tensile strength, and the tensile modulus increased with the fiber content. There was an improvement in the flexural strength and modulus as the fiber content increased; increasing the fiber content caused a decrease in the impact strength. Ismail & Sahrom [35] examined the lateral crushing energy absorption of a cylindrical kenaf fiber-epoxy composite material. Two important parameters are taken into consideration when preparing the composite as a number of layers and fiber orientation. It has been found that fiber orientation is not an important factor to increase specific energy absorption rate and power factor.

A crash deceleration beat with an early peak in time and a steady rot is more advantageous for assurance of a controlled inhabitant. In this manner, the goal of crashworthiness is an advanced vehicle structure that can retain the accident vitality by controlled vehicle distortions while keeping up satisfactory space so that the leftover accident vitality can be overseen by the limitation systems to minimize crash loads exchange to the vehicle occupants, hybridization has an effect on crashworthiness goal on crushing behaviour kenaf bast fiber is suitable for hybrid natural fiber/kevlar reinforced epoxy composites with using the analytical hierarchy process(AHP) [36]. The vehicle should also be designed to possess high over strength and high retention strength for large mass components. To minimize the effects of post-crash hazards, Ataollalhi et al. [37] studied the influence of the wall lengths on the compressive response and failure mode of natural silk/epoxy composite square tubes. They found that decrease in the length of tubes leads to increase the specific energy.

Materials Research Forum LLC

doi: http://dx.doi.org/10.21741/9781945291876

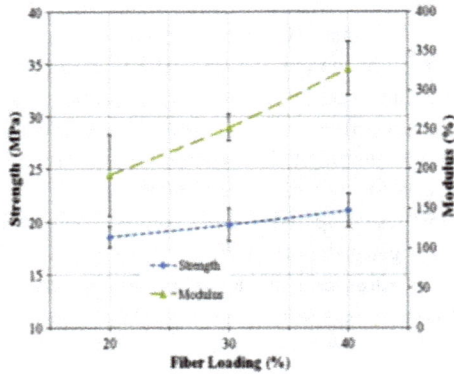

Figure 2 The effect of the fiber content on the modulus and tensile strength of epoxy composites [34].

1.7 Crashworthiness prerequisites

The structure should yield a deceleration pulse that satisfies the following requirements for a range of occupant sizes, ages, and crash speeds for both. It should minimize high-frequency fore-aft vibrations that give rise to harshness. In addition, the vehicle structure should be sufficiently stiff in bending and torsion for proper ride and handling [38].

Crashworthiness features are required to prevent excessive and injurious acceleration forces from being transmitted to the occupants under crash impact conditions. The forward vehicle structure should be designed to absorb energy during the incident. Anyway, energy-absorbing equipment should be provided for survivability and minimize the severity of injuries. In order, to satisfy the crashworthiness design criteria discussed a total system approach is needed which includes a strong protective shell to protect the occupants from crushing as well as energy-absorbing components to minimize the severity of injuries. It is important to demonstrate that in replacing metals with composite materials in crashworthiness structures, the capability to absorb energy and to maintain post-crash integrity is not compromised. The load-displacement relationship for epoxy composite tubes, which includes the essential factors and the primary regions, is shown in Figure 3.

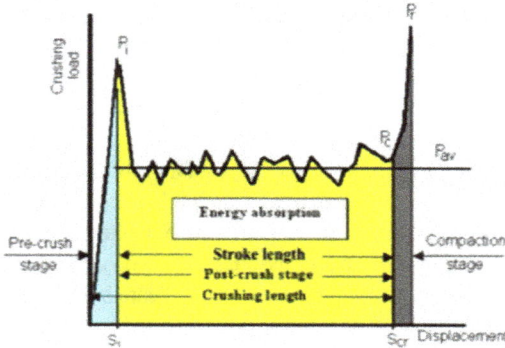

Figure 3. A schematic of the load-displacement relationship for after 40 mm [39].

1.8 Energy-absorbing thermoset composite structures

Polymer-based fiber composite materials offer significant benefits over metallic structures in reducing weight and cost as well as improving fatigue and corrosion resistance. However, these materials exhibit lower rate and behaviour of collapse compared to such metals as aluminium, a ductile metal that can tolerate rather large strains, deform plastically, and absorb a considerable amount of energy in the nonlinear region without fracture. Because of this difference between metals and composites, crash energy absorption with composites must come from innovative designs to enhance stress-strain behaviour.

Assembly operations such as mechanical joining, adhesive bonding, other kinds of attachments contribute and there are other parameters affecting (e.g., specimen geometry, processing conditions, fiber volume fraction) to crashworthiness. Effects of fiber volume fraction on absorbed energy and impact toughness of kenaf/epoxy composite, is shown in Figure 4. For crashworthiness considerations, the empirical relationship between these parameters and energy absorption are also required. Thermoplastic/thermoset multilayer composites can improve the impact damage tolerance of thermosetting resin matrix composites and thus will contribute to the ability to absorb most of the energy. A further enhancement in energy absorption and post-crash integrity can be achieved by optimizing the design of parameters.

High energy absorption, high strength, and rigidity are obtained primarily due to the property of mass reduction by the composite materials widely used in the automotive and motor sports industry [41]. The potential of NFPCs for application in providing

11

sustainable energy absorption was investigated by Meredith et al. while focusing on motor sports. The used vacuum assisted resin transfer molding technology, test conical specimens of jute, hemp and flax fabric reinforced polypropylene composites for their properties and features and recorded various values by different kinds of materials to analyze specific energy absorption (SEA). Improvement in energy absorption is evident from the increase in volume fraction that is possible only in the presence of low speed such as 2.5 m/s [42]. On the other hand, at high speeds such as 300 m/s, similar performance is shown with jute, hemp, and flax, but jute showed the low strength of fibers and brittleness [43]. The use of NFPCs has expanded considerably in products developing industry fields in recent years. As indicated by current pointers are that interest NFPCs in the industry will continue to growing quickly around the world. Over 5 years (2011–2016), the NFPCs industry is estimated to grow 10% worldwide [44].

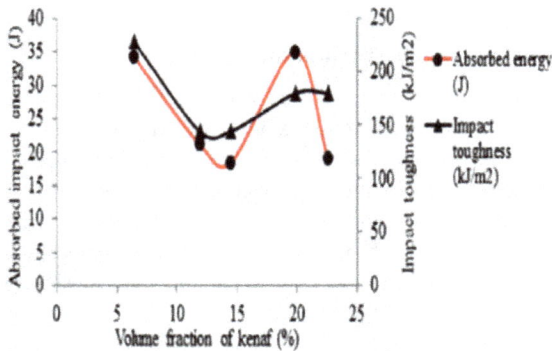

Figure 4 Shows the effects of volume fraction on absorbed energy and impact toughness of epoxy composite[40].

1.9 Assessing factors of energy absorption capability

1.9.1 Crush force efficiency (CFE)

Crush force efficiency is the ratio between the average crush load and the initial crush failure load. It is useful in measuring the performance of an absorber. It is calculated as Eq. (1).

$$P_m \qquad\qquad (1)$$

where, P_i and P_m are the initial and the average crushing loads, respectively. This ratio should be as close to 100% as possible, which is difficult to achieve in practice, but an ideal absorber is said to exhibit a crush force efficiency of 100%.

1.9.2 Stroke efficiency (SE)

The relative deformation of the absorber is referred to as the stroke efficiency (SE) of the absorber. This can be calculated as:

$$SE = \frac{u}{H} \qquad\qquad (2)$$

where u and H represent the stroke and the total height of the structure, respectively.

1.9.3 Initial failure indictor (IFI)

The ratio between initial crush load and critical crush load are calculated as:

$$IFI = \frac{P_i}{P_{cr}} \qquad (3)$$

where P_i is the initial crushing load, and P_{cr} is the critical crushing load.

1.9.4 Specific energy absorption E_S

Specific energy absorption (E_s) is defined as the energy absorbed per unit mass of material. Figure 5 shows a typical load-displacement curve obtained from the progressive crushing of a composite tube specimen(Alkateb et al [45]). The total work is done, or energy absorbed W_t, in the crushing of composite specimens for the area under the load-displacement curve is:

$$W_t = \int_{S_i}^{S_b} pds \qquad\qquad (4)$$

where W_t is the total energy absorbed in the crushing of the composite tube specimen, and the more characteristic property of progressive crushing mode is:

$$W_t = \int_{si}^{sb} P_m ds = P_m (S_b - S_i) \qquad\qquad (5)$$

where S_b and S_i are the crush distances, and P_m is the mean crush load as indicated.

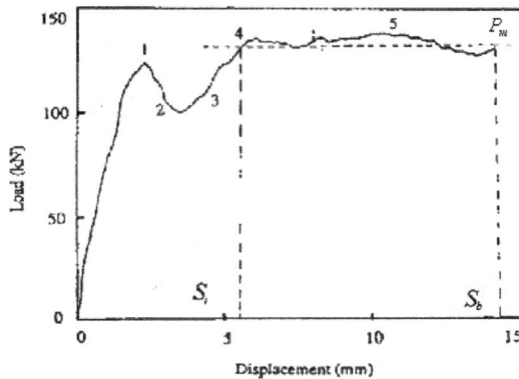

Figure 5 Typical Load-displacement curves for a progressively crushed epoxy composite tube [45].

1.10 Volumetric Energy absorption capability

The volumetric energy absorption capability (i.e., energy absorbed per unit volume) is also an essential parameter for an energy absorbing system design, where space is a restraint factor. The volume occupied by kenaf fiber reinforced elliptical epoxy composite cones before crushing can be calculated as:

$$V = \pi \times a \times b \times \frac{h}{3}$$

(6)

The energy absorbed per unit volume E_V can be calculated as:

$$E_V = \frac{E}{V} = \frac{1}{V} \sum_{i=1}^{N} P_i \bullet \frac{\left(s_{i+1} - s_{i-1}\right)}{2} = \frac{3}{\pi \times a \times b \times h} \sum_{i=1}^{N} P_i \bullet \frac{\left(s_{i+1} - s_{i-1}\right)}{2}$$

(7)

1.11 Energy absorption

The energy absorption capability of a composite material is critical to developing improved human safety in an automotive crash. The capacity of energy absorption is considered one of the general characteristics of thermoset epoxy composites such as biodegradability, mechanical properties, viscoelastic behavior and flame retardant.

The energy absorption capacity of epoxy composite materials is important for improving the safety of human beings at the time of an automotive crash. Energy absorption depends on many parameters such as sample shape, fiber type, fiber structure, matrix type, fiber volume fraction, processing conditions, test speed etc. When designing energy absorption composite structures, one of the aims, at absorbing most of the kinetic energy of impact within the device itself in an irreversible manner, this contributed towards a better understanding of the modes of failure and the energy dissipation patterns during impact in such safer structures and in evaluating existing ones for specific uses. Energy absorption is the concept of absorbing energy by converting the kinetic energy into another form of energy, therefore reduce losses in human and material resources [46]. The conversion of kinetic energy into plastic deformation depends on the method and magnitude of application of loads, material properties, displacement patterns or the deformation and transmission rates [47]. On the other hand, the total energy absorbed (E) is the area under the load/displacement curve, which is a function of the specimen cross-sectional area and the material density and is a load-displacement curve obtainable by numerical integration.

An attempt in this chapter is made to display some related works completed in the field of epoxy composite energy capability absorption that has been published in the literature to better understand the effect of parameters on the energy absorption capability of composite materials. In the following section, the review is focused on the research activities in the development of energy-absorbing composite structures which demonstrate compliance with the crashworthiness design criteria.

1.12 Literature survey

Many researchers have done many studies on the energy absorption capability of composite materials. Axially symmetric tubes have been used to perform many experimental studies on the energy absorption of composites because they are easy to manufacture and close to the geometry of the actual impact structure. In addition, composite tubes can be easily designed for stable crushing.

Several studies on the energy absorption capability of epoxy composite materials carried out by many researchers. Axially symmetric tubes have been used to perform many experimental studies on the energy absorption of composites because they are easy to manufacture and close to the geometry of the actual impact structure. In addition, composite tubes can be easily designed for stable crushing.

Alkbir et al. [48] study the energy absorption performances of hexagonal shaped tubes fabricated using non-woven kenaf fiber reinforced epoxy composites. Their work investigated the effect of hexagonal geometries on the crushing performances. It was found that the values of the specific energy absorption depend on the tube geometry.

Thermoset Composite

Materials Research Forum LLC

Materials Research Foundations 38 (2018)

doi: http://dx.doi.org/10.21741/9781945291876

Eshkoor et al. [49] found similar values of the SEA for non- triggered and triggered woven natural silk/epoxy composites. Oshkovr et al. [50] reported that the specific energy in short and mid-lengths of silk/epoxy composite square tubes was increased as the number of specimen's layers increases. On the other hand, the specific energy of the silk/epoxy composite square tube with 30 layers decreased due to the increasing of the specimen's weight. Yan et al. [51] concluded that the specific absorbed energy of the natural flax/epoxy composite tube with triggered and foam-filled tubes had greater value than that of the triggering tubes. But these energy absorption values of triggered and foam-filled tubes can be higher or smaller than that of foam-filler tubes. Ataollahi et al. [52] manufactured and tested square tubes based on silk fiber reinforced epoxy resins and studied the effect of tube lengthen SEA. The authors found that decrease in the length of tubes leads to increase of the specific energy. Mahdi et al. [53] investigated experimentally the crushing behavior of hybrid and non-hybrid natural fiber/polyester composite solid cones. Two types of natural fiber used oil palm and coir fiber to fabricate solid cones with vertex angles varied from 0° to 60°. They pointed out that solid cones greatly affects the crashworthiness for the natural fiber. Yan et al. [54] concluded that the specific absorbed energy of the natural flax/epoxy composite tube with triggered and foam-filled tubes had greater value than that of the triggering tube as depicted in Figures 6 and 7. But these energy absorption values of triggered and foam-filled tubes can be higher or smaller than that of triggering tube as depicted in figures 8 and 9 also the figures indicate that the presence of triggering on the total absorbed energy and the specific absorbed energy is insignificant. Yan and Chouw [55] studied woven flax/epoxy composite tubes with different tube inner diameters, cell wall thickness, and length to diameter ratios under quasi-static axial crushing were tested. The study showed the clear majority of empty flax/epoxy composite tubes were crushed in a brittle manner and commented in a progressive crushing pattern with advantageous specific energy absorption capability. Therefore, these subsequent studies (e.g. [56]) indicate that the natural fiber-reinforced polymer composite has potential to be an energy absorber in axial crushing. Eshkoor, et al, [57], studies investigate the energy absorption response of rectangular woven natural silk/epoxy composite tubes when subjected to an axial quasi-static crushing test using a trigger mechanism. Results showed the specific energy absorption values decreased with increased length of the composite specimen, whereas total energy absorption increased with the increased length of the epoxy composite specimen.

Hamada and Ramakrishna [58], shows the epoxy composite tubes with t/D ratios of less than 0.015 fail by brittle fracture whereas tubes with t/D ratios in the range 0.015–0.25 crush progressively. Specific energy absorption capability is dependent on the absolute value of t, rather than the t/D ratio, and it increases with increasing t up to a certain value above which it decreases. Highest energy absorption capability was displayed by tubes with values of t in the range of 2–3 nm. This variation in specific energy is attributed to changes in the crush zone morphology. Also, the changes in fiber orientation for a single layer fiber do not significantly affect energy absorption performance. As expected, as the number of layers increases, the energy absorbed also increases [59]. Warrior et al. [60] studied the effect of tube geometry on the crashworthiness performances of kenaf reinforced epoxy composite tubes under quasi-static uniaxial compressive load. In their study, the randomly oriented non-woven kenaf fibers were cured into hexagonal tubes. They found that hexagonal tube angle changes affect (as depicted in Figure 9) crashworthiness parameters with different distinct failure modes.

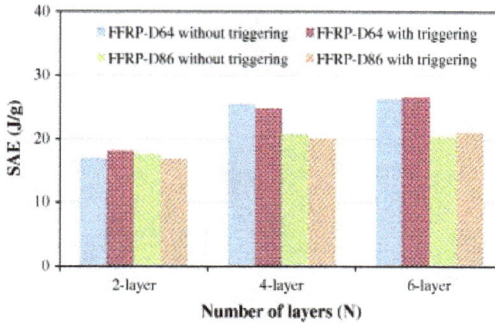

Figure 6 Effect of triggering on the specific absorbed energy of all the empty flax/epoxy composite tubes [37].

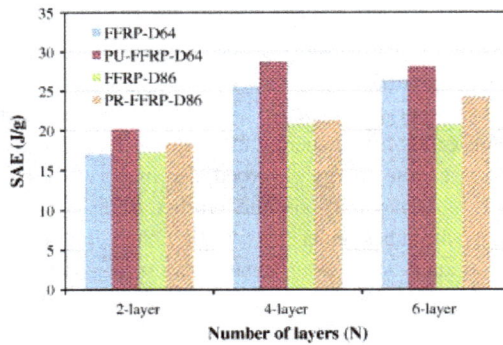

Figure 7   Effect of triggering on the total absorbed energy of all the empty flax/epoxy composite tubes [37].

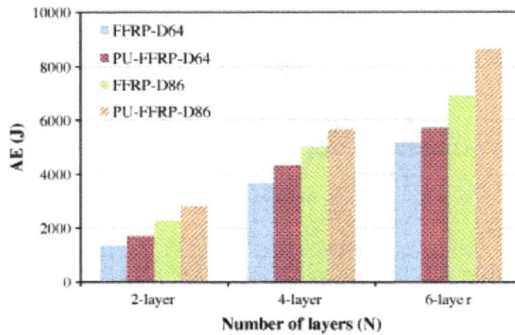

Figure 8   Effect of foam-filled on the total absorbed energy of all the empty flax/epoxy composite tubes [37].

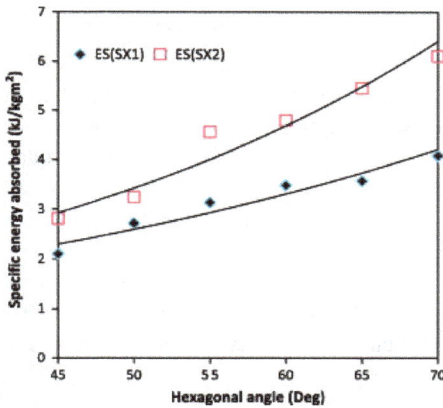

Figure 9   The effect of hexagonal angle on energy absorption capability of epoxy composite [60].

Research also took care of the use of thermoset polymer resins composites because they have a role in improving performance and the mechanical characteristics of natural fibers such as stiffness, toughness, and impact resistant.

Jeyanthi, et al, [61], focused on partially eco-friendly hybrid long fiber reinforced thermo plastics with natural kenaf fiber to enhance the desired mechanical properties for car bumper beams as automotive structural components. A specimen without any modifier is tested and compared with a typical bumper beam material called LFRT, the results indicate that some mechanical properties such as tensile strength, Young's modulus, flexural strength of a flexural modulus are more advantageous to LFRT, the new material also must improve the ability to absorb more impact load and increase the protection of the front car component. Saijad et el. [62] manufactured and tested a range of tubes with different shapes and geometric design in an axial crushing of epoxy composite materials, progressive failure absorbs more energy due to the presence of multiple failure modes and geometric design can improve energy absorption capability. On the other hand, using natural fiber reinforced a plastic composite material (NFPCs), since the ratio of the volume of the tube due to the natural fiber is lower density, environmentally friendly, bio degradable and lower the cost. Ataollahi et al. [63] manufactured and tested square tubes based on silk fiber reinforced epoxy resins and studied the effect of tube lengthen SEA. The authors reported relatively low values of the SEA, typically in the range of 4 to 5 kJ/kg. Energy Absorption of the NFPCs. Each contribution of the energy absorption mechanisms depending upon the overall toughness and the properties of the composite varies.

RĤžiþka el at. [64] presented experimental investigation and numerical prediction of the deformation behavior of several types of composite deformation elements. Experiments were conducted on two types of samples. First, on filament wound composite tubes with thermoset polymer matrix and on moulded thermoplastic corrugated plates. Second, deformation elements made from moulded thermoplastic carbon sheets are also promising for parts of composite absorbers. The authors found that the absorbed deformation energy of elements made from filament wound thermoset composite tubes depends on a laminate layup. Accordingly, simulation tests showed that the deformation boxes applied to the front bumper can effectively absorb most of the impact energy at the initial stage of the crash.

Salman S.D., et al, [65], discussed the influence of resin system on the energy absorption capability and morphological properties of plain woven kenaf composites. The results show that a significant increase in energy absorption, strength, and toughness of the kenaf/PVB composite was found at different energy levels. The impact strength and toughness of the kenaf/PVB film was 6 times that of the kenaf/epoxy composite material, which was particularly at different energy levels. The SEM examinations of the Charpy impact test specimens show that there is no delamination between the layers and the interfacial bond between kenaf fiber and the two resins but the PVB film is the highest. kenaf/PVB composite failed by fiber fracture while kenaf/epoxy composite failed by a combination of fiber pull-out and although the kenaf/PVB composite material failed due to fiber fracture while the kenaf/epoxy composite material was failed by a combination of fiber pull-out and fiber fracture as well as crack propagations through the matrix.

Bakar el at. [66] investigated the tensile properties and low-velocity impact behavior of long kenaf with two different thermosetting polymer resins (epoxy and polyester) Tensile properties of kenaf/epoxy and kenaf/polyester composites were studied experimentally. Tensile tests of kenaf fibers were carried out by different fiber weight percentages (10, 15, 20 and 25%) and the impact test was conducted using an instrumented drop tower device. The results of this study indicated that kenaf/polyester composites have better impact absorption energy than kenaf/epoxy composite material. However, kenaf/epoxy composites have higher tensile properties than kenaf/polyester composites. This is a good indication that epoxy bonding is superior to other thermoset polymer resins. For that reason, appropriate process techniques and parameters should be rigorously chosen in order to get the best characteristics of producing a composite. Dehkordi et al. [67] evaluated the damage tolerance is assessed through compression after Impact (CAI) tests and response of thermoset composites containing basalt/Nylon hybrid yarns. Their results indicated that hybrid laminates have 41–82% lower compressive strength than the basalt laminates. However, hybrid laminates had higher residual strength with increasing impact energy level.

A similar finding was also observed in the average impact strength of the two types of composites. It was clear that for the composite specimens fabricated using plain woven kenaf/PVB composite recorded the higher values of impact strength compared with the composite specimens fabricated using epoxy resin, at different energy levels. This may be explained by that PVB film has higher elongation properties compared to epoxy resin due to its low viscosity which helps in penetrating fabrics easily. High strength and viscosity of the epoxy resin have increased the specimen stiffness and decreased from its ability to absorb impacts. Furthermore, using a hot press technique to fabricate the kenaf/PVB composite has affected the impact properties of the composites which lead to good resin/fabrics penetration. As research studies have reported, the type of polymer is a critical factor which has an important influence on the impact strength of the natural fibers composites [68-70].

Main practice of natural fiber thermoplastic resin was limited because it cannot withstand high temperatures during processing and moisture absorption. These limitations can be controlled by some of the components such as proper selection of natural fibers, selection of fillers, methods of separating fibers from bast, fiber processing techniques, fiber-matrix bonding, injection molding, extrusion, compression molding, etc. [71].

Thermal composite materials are easier to recycle than mineral-based fibers. Thermoplastic is more recyclable than thermoset plastics composite. Grinding reprocessing is a common method for recycling thermoplastic composite materials. This recycling process is more efficient and more economical than other recycling methods such as chemical recycling, particle recycling, energy recycling [72].

Natural fiber reinforced composite efficiency depends on the fiber to the matrix interface and the capability of adhesion over the matrix to the fiber. This can be maximized by increasing the bond between the fibers and the matrix and changes the length of the short fibers to the length of the long fibers. Influence of fiber length and fiber distribution give greater impact while developing natural fiber thermal composite using extrusion process or injection moulding [73]. Efficiency and performance are less for using short fibers compared with long fiber composites due to fiber distribution and fiber orientation while using short fibers for the composite. Through previous studies carried out to investigate the potential use of natural fibers as reinforcing agents in polymers. Table 2 shows the application of natural fiber composite applications in the industry. Finally, most of the results show that natural fibers can be used as a reinforcing tool for plastics. However, its mechanical strength is relatively lower than that of synthetic fiber epoxy composite material [74].

Table 2 Natural fiber composite applications in industry [75, 76–78].

Fiber	Application in building, construction, and others
Hemp fiber	Construction products, textiles, cordage, geotextiles, paper & packaging, furniture, electrical, manufacture bank notes, and manufacture of pipes
Oil palm fiber	Building materials such as windows, door frames, structural insulated panel building systems, siding, fencing, roofing, decking, and other building materials [14]
Wood fiber	Window frame, panels, door shutters, decking, railing systems, and fencing
Flax fiber	Window frame, panels, decking, railing systems, fencing, tennis racket, bicycle frame, fork, seat post, snowboarding, and laptop cases
Rice husk fiber	Building materials such as building panels, bricks, window frame, panels, decking, railing systems, and fencing
Bagasse fiber	Window frame, panels, decking, railing systems, and fencing
Sisal fiber	In construction industry such as panels, doors, shutting plate, and roofing sheets; also, manufacturing of paper and pulp
Stalk	Building panel, furniture panels, bricks, and constructing drains and

fiber	pipelines
Kenaf fiber	Packing material, mobile cases, bags, insulations, clothing-grade cloth, soilless potting mixes, animal bedding, and material that absorbs oil and liquids
Cotton fiber	The furniture industry, textile and yarn, goods, and cordage
Coir fibers	Building panels, flush door shutters, roofing sheets, storage tank, packing material, helmets and postboxes, mirror casing, paper weights, projector cover, voltage stabilizer cover, a filling material for the seat upholstery, brushes and brooms, ropes and yarns for nets, bags, and mats, as well as padding for mattresses, seat cushions
Ramie fiber	Use in products as industrial sewing thread, packing materials, fishing nets, and filter cloths. It is also made into fabrics for household furnishings (upholstery, canvas) and clothing, paper manufacture.
Jute fiber	Building panels, roofing sheets, door frames, door shutters, transport, packaging, geotextiles, and chip boards.

1.13 Conclusions

The use of natural fibers as reinforcement for polymer and thermoset polymer composites has a positive effect on the mechanical behavior of the polymer. High strength, energy absorption, and rigidity are obtained by the composite materials, which are widely used in automobiles.

In recent years, applied research on natural fiber composite materials specifically on energy absorbing has grown rapidly considering encouragement from many researchers who emphasize sustainability and the commercial industry and using green materials. Natural fiber composite has good properties and potential to be developed as a component of engineering.

Thermoset matrix systems dominate the composites industry because of their reactive nature and ease of impregnation, characterized by low viscosity. The use of natural fiber to reinforced composite materials can add the structural strength of the material. In addition, it increases the energy absorption capacity of the material. There are several conclusions that can be drawn:

Failure modes have been studied where these different failure modes affected peak load and specific energy absorption significantly on the failure mechanisms of fiber reinforced epoxy thermoset composites.

Geometry is considered an important factor in determining the ability of a substance to absorb energy. The thin-walled tube is geometrically superior for energy absorption capacity.

Several factors have direct effects on specific energy absorption such as shapes, triggering, fiber orientations, fiber types and material hybridization. And others, have Indirect effects on specific energy absorption such as technician skill, quality of materials and manufacturing conditions, etc.

The mechanical properties of the materials play a large role in the mechanism and behavior of the failure and consequently to withstand the shocks by the vehicle.

The previous studies show the possibility that natural fiber reinforced epoxy composite tubes can be applied as energy absorbing devices. However, the energy absorption behavior of the epoxy composite structure is not easily predicted because the destruction mechanism occurring in the material is very complicated. Finally, further studies should be continued to improve the crashworthiness performance to demonstrate maximum performance.

Acknowledgments

The authors want to thank the Universiti Putra Malaysia for the financial support for this research programme using HiCoE Grant, Ministry of Higher Education, Malaysia.

References

[1] D.N. Saheb, J. Jog, Natural fiber polymer composites: a review, Advanced Polymer Technology. Vol.18 (1999) 351–363. https://doi.org/10.1002/(SICI)1098-2329(199924)18:4<351::AID-ADV6>3.0.CO;2-X

[2] B.A. Cheeseman,T.A. Bogetti, Ballistic impact into fabric and compliant composite laminates, Composite Structures.Vol. 61 (2003) 161-173. https://doi.org/10.1016/S0263-8223(03)00029-1

Materials Research Forum LLC
doi: http://dx.doi.org/10.21741/9781945291876

[3]   E. Gooding, energy absorbing structure US Patent Specification Adaptive, (1999) 581-591.

[4]   O. Faruk, A. K. Bledzki, H.P. Fink, M. Sain, Biocomposites reinforced with natural fibers: 2000–2010 (Progress in Polymer Science. Vol.37(11)(2012) 1552–1596. https://doi.org/10.1016/j.progpolymsci.2012.04.003

[5]   H. Ku, H. Wang, N. Pattarachaiyakoop, M. Trada, A review on the tensile properties of natural fiber reinforced polymer composites, Composites Part B: Engineering. Vol.42 (2011) 856–873. https://doi.org/10.1016/j.compositesb.2011.01.010

[6]   A. Ticoalu, T. Aravinthan, F. Cardona, A review of current development in natural fiber composites for structural and infrastructure applications, in Proceedings of the Southern Region Engineering Conference (SREC), Toowoomba, Australia. Vol.10 (2010) 113–117.

[7]   O. Faruk, A. K. Bledzki, H.P. Fink, M. Sain, Biocomposites reinforced with natural fibers: 2000–2010, Progress in Polymer Science .Vol.37 (2012) 1552–1596. https://doi.org/10.1016/j.progpolymsci.2012.04.003

[8]   F.Z. Arrakhiz, M. El Achaby, M. Malha, Mechanical and thermal properties of natural fibers reinforced polymer composites: doum/low density polyethylene, Materials & Design.Vol.43 (2013) 200–205. https://doi.org/10.1016/j.matdes.2012.06.056

[9]   G. Di Bella, V. Fiore, G. Galtieri, C. Borsellino, A. Valenza, Effects of natural fibers reinforcement in lime plasters (kenaf and sisal vs. Polypropylene), Construction and Building Materials. Vol.58 (2014) 159–165. https://doi.org/10.1016/j.conbuildmat.2014.02.026

[10]   A. Shalwan, B.F. Yousif, In state of art: mechanical and tribological behaviour of polymeric composites based on natural fibers, Materials & Design. Vol.48 (2013) 14–24. https://doi.org/10.1016/j.matdes.2012.07.014

[11]   M.A. Norul Izani, M. T. Paridah, U. M. K. Anwar, M. Y. Mohd Nor, P.S. H'Ng, Effects of fiber treatment on morphology, tensile and thermogravimetric analysis of oil palm empty fruit bunches fibers, Composites Part B: Engineering. Vol.45 (2013) 1251–1257. https://doi.org/10.1016/j.compositesb.2012.07.027

[12]   I.S.M. A. Tawakkal, M.J. Cran, S.W. Bigger, Effect of kenaf fiber loading and thymol concentration on the mechanical and thermal properties of

PLA/kenaf/thymol composites, Industrial Crops and Products.Vol.61 (2014) 74–83. https://doi.org/10.1016/j.indcrop.2014.06.032

[13]   S. Shinoj, R. Visvanathan, S. Panigrahi, M. Kochubabu, Oil palm fiber (OPF) and its composites: a review, Industrial Crops and Products. Vol.33 (2011) 7–22. https://doi.org/10.1016/j.indcrop.2010.09.009

[14]   K.S. Sanjay, M. d. Abdul Sami, Sai Anoop, B.V. Gudipudi, S.V. Lahari and A.U. Syed, Advanced Composite Materials in Typical Aerospace Applications,Global Journals Inc (USA). Vol.14 (2014) 0975-5861.

[15]   I.Y. Chang, J.K. Lees, Recent development in thermoplastic composites: a review of matrix systems and processing methods, J Thermoplast Compos. Vol.1 (3) (1988) 277-296. https://doi.org/10.1177/089270578800100305

[16]   R.J. Lee, Compression strength of aligned carbon fiber-reinforced thermoplastic laminates Composites. Vol.18 (1) (1987) 35-39. https://doi.org/10.1016/0010-4361(87)90005-X

[17]   B. Vieille, V.M. Casado, C. Bouvet, About the impact behavior of woven-ply carbon fiber-reinforced thermoplastic- and thermosetting-composites: a comparative study, Compos Struct. Vol.101 (2013) 9-21. https://doi.org/10.1016/j.compstruct.2013.01.025

[18]   W. Hufenbach, R. Böhm, M. Thieme, A. Winkler, E. Mäder, J. Rausch, Polypropylene/glass fiber 3D-textile reinforced composites for automotive applications, Mater Desig. Vol.32 (3) (2011) 1468-1476. https://doi.org/10.1016/j.matdes.2010.08.049

[19]   N. Bernet, V. Michaud, P.E. Bourban, J. A.E. MånsonCommingled yarn composites for rapid processing of complex shapes, Compos Part A-Appl S. Vol.32 (11). (2001) 1613-1626. https://doi.org/10.1016/S1359-835X(00)00180-9

[20]   M.D. Wakeman, T.A. Cain, C.D. Rudd, R. Brooks, A.C. Long,Compression moulding of glass and polypropylene composites for optimised macro- and micro-mechanical properties-1 commingled glass and polypropylene, Compos Sci Technol. Vol.58 (12) (1998) 1879-1898. https://doi.org/10.1016/S0266-3538(98)00011-6

[21]   D. Trudel-Boucher, B. Fisa, J. Denault, P. Gagnon, Experimental investigation of stamp forming of unconsolidated commingled E-glass/polypropylene fabrics, Compos Sci Technol. Vol. 66 (2006) 555-570. https://doi.org/10.1016/j.compscitech.2005.05.036

Thermoset Composite

Materials Research Forum LLC

Materials Research Foundations **38** (2018)

doi: http://dx.doi.org/10.21741/9781945291876

[22] S. Mazumdar, Composites manufacturing: materials, product and process engineering, CRC Press, Florida (2001). https://doi.org/10.1201/9781420041989

[23] S. Ramakrishna, H. Hamada, Energy absorption characteristics of crash worthy structural composite materials, Key Engineering Materials. Vol.141–143 (1998) 585-620.

[24] T. Hou, G. Pearce, B. Prusty, D.Kelly , R. Thomson, Pressurised composite tubes as variable load energy absorbers, Composite structures.Vol.120 (2015) 346-357. https://doi.org/10.1016/j.compstruct.2014.09.060

[25] C. Priem, R. Othman, P.D. Rozycki, Experimental investigation of the crash energy absorption of 2.5 D-braided thermoplastic composite tubes, Composite structures.Vol.116 (2014) 814-26. https://doi.org/10.1016/j.compstruct.2014.05.037

[26] J.S.Kim, H. J. Yoon, K.B. Shin, A study on crashing behaviours of composite circular tubes with different reinforcing fibers, International Journal of Impact Engineering. Vol.38 (2011) 198-207. https://doi.org/10.1016/j.ijimpeng.2010.11.007

[27] A. Othman, S. Abdullah, A. Ariffin, N. Mohamed, Investigating the   quasi-static axial crashing behaviour of polymeric foam-filled composite pultrusion square tubes, Materials & Design. Vol.63(2014) 446-59. https://doi.org/10.1016/j.matdes.2014.06.020

[28] Mohamed Alkateb, S.M. Sapuan, Z. Leman, M. R. Ishak, Mohammad Jawaid, Energy absorption capacities of kenaf fiber-reinforced epoxy composite elliptical cones with circumferential holes, Fibers and Polymers.Vol.18 (2017) 1187-1192. https://doi.org/10.1007/s12221-017-1244-0

[29] A. Elgalai, E. Mahdi, A. Hamouda, B.Sahari, Crashing response of composite corrugated tubes to quasi -static axial loading, Composite Structures. Vol.66 (2004) 66 -671. https://doi.org/10.1016/j.compstruct.2004.06.002

[30] E. Mahdi, A. Hamouda, Energy absorption capability of composite hexagonal ring systems, Materials & Design. Vol.34 (2012) 201-10. https://doi.org/10.1016/j.matdes.2011.07.070

[31] E. Mahdi, A. Hamouda, B. Sahari,Y. Khalid, Experimental quasi-static axial crashing of cone–tube–cone composite system, Composites PartB: Engineering. Vol.34 (2003) 285 -302. https://doi.org/10.1016/S1359-8368(02)00102-6

[32]   M. Jawaid, H. A. Khalil, A .Hassan, R. Dungani, A. Hadiyane, Effect of   jute fiber loading on tensile and dynamic mechanical properties of oil palm epoxy composites, CompositesPart B: Engineering.Vol.45 (2013) 619- 624. https://doi.org/10.1016/j.compositesb.2012.04.068

[33]   M.M. Davoodi, S.M. Sapuan, D. Ahmad, Aidy Ali, A. Khalina , Mehdi Jonoobi, Mechanical properties of hybrid kenaf/glass reinforced epoxy composite for passenger car bumper beam, Materials and Design. Vol.31 (2010) 4927-4932. https://doi.org/10.1016/j.matdes.2010.05.021

[34]   Y. El-Shekeil, S. M. Sapuan, M. Algrafi, Effect of fiber loading on mechanical and morphological properties of cocoa pod husk fibers reinforced thermoplastic polyurethane composites, Materials & Design.Vol.64 (2014) 330-3. https://doi.org/10.1016/j.matdes.2014.07.034

[35]   A.E. Ismail, M.F. Sahrom, Lateral crushing energy absorption of cylindricalkenaf fiber reinforced composites, International Journal of Applied Engineering Research. Vol.10(8) (2015) 19277-19288.

[36]   R. Yahaya, S.M. Sapuan, Z. Leman, Selection of natural fiber for hybrid laminated composites vehicle spall liners using analytical hierarchy process (AHP), Appl. Mech. Mater. Vol.564 (2014) 400–405. https://doi.org/10.4028/www.scientific.net/AMM.564.400

[37]   S. Ataollahi, S. Taher, R.A. Eshkoor, A.K. Ariffin, C.H. Azhari, Energy absorption and failure response of silk/epoxy composite square tubes: experimental, Compos B: Eng.Vol.43 (2012) 42–548. https://doi.org/10.1016/j.compositesb.2011.08.019

[38]   R. Mahjoub, J. M. Yatim, A. R. Mohd Sam, M. Raftari, Characteristics of continuous unidirectional kenaf fiber reinforced epoxy composites Materials & Design. Vol.64 (2014) 640–649. https://doi.org/10.1016/j.matdes.2014.08.010

[39]   M.F.M.Alkbir, S.M. Sapuan, A.A. Nuraini, M.R. Ishak, The Effect of Fiber Content on the Crashworthiness Parameters of Natural Kenaf Fiber-Reinforced Hexagonal Composite Tubes, Journal of Engineered Fibers and Fabric.Vol. 11 (2016) 75 –86

[40]   R. Yahaya, S.M. Sapuan, M. Jawaid, Z. Leman, E.S. Zainudin, Mechanical performance of woven kenaf-Kevlar hybrid composites, Journal of Reinforced Plastics and Composites.Vol.33 (2014) 2242–2254. https://doi.org/10.1177/0731684414559864

Materials Research Forum LLC
doi: http://dx.doi.org/10.21741/9781945291876

[41]   G. Savage, Development of penetration resistance in the survival cell of a Formula 1 racing car, Engineering Failure Analysis.Vol. 17 (2010) 116–127. https://doi.org/10.1016/j.engfailanal.2009.04.015

[42]   J. Meredith, R. Ebsworth, S. R. Coles, B. M. Wood, K. Kirwan, Natural fiber composite energy absorption structures, Composites Science and Technology. Vol. 72 (2012) 211–217. https://doi.org/10.1016/j.compscitech.2011.11.004

[43]   P. Wambua, B. Vangrimde, S. Lomov, I. Verpoest, The response of natural fiber composites to ballistic impact by fragment simulating projectiles, Composite Structures. Vol. 77 (2007) 232–240. https://doi.org/10.1016/j.compstruct.2005.07.006

[44]   N. Uddin, E. d. Developments in Fiber-Reinforced Polymer (FRP) Composites for Civil Engineering, Elsevier, (2013). https://doi.org/10.1533/9780857098955

[45]   M. Alkateb, E. Mahdi, A. Hamouda and M. Hamdan, On the energy absorption capability of axially crashed composite elliptical cones, Composite structures. Vol. 66 (2004) 495-501. https://doi.org/10.1016/j.compstruct.2004.04.078

[46]   F.L. Matthews, R. D. Rawlins, Composite Materials, Engineering & Science, Chapman & Hall, London. (1994).

[47]   D. A. Bahwgan, J. B. Lawrence, Analysis and performance of fiber composites, New York: John Wiley & Sons, Inc. (1990) 282-322.

[48]   M.F.M Alkbir, S. M. Sapuan, A. A. Nuraini, M. R. Ishak, Effect of geometry on crashworthiness parameters of natural kenaf fiber reinforced composite hexagonal tube, Materials and Design. Vol.60 (2014) 85–93. https://doi.org/10.1016/j.matdes.2014.02.031

[49]   R. A. Eshkoor, S. A. Oshkovr , A. B.Sulong, R. Zulkifli, A. K. Ariffin, C. H. Azhari, Comparative research on the crashworthiness characteristics of woven natural silk/epoxy composite tubes, Materials & Design.Vol.47 (2013) 248–257. https://doi.org/10.1016/j.matdes.2012.11.030

[50]   S.A. Oshkovr, R. A. Eshko, S. T. Taher, A. K. Ariffina, C. H.Azhari, Crashworthiness characteristics investigation of silk/epoxy composite square tubes, Compos Struct.Vol.94 (2012) 2337–42. https://doi.org/10.1016/j.compstruct.2012.03.031

[51]   L.B. Yana, N. Chouw, K. Jayaraman, Effect of triggering and polyurethane foam-filler on axial crushing of natural flax/epoxy composite tubes, Mater Des. Vol. 56 (2014) 528–41. https://doi.org/10.1016/j.matdes.2013.11.068

[52]   S. Ataollahi, S.T. Taher, R.A. Eshkoor, A.K. Ariffin, C.H. Azhari, Energy absorption and failure response of silk/epoxy composite square tubes: experimental, CompoB: Eng. Vol.43 (2012) 542–8. https://doi.org/10.1016/j.compositesb.2011.08.019

[53]   E.Mahdi, A.S.M. Hamouda, A.C. Sen, Quasi-static crushing behaviour of hybrid and non-hybrid natural fibre composite solid cones, Compos Struct. Vol.66 (2004) 647–63. https://doi.org/10.1016/j.compstruct.2004.06.001

[54]   L.B. Yana, N. Chou, K. Jayaraman, Effect of triggering and polyurethane foam-filler on axial crushing of natural flax/epoxy composite tubes, Mater Des. Vol. 56 (2014) 528–41. https://doi.org/10.1016/j.matdes.2013.11.068

[55]   L.B.Yan, N. Chouw, Crashworthiness characteristics of flax fibre reinforced epoxy tubes for energy absorption application, Mater & Des. Vol.51 (2013) 629–64. https://doi.org/10.1016/j.matdes.2013.04.014

[56]   J. Meredith, R. Ebsworth, S.R. Coles, B.M. Wood and K.Kirwan, Natural fibre composites energy absorption structures, Compos Sci Technol. Vol.72 (2012) 211–217. https://doi.org/10.1016/j.compscitech.2011.11.004

[57]   R.A. Eshkoor, S. Ude, A. Oshkovr, R. Sulong, R. Zulkifl, A. Ariffin, Failure mechanism of woven natural silk/epoxy rectangular composite tubes underaxial quasi -static crashing test using trigger mechanism, International Journal of Impact Engineering. Vol. (2014) 53 -61.

[58]   H. Hamada, S. Ramakrishna, Energy Absorption Capabilities of Fiber Reinforced Thermoplastic Composite Materials, Proceedings of Japan-US VIIth Conference on Composite Materials. (1995) 609-616.

[59]   A.E. Ismail, M.K. Awang, M.H. Sa'at, Tensile strength of natural fiber reinforced polyester composite, AIP Conf. Proc. Vol. 909 (2007) 174-179.

[60]   A.E. Ismail, Energy Absorption Performances of Square Winding Kenaf Fiber Reinforced Composite Tubes, International Journal of Engineering and Technology. Vol.6 (2015) 0975-4024.

[61]   S. Jeyanthi, J.J Rani, Improving mechanical properties by kenaf natural long fiber reinforced composite for automotiv e structures, India Journal of Applied Science and Engineering. Vol.15 (2012) 275-280.

[62]   T.W. lau Saijod , M.R. said, mohd yuhazri Yaakob, On the effect of geometricl designs and failure modes in composite axial crushing: A literature review, Composite sturctures designs. Vol.94 (2012) 803-812.

[63]   S. Ataollahi, S.T. Taher, R.A. Eshkoor, A.K. Ariffin and C.H. Azhari, Energy
absorption and failure response of silk/epoxy composite square tubes:
experimental, CompoB: Eng. Vol. 43(2) (2012) 542–8.
https://doi.org/10.1016/j.compositesb.2011.08.019

[64]   RĤžiþka Milana, Kulíšek Viktora, Bogomolov Sergiia, ShánČl Víta, Development
of composite energy absorber, Modelling of Mechanical and Mechatronic
Systems. Vol. 96 (2014) 392 – 399.

[65]   S.D. Salman, Z. Leman, M.T.H. Sultan, M. R. Ishak, F. Cardon, Influence of resin
system on the energy absorption capability and morphological properties of plain
woven kenaf composites, IOP Conference Series: Materials Science and
Engineering. Vol. 100 (2015) 18–19. https://doi.org/10.1088/1757-
899X/100/1/012053

[66]   N.H. Bakar, K.M. Hyie, A.F. Mohamed, Z. Salleh, A. Kalam, Kenaf fiber
composites using thermoset epoxy and polyester polymer resins: energy absorbed
versus tensile properties, Materials Research Innovations. Vol. 18 (2014) 505-50.
https://doi.org/10.1179/1432891714Z.0000000001037

[67]   M. Tehrani Dehkordi, H. Nosraty , M.M. Shokrieh, G. Minak, D. Ghelli, The
influence of hybridization on impact damage behavior and residual compression
strength of intraply basalt/nylon hybrid composites, Materials & Design.
Vol..43(0) (2013) 283-290. https://doi.org/10.1016/j.matdes.2012.07.005

[68]   M. Aktaş, C. Atas, B.M. İçten, R. Karakuzu, An experimental investigation of the
impact response of composite laminates, Composite Structures. Vol.87 (2009)
307-13. https://doi.org/10.1016/j.compstruct.2008.02.003

[69]   S.M. Sapuan, N. Harun , K. Abbas, Design and fabrication of a multipurpose table
using a composite of epoxy and banana pseudostem fibres, Journal of Tropical
Agriculture, Vol.45 (2007) 66-8.

[70]   M.F. Omar, H.M.d. Akil, Z.A. Ahmad, A.A.M. Mazuki Yokoyama T, Dynamic
properties of pultruded natural fibre reinforced composites using Split Hopkinson
Pressure Bar technique Materials & Design.Vol.31 (2010) 4209- 18.
https://doi.org/10.1016/j.matdes.2010.04.036

[71]   M. Ramakrishna, V. kumar, N. Yuvrajsingh, Novel treated Pine Needle fiber
reinforced polypropylene composites and their characterizatio,J. Reinf. Plast.
Comp. Vol.29 (2010) 2343-2355. https://doi.org/10.1177/0731684409348969

[72]   J. Chu , J. Sullivan . Recyclability of a fiber reinforced cyclic polycarbonat composite, Polym. Comp. Vol.17(4) (1996) 556-567. https://doi.org/10.1002/pc.10646

[73]   M.M. Davoodi, S.M. Sapuan, R Yunus, Conceptual design of a polymer Composite automotive bumper energy absorber, Mater. Des. Vol.29 (2008) 1447-1452. https://doi.org/10.1016/j.matdes.2007.07.011

[74]   T.A. Warrior, E.Turner and M. Ribeaux Cooper, Effects of boundary conditions on the energy absorption of thin-walled polymer composite tubes under axial crushing, Thin-Walled Structures.Vol.46 (2008) 905–918. https://doi.org/10.1016/j.tws.2008.01.023

[75]   A. Ticoalu, T. Aravinthan, and F. Cardona, A review of current development in natural fiber composites for structural and infrastructure applications, (Toowoomba, Australia, in Proceedings of the Southern Region Engineering Conference. (SREC)-F1-5 (2010) 113–11.

[76]   T. Sen, H.N. Reddy, Various industrial applications of hemp, kinaf, flax and ramie natural fibers, International Journal of Innovation, Management and Technology. Vol.2 (2011) 192–198.

[77]   U.S. Bongarde, V.D. Shinde, Review on natural fiber reinforcement polymer composite, International Journal of Engineering Science and Innovative Technology. Vol.3 (2014) 431–436.

[78]   L. Mwaikambo, Review of the history, properties and application of plant fibers, African Journal of Science and Technology.Vol.7 (2006) 120-133.

Materials Research Forum LLC
doi: http://dx.doi.org/10.21741/9781945291876

Chapter 2

Wood Flour Filled Thermoset Composites

M. Ramesh[1,*], L. Rajeshkumar[2]

[1]Department of Mechanical Engineering, KIT-Kalaignarkarunanidhi Institute of Technology, Coimbatore-641402, Tamil Nadu, India.

[2]Department of Mechanical Engineering, KPR Institute of Engineering and Technology, Coimbatore-641407, Tamil Nadu, India.

mramesh97@gmail.com*, lrkln27@gmail.com

Abstract

The increasing environmental concerns, health related issues and depletion of fossil fuels have led to an increased interest in the development of eco-friendly bio-degradable and sustainable materials. The development of sustainable and renewable materials from natural resources has been witnessing a tremendous world-wide attention and importance. In recent years, there has been a continuous and emerging research in the preparation of novel materials from wood based resources due to their great potential as alternatives to petroleum-based products. Wood flour based composite is recognized as sustainable materials owing to its eco-friendly attributes derived from reinforcement of fully degradable wood fibers etc. which leads toward minimization of the generation of polluting agents. The aim of this chapter was to develop sustainable composites by complete replacement of the synthetic fibers by sustainable wood flour and substituting thermoset matrices. This chapter discusses the fabrication methods of wood flour based composites, and its properties in detail. From the analysis it is observed that physical and mechanical properties of the composites are greatly influenced by the fibre loading and the concentration of wood flour. The practical applications of these composites are also discussed at the end of the chapter.

Keywords

Wood Polymer Composites, Wood Flour, Eco-Friendly, Bio-Degradability, Thermoset Composites

Contents

2.1   Introduction

In our day-to-day life wood plays a significant role and it has historically been, and indeed remains today, one of the most widely used structural materials. Wood is a naturally occurring, renewable, and bio-degradable material with excellent axial stiffness-to-weight and strength-to-weight ratios [1-3]. However wood resources are getting depleted continuously while the demand for the material is ever increasing. According to the statistics, by the beginning of the next century wood will be scarce for the whole world [4]. This situation has urged researchers to develop alternative materials. Among the various synthetic materials that have been explored, polymer based materials claim a major share as wood substitutes [5].One more ancillary materials were plant fibres and a thorough investigation of their mechanical and thermal properties, eco-friendly nature and prospects of reinforcing were made. Wood as a raw material has taken on increasing importance as one of the most essential renewable resources for meeting the growing demand for bio-energy, construction material, composites and paper [6]. There are many literatures which describe the potential of the plant fibres used as reinforcing materials [7-11]. The usage of plant fibres especially bast and leaf fibres, its properties and technical applications were discussed by several authors previously [5, 12-32]. The investigation of plant fibre and its composites was done by many researchers on a broad spectrum [33-44]. Ramesh et al. have reviewed the potential of plant fibres and its composites in detail [45]. The preparation and properties of kenaf fibre and its reinforcement in composites were reviewed [46].

Contrary to plant fibres in recent years, the use of wood flour (WF), the most common wood-based filler, in the manufacture of wood flour composites (WFCs) has been of great interest to many researchers. WF is cheaper than wood fibre and is used as filler for polymer, which tends to increase the stiffness of the composite materials. Wood and other ligno-cellulosic fibres typically have higher aspect ratio than that of WF [47]. WF is obtained from natural resources, it is available in large quantities, light, cheap, and it can be added to commercial matrices in particular amounts thus offering advantageous economically [48-50]. Although most WFC products are considerably less stiff than solid wood, adding WF can greatly stiffen, but results in a more brittle nature, compared with pure polymeric materials [51]. WF is a leftover material that has to be removed from timber mills during the earlier times. Previously some notable waste management methods like bedding, composting, combustion, gas generation and feedstock in chemical industries etc. were employed. However, the utilization of WF as filler and support in composites has picked up. The benefits of utilizing WF in composites are that the bio-based asset is easily available, cheap, sustainable, smooth, possibility of rich filling, adaptable and recyclable [52]. A drawback is the moisture absorption of the WF which has undesirable effects on the properties of composites [53]. The use of WF as raw material for making new solid products is the most favorable due to its straightforwardness in application and low costs. These advantages make this one of the most abundant yet valuable materials [54].

Tragically, the inconsistency between polar WF and non-polar polymeric grids influences the level of dispersion of fibres in the matrix and the homogeneousness of the composite material. In order to eradicate these drawbacks, various authors recommended a wide range of solutions [55-59]. The accentuation being developed of these products is generally put on the sustainability issue. Nevertheless, the current trend is, rather than just replacing the expensive material with a cheaper alternative which has scarcely adequate properties, it is highly desirable to mend the composite performance [54]. The main drawbacks of WFCs are their water sensitivity and relatively poor dimensional stability, changing characteristics with origin and the time of the harvest, and poor adhesion towards all polymers [31, 49].

2.2 Wood polymer composites

Wood polymer composites (WPCs) are wood fibre reinforced bio-composites mainly used in construction, outdoor gardening and automotive products [60]. It is evident that the overall performance of WPCs improves through careful selection of materials and manufacturing processes [61] although the decision of selecting optimized materials can be a complicated process [62]. Major types of wood reinforcements are hardwood and softwood. Hardwood trees are from deciduous and they shed their leaves once a year whereas softwood trees are from conifers and they are evergreens. Hardwoods are sturdier and denser than softwoods, but balsa wood, being less dense, is considered a hardwood. Softwood grows faster than hardwood. Some other wood byproducts like flakes, fiber and flour are also agro-squanders which are broadly used as reinforcements in composites [63]. Studies reveal that the properties of WPCs are governed by several parameters like filler and matrix characteristics, chemical interaction between fibers and matrices, the volume fraction of wood fillers, treating conditions and moisture absorption [33, 64-71]. Some note-worthy solutions like the usage of coupling agents, initial treatment of fibers and/or polymers by means of physical and chemical treatment methods were proposed by many researchers to overcome the difficulties such as compatibility between fibres and matrix, enhance mechanical properties, fibre dispersion and resistance to moisture absorption. Low thermal resistance of the wood fibres is another solid issue and their thermal stability ruptures around 200°C which is wood species and composition dependent. In order to overcome thermo-degradation a fairly low melting point matrix may be used [71-76].

Thermoplastic matrices were generally reinforced with wood fibres to form wood composites [50, 66, 77-80]. Attributes like higher lignin content in softwood composites made them to exhibit enhanced stiffness than hardwood composites [50, 80] and higher cellulose content in hardwood composites produced good tensile and impact strength and elongation comparatively [50]. A comparative study of WPCs with poly-lactic acid (PLA) and polypropylene (PP) disclosed that PLA has superior dispersion of fibers than in PP [79]. The wood fibres were reinforced in high density polyethylene (HDPE) and the composite properties were studied [78]. Detailed review of WPCs was done by several authors [31, 80-82]. WPCs were fabricated with different fibre content and the fibres were treated with an agent to enhance the compatibility. This was made to provide a better adhesion between the fibres and the matrix at the interface, improved homogeneity of the particles and to reduce the moisture absorption of the final composite. Test results pointed out that properties were influenced by the moisture content and ultimately the strength decreased with the increase in moisture content. With the addition of compatibiliser, damping index decreased for hard WPCs. Long WPCs showed more impact resistance than hard WPCs. Short term flexural creep tests were conducted to investigate the creep behaviour of WPCs [83].

A comparison study between coconut shell, barley husk and WPCs stated that the tensile strength, elongation at break and impact strength of the coconut shell and barley husk reinforced composites was superior to that of the WFCs [84]. Investigation on the effect of oil palm shell (OPS) powder size on the mechanical properties of the composites containing 75-150 mm sized OPS exhibited the highest strength with the removal of impurities during treatment playing a major role in the presence of strong interfacial bonding between filler and matrix [85]. In addition a study on the effect of different filler content of WF and wood fibres reinforced in soybean and linseed oil-based composites revealed that the maximum tensile strength was achieved for a relatively high filler load of WF [86].

2.3 Wood flour composites (WFCs)

WF is becoming increasingly important since it offers several benefits which include low density, renewability, bio-degradability, noise absorption, minimum damage during processing, etc. [87]. Like other plant fibres, WF is extremely hygroscopic, meaning it absorbs a fair amount of moisture when being exposed to a wet environment. Thus such degraded WFCs experience a long term effect of debonding due to many internal defects like microbial attack, swelling and oxidation. All these factors have a strong unfavorable effect on the material integrity and its mechanical properties [69, 88-90]. Recently, a composite manufactured with WF has been developed for effective utilization of waste wood [91-93]. The WFC does not affect the surrounding because it is composed of only wood particles which are a natural resource. The WF products with complicated shapes can be formed by compression molding at the appropriate temperature. However, the composite which is produced by solidifying only the WF without the binder does not have high strength, and it is brittle and has poor water resistance. Therefore, to improve these defects, a bio-degradable resin is used as an adhesive [94]. The degradation of WFCs was studied by several authors and reported a reduction of tensile strength and modulus and impact strength and an increase of the elongation and crystallinity rate according to the successive processing cycles of composites [95].

There are many advantages of using WF rather than synthetic fillers in composites. WFs have greater deformability when compared to synthetic fillers, which result in less damage during processing and less abrasiveness to equipment [96]. Since, WF can contain a lot of moisture it needs to be dried before it is reinforced with a matrix. The rollers and brushers are used to produce smooth surface finish for good appearance and anti-slip purposes. One advantage of WF over that of conventional powders, such as talc, is the lower density. There is also a high level of interest in using non-wood cellulose fibres derived from agricultural feed stocks such as flax, hemp, sugarcane fibres, kenaf, and peanut shells. Although some of these fillers are more expensive than WF, the longer aspect ratio can lead better properties in the finished product. From an economical point of view, wood-based particles come from a renewable resource and are typically less expensive than synthetic or conventional powders [97].

2.3.1 Processing of WFCs

WFCs were prepared and the effects of various factors such as WF particle size and its concentrations, effect of coupling agent and impact modifier on the properties of the composites were evaluated [98]. First WF was dried in an air circulated oven over 50°C around 12 h to moisture content of 20-30% based on the oven-dry weight of the particles. Following the drying, the WFs were then processed by a rotary grinder without adding additional water. Ultimately the WF stays back in a 60-mesh screen after being ground to fine powder. Then the moisture content (1-2%) was discarded from the WF by desiccating it in an oven at 100°C for one day before using it in fabrication of composites [99]. The WF, matrix, and coupling agents were processed in a 30-mm co-rotating twin screw extruder with a length-to-diameter ratio of 30:1. The barrel temperatures were controlled at 170, 180, 185, and 190°C for different zones. The temperature of the extruder die was held at 200°C. The extruded samples passed through a water bath and were subsequently pelletized. The pellets were stored in a sealed container and then dried to the moisture content of 1-2% in an oven before processing. The temperature used for processing the samples was 180-200°C from feed zone to die zone. The samples were processed at the pressure between 4-5MPa with cooling time of about 20 s. Then, the samples were conditioned at 23°C and relative humidity of 50% according to ASTM D 618 standards [99, 100].

WF were blended with coupling agent and matrix in a high velocity mixer after being oven dried at 80°C for 24 h. Various quantities of WFs, required quantity of polypropylene (PP) along with the coupling agent were mixed together to prepare WPCs. Then the granules of WPCs were again oven dried, before moulding, at 80°C for 24 h [83]. For preparation of WFCs, a twin-screw extruder with aspect ratio of 40, diameter of screw as 2 mm, a screw speed of 150 rpm and a material input (Q) of 1 kg/h was set and used. A temperature of 180°C was set between hopper and die as process temperature. Before processing the PP, WF and fire retardant in an extruder, they were homogenously mixed. As per JIS K7113 standards an end-gated mould was used in a screw injection-moulding machine to prepare the tensile bars which were dumbbell-shaped at a temperature of 210°C [101].

The addition of WF was started 5 min after reaching a constant torque. The WF was added in three equal portions in 2 min intervals. Kneading was continued until a constant torque was reached, in a time interval of 10 to 25 min. The temperature was increased to 150°C after few min and the mixture was allowed to react for 20 min. The WPC were removed from the kneader and after cooling cut into granules of about three mm in radius [102]. The compounding of WFC was carried out in an extruder in the processing temperature range of 190-230°C. The compounded samples were prepared by an injection molding machine [63]. Gravimetric-type material feeder fitted extruders were used to prepare the composites. A twin-screw slide feeder was to enforce the WF into a polymer melt while a peristaltic pump was used to feed the silane solution. In order to prevent the evaporation of water and other reactants in the melt, all atmospheric-pressure ventilations and vacuum ventilations were made impassable. Peroxides disintegrate as radicles when the silane solution passed in through the extruder thus splicing the silane into the composites. Apart from the rate of decomposition of dicumyl-peroxide the extruder temperature also depended upon the ability to attain sufficient compounding of the material. For a residence time of 55-60 s, the speed of the screw was maintained at 155 rpm and the range of temperature was 180°C-200°C. Almost 97% of dicumyl-peroxide was decomposed, which can be approximated theoretically to five half-life times if an actual process temperature for melting was kept around 195°C. The composites, immediately after processing, were tested for their ability of cross-linking. Cross-linking was prevented by the low temperature in the freezer as hydrolysis was slowed down [103].

2.3.2 Properties of WFCs

Injection moulding process and extrusion was used to prepare the WFCs. Investigation of mechanical and thermos-dynamic properties were done with WF loading as a function. Characterization techniques like scanning electron microscopy (SEM) and differential scanning calorimetry (DSC) were used to inter-relate the morphology of the fiber/matrix interface with the effects of fibre-silane thermo-chemical vapor deposition treatment for WF and a maleic anhydride (MA) graft co-polymerization technique. Theoretical models for the stiffness of the composites like Halpin–Tsai/Tsai–Pagano micro-mechanical model was used to compare the results in order to investigate the MA enforcement into fibre/matrix interface and the induced phenomenological effects of silane coupling agent [104].

2.3.2.1 Mechanical properties

A possible solution to enhance the mechanical properties of natural fibre reinforced composites, suggested by several researchers is the hybridization of fibres with inorganic fillers [105–111]. The mechanical properties of WPCs may not be significantly affected by fibre type [112], but it has also been reported that the type of fibre and the lignin, cellulose and hemicelluloses content can have a strong influence on the mechanical properties [113]. Addition of WF enhanced the mechanical properties of composites, but simultaneously it increased the burning speed of the materials [101]. The mechanical properties such as tensile strength, elongation at break, toughness, fracture energy, and Young's modulus of the WPCs were measured with an Instron tester (Model:4201) [98], Zwick/Roell model Z010 [100], universal testing machine (UTM) [63, 114] following ASTM D638 [63, 98, 100], ISO 527-2 [114] standards with the speed of cross head as 10 mm/min [63, 100]. The standard dimension is 20 mm length, 12.5 mm width and 3 mm thickness [63]. The impact strength of the specimen was tested by using an impact tester (Model: TMI 43-01) [98] following ASTM D 256 [100] standards. According to ASTM D 790 [100, 115] and ASTM D 7264 [116] standards, the flexural tests were carried out at room temperature with a load capacity of 10 kN in Zwick/Roell model Z010 UTM. According to ASTM D 790-86 standard, a specimen of dimension 70 × 12 × 3 mm were prepared for a three point bending test and compression test conducted at room temperature in UTM with a cross head speed of 1 mm/min and 0.5 mm/min respectively [115]. A cross head speed of 2 mm/min, support span of 140 mm, a square plate specimen of dimensions 200 mm × 30 mm × 10 mm were the requirements and parameters for a four point flexural test. For evaluating Young's modulus through rate of strain, strain gauges were used. A constant temperature of 23±1°C and a relative humidity of 50 ± 5% were considered while taking all mechanical measurements [114].

Marcovich et al. [69] investigated the mechanical properties of WFCs on the moisture content at 60 and 90% relative humidity. Results revealed that moisture content drastically reduced them but at the same time as the actual values of the compression recuperated after drying the effects of moisture content was reversible. To measure the impact characteristics values, the specimens were tested by using a low-velocity falling weight impact tester at room temperature in non-penetration mode [83]. WFCs were prepared and tested for its mechanical and dynamic mechanical properties like elasticity modulus, ultimate tensile strength and maximum strain [115]. The longitudinal and transverse strains were measured using a 50 mm gauge length extensometer and a strain gauge, respectively [104]. Improved interfacial adhesion was the convergence point for all results.

The tensile strength, tensile modulus and strain at break of the composites were analyzed. The product of cross head speed and time gave displacement which was used to calculate the rate of strain [103]. The prepared specimens had an approximate aspect ratio of 2 and were square in shape. Molybdenum sulfide wax was used to lubricate the precisely

Thermoset Composite

Materials Research Forum LLC

Materials Research Foundations 38 (2018)

doi: http://dx.doi.org/10.21741/9781945291876

machined parallel faces of the specimen [115]. Fig. 1 shows the variation of flexural properties of WF composites with respect to time of exposure. Graph depicts the huge fall of strength and stiffness with respect to time for all compositions of WF. The interfacial adhesion between the fibre and matrix was majorly reduced due to hydrophilic nature of the composites. This indirectly affects the stress transfer between the fibre and the matrix due to poor interfacial bonding. When a newer bond forms between cellulose and water molecules, the intra-molecular hydrogen bonding was diluted [117].

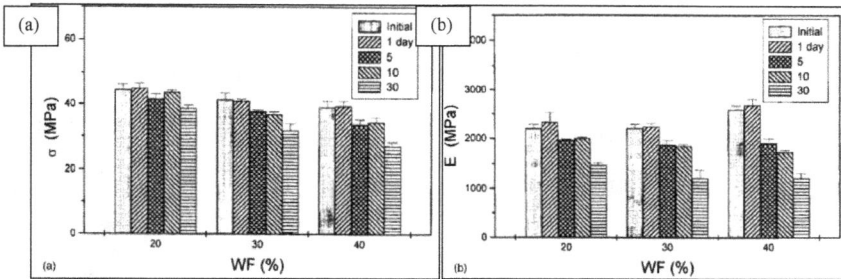

Fig. 1 Variation of (a) flexural strength and (b) flexural modulus of WFCs.

2.3.2.2 Surface roughness and wettability

The surface roughness and wettability of composites filled with WF have been evaluated [99]. It could be observed that the surface roughness values of WFCs were inversely proportional to the polymer content. The composites without the MA grafted PP were found to have higher surface roughness but better wettability as compared with the ones with the MA grafted PP. The wettability of the composites increased with increasing WF content. The incorporation of the coupling agent decreased the wettability of the composite specimens compared with untreated ones. The test result showed that WF could be utilized in the production of the filled composites because of the surface properties of the composites. The effect of WF loading on the surface roughness of the WF filled composites is presented in Fig. 2 [99]. It was found that WFCs had higher surface roughness in the absence of MAPP and PP when compared with the ones having them. The wettability of the composites increased with increasing content of the WF is shown in Fig. 3 [99].

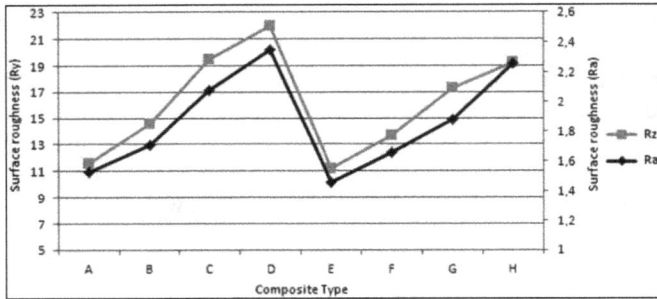

Fig. 2 Effect of the WF loading on the surface roughness of the unfilled and WF filled composites.

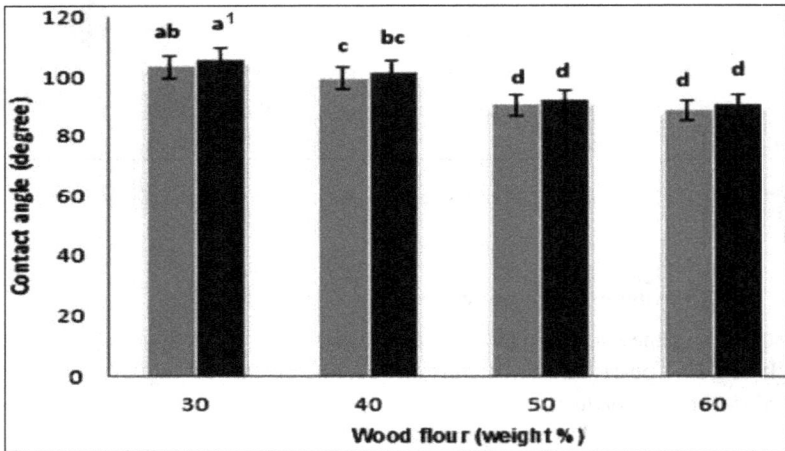

Fig. 3 Effect of the WF loading and coupling agent on the wettability of the unfilled and WF filled composites.

2.3.2.3 Water absorption tests

ASTM D 570 standard was followed for water absorption tests according to which the specimens were oven dried at 105°C for 24 h. At a normal temperature of 23±2°C the specimen were submerged in water for 2 h after which they were wiped off with a cloth to remove the superficial water and then weighed. This process was repeated after 2 h again and weighed after 24 h [116].The results of water immersion tests of MAPP and PP WF composites for 2 h and 24 h followed same pattern of increase in water absorption as the content of WF increases due to their hydrophilic nature. But the results were just opposite for the wood samples because of its hydrophobic nature whereas the water absorption could be more due to the presence of more spots as wood content increases. Attributes like presence of lumens, hydrogen bonding sites and the interfacial gap between reinforcement and matrix were the main reason for water absorption in WF composites [118].

Immersion tests were conducted on rectangular specimen of dimension 35 × 12 × 3 mm to determine the moisture content as per DIN 52375 [83], ASTM D 570-81 [115] standards. At least three specimens for each sample were taken for the test and their volume and moisture content was determined [83]. Within a time of 24 h different readings of weight gain were taken from the specimen soaked in distilled water at room temperature before which they were conditioned to reach a constant weight. When the readings reach saturation by giving the same value in three consecutive measurements the final weight gain was calculated [115]. Fig. 4 shows the relationship between the water absorbed and percentage of untreated and MAPP treated WFCs with respect to the time of their presence in hygro-thermal chamber. It could be observed that for higher fiber content the water absorption is more articulated [117].

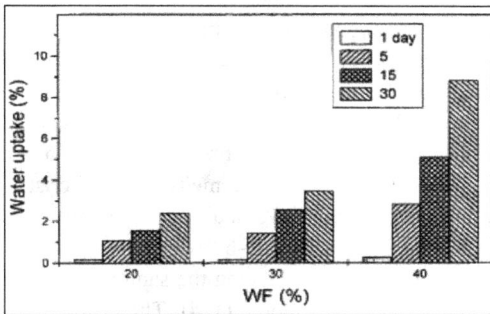

Fig. 4 Water uptake of WFCs.

2.3.2.4 Thermo-gravimetric analysis (TGA)

TGA is used to investigate the thermal behavior of the WF, matrix and their resulting composites. The tests were conducted with 15 mg mass of sample under a flowing nitrogen atmosphere with a heating rate of 10°C/min between 35°C and 700°C according to ASTME1131-08 standards [63, 100, 102, 119]. TGA/DTA measurements were performed on a Netzsch STA 409 TG analyzer. A sample was cut into small pieces and conditioned in vacuum at room temperature for 24 h. The thermal degradation of WF occurs over a wide temperature range in both oxygen and nitrogen atmosphere. But in contrast to the polymer matrix, some mass loss of WF in nitrogen occurs in the temperature range between 25°C and 150°C, 150°C and 250°C and between 400°C and 600°C while the mass loss of the polymer matrix occurs mainly between 300°C and 500°C. Wood degradation in the temperature range from 200°C until 350°C is assigned to hemicellulose and cellulose degradation and from 250°C up to 500°C to lignin degradation [53]. Wood and polymer matrix degradation overlap between 200°C and 350°C which means that step separation for wood and matrix could not be achieved. However, when WF was degraded between 200°C and 390°C, then the matrix was degraded between 390°C and 500°C, and finally, by changing the inert atmosphere to an oxidative atmosphere, the char of WF burnt residue-free. With the atmosphere change from nitrogen to oxygen for both ingredients WF and polymer, a complete thermal degradation is ensured [120].

2.3.2.5 Differential scanning calorimetry (DSC)

The melting and crystallization behaviors of WFCs were assessed through DSC using a Perkin Elmer [104, 121], Mettler-Toledo DSC 821 calorimeter [102] apparatus equipped with a cooling attachment, under a nitrogen atmosphere. The composite samples were heated from -50°C to 185°C at the rate of 3°C/min before which 10-15 mg of the samples were preheated to 50°C for 5 min. Afterwards the samples were glazed 20°C/min to room temperature [104]. Two heating steps interspersed with a cooling step from 20°C to 210°C at a constant rate of 10°C/min were carried out. The samples were analyzed in standard aluminium DSC pans [121]. Thermal properties of the composites like fusion/crystallization enthalpies H_c and H_f and melty (T_f) and cristally (T_c) temperatures were examined using a non-isothermal DSC analyzer. Alternative heating and cooling of the samples were done between 210°C and -60°C in two stages; heating to 210°C and cooling them to -60°C as first stage and heating the samples again from -60°C to 210°C as second stage, at a rate of 10°C per minute [114]. The heating process was performed twice; glass transition temperatures were obtained from the second run [102].

Fig. 5 shows the DSC thermograms in which the development of double melting peaks, transference of the melting temperature and abridged heat of melting depicts the imperfect polymer crystal structure when WF is impregnated into the matrix. During the transition from molten to crystalline state WF enables the fast formation of transcrystalline structures along the fibre surfaces which portrays WF as good nucleating agents. Due to WF impingement on bulk formed crystals and disordered crystal growth at the interface of matrix/fibre, some discontinuities originate in the matrix crystal structure which led to the reduction in whole polymer crystallinity [104].

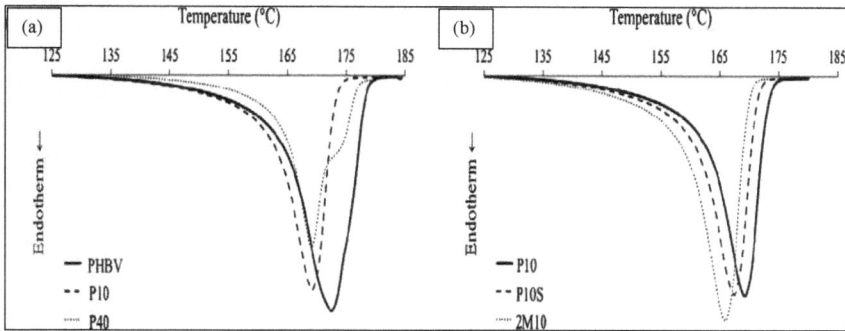

Fig. 5 DSC thermograms of WFCs.

2.3.2.6 Dynamic mechanical tests (DMA)

Dynamic mechanical measurements were performed by means of a Perkin-Elmer dynamic thermal analyzer. DMA were performed using a rheometer in torsion mode according to DIN EN ISO 6721-1 standards. A temperature ramp was driven from 0 to 60°C at a heating rate of 1°C/min. For determination of flow activation energies the measurement was performed on five different frequencies of 0.10, 0.316, 1.00, 3.16 and 10 Hz respectively [102]. The test parameters were 15 mm length of specimen platform, static and dynamic stress of 3×10^5 and 5×10^5 Pa respectively, forced oscillation frequency of 1 Hz [115]. A three point bending fixture was used testing in temperature scan mode. The effect of the addition of sisal fibres with bisphenolic matrix is shown in Fig. 6. Storage modulus of the material mostly goes in hand cohesion density at room temperatures which is due to the effect of washing the fibres with acetone. Material cohesion improvement and high storage modulus at room temperature were the major advantages of the mechanical mooring of the resin on fiber surface after washing which also coincides with the results of water absorption by the composites at room temperature [115].

2.3.2.7 Creep test

The creep tests were performed according to EN ISO 899-2 standards. Injection molded flexural specimen from the composites with dimensions 60 x 12 x 2.5 mm were subjected to creep test in a dynamic mechanical analyzer at constant stress in dual cantilever mode [103]. An insulated chamber with a mini fan and light bulb setup, for heating and air circulation, was used for the temperature testing of the creep devices. The temperatures used for testing were 23, 40 and 60°C and these are controlled by a temperature controller. By rising the time of exposure to 180 min, the creep strength and modulus were measured in terms of the deformation of the composites [83]. By application of a static stress of 5 MPa at 30°C for 300 min the initial stage of test was done then as a next stage the creep cycling test with a stress of 2 MPa at 60°C for 60 min is carried out to evaluate the creep strain with respect to time [103]. These stages of tests were repeated and after each test, a recovering time of 60 min was provided to the composites.

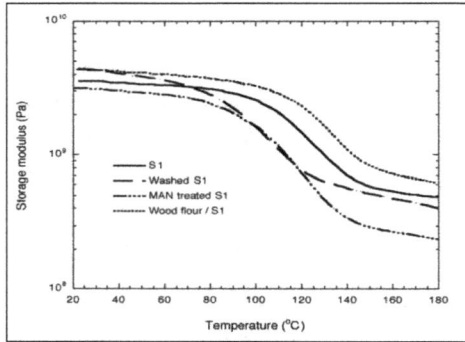

Fig. 6 Storage modulus vs. temperature of WFCs.

Cross-linked composites shows better results in primary and steady state creep regions in the first short term creep test results as shown in Fig. 2.7a. The primary stage of the composites contains lower creep strain region in the creep strain curve of the dry WF cross-linked samples when compared with wet WF samples whereas no major variation of the creep strain for the materials that underwent same storage mode in the steady state region. But when the creep strain rate of the composites stored in a thermal bath and room temperature were compared, the latter exhibited higher strain rate in steady state. The main reason for the higher primary creep strain was the poor interaction between the polymer and wood of X-dry composites due to which the tensile strength of X-wet composites were higher than the counterparts. The main indication of the primary creep strain and tensile strength was that the X-wet composites experienced cross linking majorly in matrix and lesser at the interface. Besides this the same property was equally distributed among matrix and interface for the X-dry composites. Results obtained from the primary creep test at 30°C and creep cycling test at 60°C of cross-linked samples were coinciding with respect to low creep response as shown in Fig. 7b [103].

Identical degrees of cross-linking was observed in X-wet SA and X-dry RT composites but they stand apart in creep strain rates measured by creep test modes which indicates the differences in creep cycling for various storage mode of cross linked composites. A difference in impact due to higher temperature, on the samples at ambient temperature, was observed in such a way that it was greater during primary creep strain and smaller at the beginning of steady state creep in creep cycling strain curves. So, this depicts the change in structure of the matrix at higher temperatures with respect to its storage mode. Major portion of the amorphous phase in the polymer matrix of the composites, which were placed in the thermal bath, experienced hydrolysis and high quantum of cross linking. Thus, when the composites were subjected to creep test the cross-linked samples recovered quickly after each stress cycle release while the non-cross-linked composites had residual deformations. Bengtsson and Oksman concluded that all the composites other than the non-cross-linked ones portrayed low creep responses [122-124].

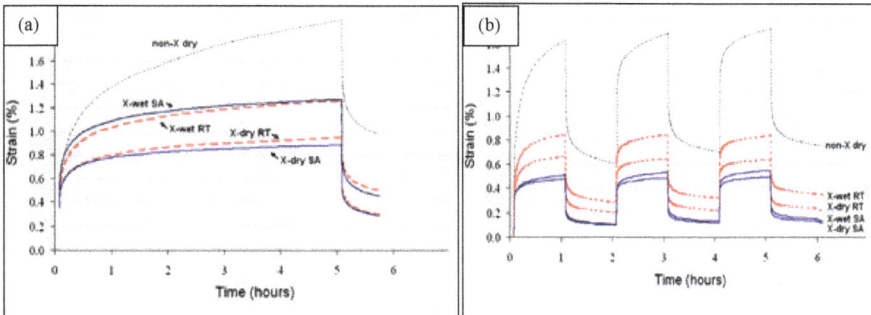

Fig. 7 Creep strain curves for the composites at a stress of (a) 5 MPa at30°C; and (b) 2 MPa at 60°C.

2.3.2.8 Flammability characteristics

The flammability characteristics of the WFCs specify the fire performance of the material that is evaluated and classified during compounding process by a flammability test [125]. Effective flame retardants like chlorine and bromine halogenated compounds may be used but as they give out toxic gases during the reaction they should be avoided. Studies by Sain et al. [125] showed that due to the addition of fire retardants, only a marginal drop in mechanical properties were found and to bring down the burning rate of WPCs to 50% without usage of flame retardant 25% magnesium hydroxide was effective. According to Gracia et al. [126] self-extinguish materials were developed from polyethylene based composites due to the addition of fire retardants when reagents of hydroxide or phosphate were used. Experiments by cone calorimetry conducted by Stark et al. [127] arrived at a result that phosphate or hydroxide and various other fire retardants could be used in WFCs.

The improvement in flammability behaviour of WFCs by the use of three types of fire retardants like melamine polyphosphate (MPP), aluminium hydroxide and phosphate was done [101]. From the investigation of WF addition upon the flammability behaviour of the composites, it was found that their addition speeded up the burning rate of the composites. Due to the synergy effect of the WF and fire retardant, the self-extinguishing property could be reached by smaller quantities of fire retardants as per the amount of WF added. In order to study the least quantity of fire retardant required and its effectiveness for improving the fire performance of WFCs, a burning test was conducted. Similarly, tensile tests and cone calorimetry tests were conducted to evaluate the effect of the fire retardant on the strength of the composites and to deduce the burning occurrence of the WFCs with fire retardant respectively [101].

The influence of fire retardants on the composites flammability was studied by two types of burning tests such as horizontal and vertical tests. Before the test, the specimen were dried at 80°C for 24 h. When the sample was positioned horizontally and when one of its ends was exposed to a natural gas fueled flame for 20 s then it is called a horizontal test. Similarly, when the sample was positioned vertically and when one end of the sample was exposed to the flame for 10 s then we called it a vertical test. The height of the flame and the angle of the flame for the horizontal and vertical tests were 10 mm, 30° and 20 mm, 30° respectively. Burning time, which was measured, indicates the time of the flame to reach the second reference point from the first which are 80 mm apart. In order to study the effect of dripping of the specimen from the fire source, a wire sheet was held at a distance of 10 mm under the specimen. In case the specimen is self-extinguished or incombustible in the horizontal burning test, then it was subjected to the vertical burning test. Class of the sample was determined by the sample dripping, its burning state and combustion time [101].

2.3.2.9 Tomography

In order to visualize WF dispersion and their size, composites samples were examined using a high-resolution computed tomography system. The dimension of examined volume of all samples were $(3.6–3.8) \times (2.8–3.8) \times (0.94–1.43)$ mm^3 [114]. Tomography is anon-destructive imaging technique which renders the visualization of WFs dispersion across the matrix. Even though the tomography technique is not very suitable for composites, many earlier studies had used it in order to understand the mechanical properties of the composites, with 20%, 30% and 50% of wood content. By using tomography the sample volume can be reconstructed in three dimensions. The tomography images with 30% of wood content are presented in Fig. 8 [114]. Homogenous dispersion of the fibers into the matrix and the matrix and particles had no voids were observed from the images. These are due to good interfacial interaction for better mechanical properties and symptoms of appreciable cohesion amongst composites respectively.

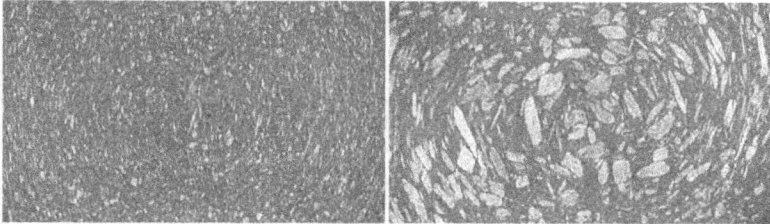

Fig. 8 Tomography pictures of WFCs.

2.3.3 Scanning electron microscopy (SEM) analysis

The fractured surface of WFC samples was examined using an environmental scanning electron microscopy (ESEM). To study the morphology of the interface and the wood-matrix interaction ESEM was used [114]. Surface morphology of the composites was examined at an accelerated voltage of 25kV without coating metallization of the composites which is considered to be the merit of ESEM. The image of the specimen were recorded using a FEI XL 30 Sirion [104], Zeiss Evo LS 10 [100], SEM-EDX Philips [63], JEOL 5500 LV [128], from which the dispersion of the particles, interfacial adhesion between fiber and matrix, and the filler morphology could be studied. The samples were fitted to a mount and coated with a DAG T-502 carbon paint using a sputter coater for 45 s operated at a current of 20 mA [104]. Then along the direction of injection mould fill flow the specimen was fractured in longitudinal and horizontal directions after freezing them with liquid nitrogen.

Fig. 9 shows cross sectional morphology of WFCs visualized through SEM micrographs. It could be seen that neither individual silver and zinc oxide micro or nano-particles nor their combination seemed to improve the poor surface adhesion of WF with the matrix. It could be observed that the process of compounding did not alter the morphology of the fillers and components and the silver and zinc oxide particles were stuck to the WF particles surface. At the same time, during compounding, antibacterial property of the material could be improved by the positive effect of non-agglomeration of silver and zinc oxide which were unwrapped from the surface and dispersed into the matrix [54].

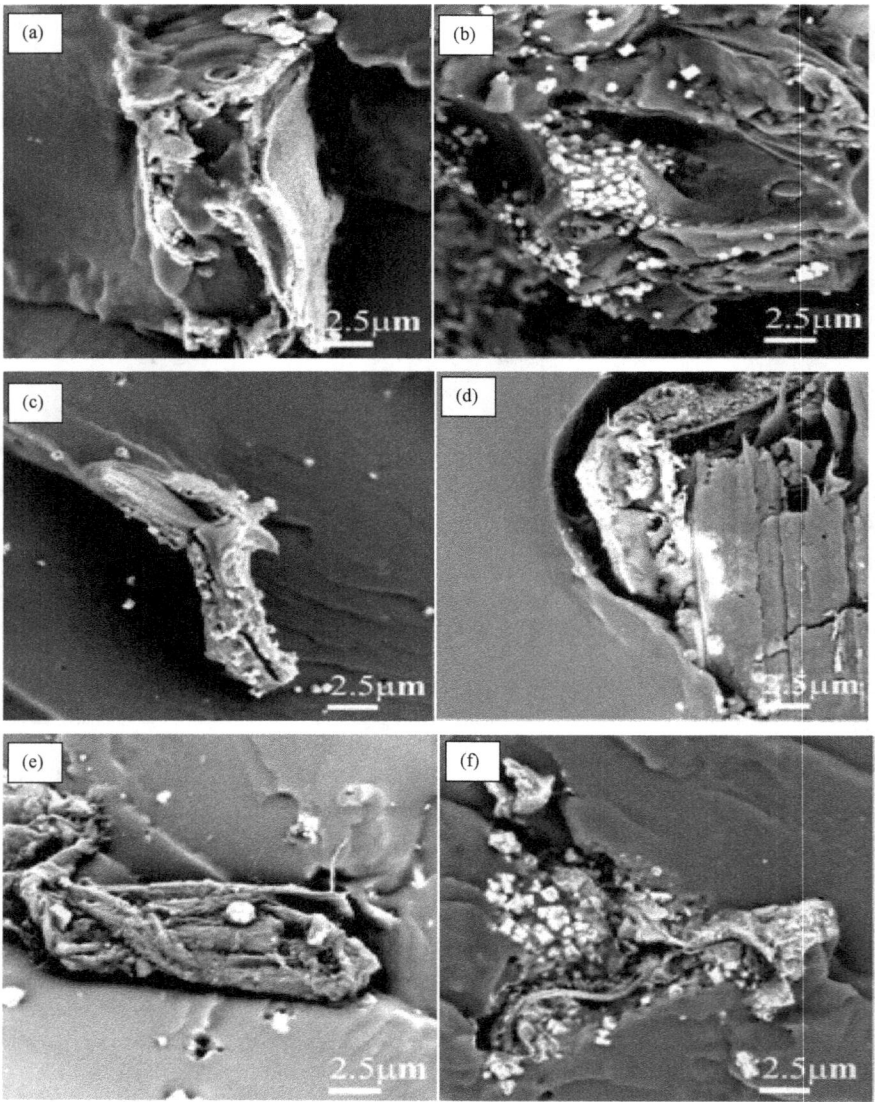

Fig. 9 SEM micrograph of WF/PVC composites (a) WF-Ag/PVC; (b) WF-ZnO/PVC; (c) WF-Ag/ZnO/PVC; (d) WF-Ag-AA/PVC; (e) WF-ZnO-AA/PVC; (f) WF-Ag/ZnO-AA/PVC.

SEM micrographs of fractured surfaces of tensile test specimens of WFCs are shown in Fig. 10. High content of WF is obvious in Fig. 10 (a) and pull out of particles is observed which indicates a relatively poor bonding between the WF and the matrix. Fig. 10 (b), on the other hand, shows that both the higher aspect ratio and better interfacial adhesion between the fibre and matrix are the reasons for improved mechanical strength of the WFCs. The fibre pull out and fracture can be easily seen in the micrograph. This is an indication of strong interfacial bonding between the fibre and the matrix [63].

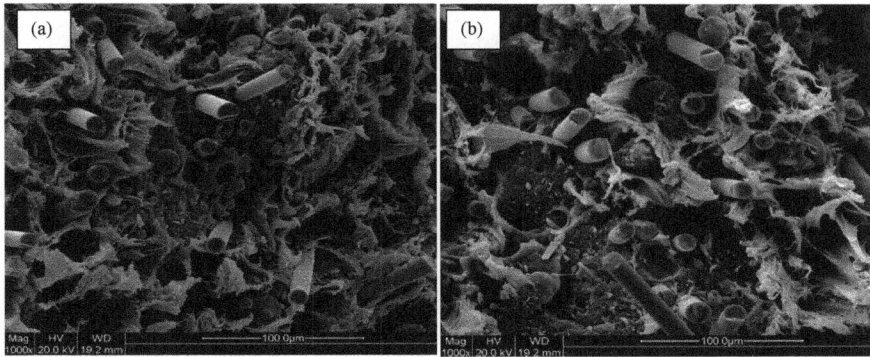

Fig. 10 SEM micrographs of cracks developed in the WFCs.

2.4 Practical applications

Practical applications of WFCs are highly depends on the hydrophilic nature of cellulose, which causes dispersion and moisture absorption problems. The major areas of applications of WFCs is the construction, aerospace and automobile industries [45, 48, 129, 130], but they are also applied for packaging, for the preparation of various household articles, furniture, office appliances and other related items [48, 129]. In most of these applications, they are used as structural materials, where the load-carrying capacity of the dispersed component is playing a major role. This is determined by the particle characteristics of the reinforcing materials and by interfacial adhesion between the reinforcement and matrix [131-133].

Conclusions

WFCs comprising of wood powder and thermoset or thermoplastic resins were fabricated by using different fabrication methods and the effect of WF loading is investigated in this chapter. The various properties of these WFCs are also discussed in an elaborative manner. From the investigation the following conclusions have been made:

- From this investigation long WPCs are significantly more hygroscopic than the hard WPCs which is close to two times more at wood fiber content of 60 wt.%.

- The impregnation of WF with thermoset compounds before compounding was successfully made and used for fabrication of the composites.

- The properties of WFCs can be improved by treating the fibres with different chemical reagents and also by using a suitable composite fabrication method.

- The techniques used to enhance compatibility improved the distribution of fibers and wettability and also increased the properties by improving interlocking between matrix and fibers.

- The interfacial quality of the matrix and filler resulted in better properties and effective load transfer occurred between matrix and filler.

- Improvement in properties and their variation were due to homogenous fibers dispersion and difference in WF particle size. This analysis was made by X-ray tomography.

- The creep strength and modulus of WFCs increased with increasing of WF and decreases with increasing temperature.

- The reinforcement of WF in composites led to a raise in material stiffness and to diminish in strength and elongation-to-break.

- The addition of WF in polymer composites produces elevated maximum stresses and retained the typical property of the non-linear stepwise breakage of fibrous composites. Thermal and fire properties are also improved.

References

[1]　J.M. Dinwoodie, Timber: It's Nature and Behaviour, second ed. E & FN Spon, New York, 2000. https://doi.org/10.4324/9780203477878

[2]　L.J. Gibson, M.F. Ashby, Cellular Solids: Structure and Properties, Cambridge University Press, 1997. https://doi.org/10.1017/CBO9781139878326

[3]   J. Bodig, B.A. Jayne, Mechanics of Wood and Wood Composites, Van Nostrand Reinhold, New York, 1982.

[4]   S.P. Singh, Agro-industrial wastes and their utilization, In: Proc of National Seminar on Building Materials and their Science and Technology, Roorkee, v.15, p.111, 1982.

[5]   K. Joseph, R.D.T. Filho, B. James, S. Thomas, L.H. de Carvalho, A review on sisal fibre reinforced polymer composites, Revista Brasileira de Engenharia Agricola e Ambiental, 3(3) (1999) 367-379. https://doi.org/10.1590/1807-1929/agriambi.v3n3p367-379

[6]   R.M. Rowell, Handbook of Wood Chemistry and Wood Composites, second ed., Boca Raton. FL: CRC Press, 2012. https://doi.org/10.1201/b12487

[7]   A.K. Mohanty, M. Misra, L.T. Drzal, Natural Fibres, Biopolymers, and Biocomposites, CRC Press: Boca Raton, FL, 2005. https://doi.org/10.1201/9780203508206

[8]   F.T. Wallenberger, N.E. Weston Natural Fibres, Plastics and Composites, Springer: NY, 2004. https://doi.org/10.1007/978-1-4419-9050-1

[9]   R.S. Blackburn, Biodegradable and Sustainable Fibres, Elsevier; Amsterdam, 2005.

[10]   R.R. Franck, Bast and other Plant Fibres; CRC Press: Boca Raton, FL, 2005. https://doi.org/10.1533/9781845690618

[11]   S. Thomas, L.A. Pothan, Natural fibre reinforced polymer composites: From macro to nanoscale. In: Archives Contemporaines. Old City Publishing: Philadelphia, PA, USA, 2009. https://doi.org/10.1504/IJMPT.2009.027839

[12]   L. Yan, N. Chouw, K. Jayaraman, Flax fibre and its composites: A review, Compos. Part B: Eng. 56 (2014) 296–317. https://doi.org/10.1016/j.compositesb.2013.08.014

[13]   A. Shahzad, Hemp fibre and its composites: A review, J. Compos. Mater. 46 (2012) 973–986. https://doi.org/10.1177/0021998311413623

[14]   A.K. Mohanty, M. Misra, Studies on jute composites: A literature review, Polym-Plast. Technol. Eng. 34 (1995) 729–792. https://doi.org/10.1080/03602559508009599

[15]   H.M. Akil, M.F. Omar, A.A.M. Mazuki, S. Safiee, Z. Ishak, A. A. Bakar, Kenaf
       fibre reinforced composites: A review, Mater. Des. 32 (2011) 4107–4121.
       https://doi.org/10.1016/j.matdes.2011.04.008

[16]   Y. Li, Y.W. Mai, L. Ye, Sisal fibre and its composites: A review of recent
       developments, Compos. Sci. Technol. 60 (2000) 2037–2055.
       https://doi.org/10.1016/S0266-3538(00)00101-9

[17]   D.S. Varma, M. Varma, I.K. Varma, Coir fibres. Part I. Effect of physical and
       chemical treatments on properties, Text. Res. J. 54 (1984) 827–832.
       https://doi.org/10.1177/004051758405401206

[18]   C.G. Jarman, Banana fibre: A review of its properties and small-scale extraction
       and processing, Trop. Sci. 19 (1977) 173–185.

[19]   M.A.S. Spinace, C.S. Lambert, K.K.G. Fermoselli, M.A. De Paoli,
       Characterization of lignocellulosic curaua fibres, Carbohyd. Polym. 77 (2009) 47–
       53. https://doi.org/10.1016/j.carbpol.2008.12.005

[20]   R.V. Silva, E.M.F. Aquino, Curaua fibre: A new alternative to polymeric
       composites, J. Reinf. Plast. Compos. 27 (2008) 103–112.
       https://doi.org/10.1177/0731684407079496

[21]   J.S. Caraschi, A.L. Leato, Characterization of curaua fibre, Molecul. Cryst. Liq.
       Cryst. 353 (2000) 149–152. https://doi.org/10.1080/10587250008025655

[22]   S. Mishra,A.K. Mohanty, L.T. Drzal, M. Misra, G. Hinrichsen, A review on
       pineapple leaf fibres, sisal fibres and their biocomposites, Macromol. Mater. Eng.
       289 (2004) 955–974. https://doi.org/10.1002/mame.200400132

[23]   D.S. Varma, M. Varma, I.K. Varma, Coir fibres II: Evaluation as a reinforcement
       in unsaturated polyester resin composites, J. Reinf. Plast. Compos. 4 (1985) 419–
       431. https://doi.org/10.1177/073168448500400406

[24]   D.H. Mueller, A. Krobjilowski, New discovery in the properties of composites
       reinforced with natural fibres, J. Ind. Text. 33 (2003) 111–130.
       https://doi.org/10.1177/152808303039248

[25]   S. Shinoj, R. Visvanathan, S. Panigrahi, M. Kochubabu, Oil palm fibre (OPF) and
       its composites: A review, Ind. Crop. Prod. 33 (2011) 7–22.
       https://doi.org/10.1016/j.indcrop.2010.09.009

[26]   D. Verma, P.C. Gope, M.K. Maheswari, R.K. Sharma, Bagasse fibre composites:
       A review, J. Mater. Environ. Sci. 3 (2012) 1079–1092.

[27] Y.R. Loh, D. Sujan, M.E. Rahman, C.A. Das, Sugarcane bagasse-The future composite material: A literature review, Resour. Conser. Recycl. 75 (2013) 14–22. https://doi.org/10.1016/j.resconrec.2013.03.002

[28] N.J. Smith, G. Junior Virgo, V.E. Buchanan, Potential of Jamaican banana, coconut coir and bagasse fibres as composite materials, Mater. Character. 59 (2008) 1273–1278. https://doi.org/10.1016/j.matchar.2007.10.011

[29] S.N. Monteiro, R.J.S. Rodriquez, M.V. De Souza, J.R.M. d'Almeida, Sugar cane bagasse waste as reinforcement in low cost composites, Adv. Perform. Mater. 5 (1998) 183–191. https://doi.org/10.1023/A:1008678314233

[30] D. Liu, J. Song, D.P. Andersen, P.R. Chang, Y. Hua, Bamboo fibre and its reinforced composites: structure and properties, Cellulose 19 (2012) 1449–1480. https://doi.org/10.1007/s10570-012-9741-1

[31] A.K. Bledzki, S. Reihmane, J. Gassan, Thermoplastics reinforced with wood fillers: A literature review, Polym-Plast. Technol. Eng. 37 (1998) 451–468. https://doi.org/10.1080/03602559808001373

[32] J. Mussig, Industrial Applications of Natural Fibres: Structure, Properties and Technical Applications; Wiley: NJ, 2010. https://doi.org/10.1002/9780470660324

[33] F.P. La Mantia, M. Morreale, Green composites: A brief review, Compos. Part A 42 (2011) 579–588. https://doi.org/10.1016/j.compositesa.2011.01.017

[34] A.K. Mohanty, M. Misra, G. Hinrichsen, Biofibres, biodegradable polymers and bio-composites: An overview, Macromol. Mater. Eng. 276-277 (2000) 1–24. https://doi.org/10.1002/(SICI)1439-2054(20000301)276:1<1::AID-MAME1>3.0.CO;2-W

[35] G. Koronis, A. Silva, M. Fontul, Green composites: A review of adequate materials for automotive applications, Composites: Part B 44 (2013) 120–127. https://doi.org/10.1016/j.compositesb.2012.07.004

[36] H. Ku, H. Wang, N. Pattarachaiyakoop, M. Trada, A review on tensile properties of natural fibre reinforced polymer composites, Composites: Part B 42 (2011) 856–873. https://doi.org/10.1016/j.compositesb.2011.01.010

[37] J. George, M.S. Sreekala, S. Thomas, A review on interface modification and characterization of natural fibre reinforced plastic composites, Polym. Eng. Sci. 41 (2001) 1471–1485. https://doi.org/10.1002/pen.10846

[38] M.M. Kabir, H. Wang, K. Lau, T. Cardona, Chemical treatments on plant-based natural fibre reinforced polymer composites: An overview, Compos. Part B: Eng. 43 (2012) 2883–2892. https://doi.org/10.1016/j.compositesb.2012.04.053

[39] Y. Xie, C.A.S. Hill, Z. Xiao, H. Militz, C. Mai, Silane coupling agents used for natural fibre/polymer composites: A review, Composites: Part A 41 (2010) 806–819. https://doi.org/10.1016/j.compositesa.2010.03.005

[40] D.B. Dittenber, H.V.S. Gangarao, Critical review of recent publications on use of natural composites in infrastructure, Composites: Part A 43 (2012) 1419–1429. https://doi.org/10.1016/j.compositesa.2011.11.019

[41] R. Malkapuram, V. Kumar, Y.S. Negi, Recent developments in natural fibre reinforced polypropylene composites, J. Reinf. Plast. Compos. 28 (2009) 1169–1189. https://doi.org/10.1177/0731684407087759

[42] Q.T. Shubhra, A.K.M.M. Alam, M.A. Quaiyyum, Mechanical properties of polypropylene composites: A review, J. Thermoplast. Compos. Mater. 26 (2013) 362–391. https://doi.org/10.1177/0892705711428659

[43] J. Summerscales, N.P.J. Dissanayake, A.S. Virk, W. Hall, A review of bast fibres and their composites. Part 1: Fibres as reinforcements, Composites: Part A 41 (2010) 1329–1335. https://doi.org/10.1016/j.compositesa.2010.06.001

[44] J. Summerscales, N.P.J. Dissanayake, A.S. Virk, W. Hall, A review of bast fibres and their composites, Part 2: Composites. Composites: Part A 41 (2010) 1336–1344. https://doi.org/10.1016/j.compositesa.2010.05.020

[45] M. Ramesh, K. Palanikumar, K. H. Reddy, Plant fibre based bio-composites: Sustainable and renewable green materials, Renew. Sustain. Energy Rev. 79 (2017) 558-584. https://doi.org/10.1016/j.rser.2017.05.094

[46] M. Ramesh, Kenaf (Hibiscus cannabinus L.) fibre based bio-materials: A review on processing and properties, Prog. Mater. Sci. 78-79 (2016) 1-92. https://doi.org/10.1016/j.pmatsci.2015.11.001

[47] S.M. Mirmehdi, F. Zeinaly, F. Dabbagh, Date palm wood flour as filler of linear low-density polyethylene, Composites: Part B 56 (2014) 137–141. https://doi.org/10.1016/j.compositesb.2013.08.008

[48] A.K. Bledzki, M. Letman, A. Viksne, L. Rence, A comparison of compounding processes and wood type for wood fibre-PP composites, Composites 36 (2005) 789–797. https://doi.org/10.1016/j.compositesa.2004.10.029

Thermoset Composite   Materials Research Forum LLC
Materials Research Foundations **38** (2018)   doi: http://dx.doi.org/10.21741/9781945291876

[49] M.N. Ichazo, C. Albano, J. Gonzalez, R. Perera, M.V. Candal,
      Polypropylene/wood flour composites: treatments and properties, Compos. Struct.
      54 (2001) 207–214. https://doi.org/10.1016/S0263-8223(01)00089-7

[50] A.K. Bledzki, O. Faruk, M. Huque, Physico-mechanical studies of wood fibre
      reinforced composites, Polym-Plast. Technol. Eng. 41 (2002) 435–451.
      https://doi.org/10.1081/PPT-120004361

[51] G. Hattotuwa, B. Premalal, H. Ismail, A. Bahrain, Comparison of the mechanical
      properties of rice husk powder filled polypropylene composites with talc filled
      polyethylene composites, Polym. Test. 21 (2002) 833–839.
      https://doi.org/10.1016/S0142-9418(02)00018-1

[52] A. Ashori, A. Nourbakhsh, Preparation and characterization of
      polypropylene/wood flour/nanoclay composites, Eur. J. Wood Prod. 69 (2011)
      663–666. https://doi.org/10.1007/s00107-010-0488-9

[53] C.M. Clemons, Wood-plastic composites in the United States: the interfacing of
      two industries, Forest Prod. J. 52 (2002) 10–18.

[54] P. Bazant, L. Munster, M. Machovsky, J. Sedlak, M. Pastorek, Z. Kozakova, I.
      Kuritka. Wood flour modified by hierarchical Ag/ZnO as potential filler for wood–
      plastic composites with enhanced surface antibacterial performance, Ind. Crop.
      Prod. 62 (2014) 179–187. https://doi.org/10.1016/j.indcrop.2014.08.028

[55] D.N.S. Hon, W.Y. Chao, Composites from benzylated wood and polystyrenes:
      their processability and viscoelastic properties, J. Appl. Polym. Sci. 50 (1993) 7–
      11. https://doi.org/10.1002/app.1993.070500102

[56] D. Maldas, B. V. Kokta, Influence of phthalic anhydride as a coupling agent on the
      mechanical behavior of wood fibre-polystyrene composites, J. Appl. Polym. Sci.
      41 (1990) 185–194. https://doi.org/10.1002/app.1990.070410116

[57] M.G.S. Yap, Y.I. Que, L.H.L. Chia, Dynamic mechanical analysis of tropical
      wood–polymer composites, J. Appl. Polym. Sci. 43 (1991) 1999–2004.
      https://doi.org/10.1002/app.1991.070431106

[58] J. George, S.S. Bhagwan, N. Prabhakaran, S. Thomas, Short pineapple leaf fibre
      reinforced low density polyethylene composites, J. Appl. Polym. Sci. 57 (1995)
      843–854. https://doi.org/10.1002/app.1995.070570708

[59] N.L. Dos, M.M. Elawday, S.H. Monsour, Impregnation of white pine wood with
      unsaturated polyesters to produce wood–plastic combinations, J. Appl. Polym. Sci.
      42 (1991) 2589–2594. https://doi.org/10.1002/app.1991.070420924

[60]   T. Gurunathan, S. Mohanty, S.K. Nayak, A review of the recent developments in biocomposites based on natural fibres and their application perspectives, Compos. Part A Appl. Sci. Manuf. 77 (2015) 1-25. https://doi.org/10.1016/j.compositesa.2015.06.007

[61]   E.O. Olakanmi, M.J. Strydom, Critical materials and processing challenges affecting the interface and functional performance of wood polymer composites (WPCs), Mater. Chem. Phys. 171 (2016) 290-302. https://doi.org/10.1016/j.matchemphys.2016.01.020

[62]   B.A.A. Ali, S.M. Sapuan, E.S. Zainudin, M. Othmand, Implementation of the expert decision system for environmental assessment in composite materials selection for automotive components, J. Clean. Prod. 107 (2015) 557-567. https://doi.org/10.1016/j.jclepro.2015.05.084

[63]   M.A. AlMaadeed, R. Kahraman, P. N. Khanam, N. Madi, Date palm wood flour/glass fibre reinforced hybrid composites of recycled polypropylene: Mechanical and thermal properties. Mater. Des. 42 (2012) 289–294. https://doi.org/10.1016/j.matdes.2012.05.055

[64]   J.Z. Lu, Q. Wu, H.S. McNabb, Chemical coupling in wood fibre and polymer composites: a review of coupling agents and treatments. Wood Fibre Sci. 32 (2000) 88-104.

[65]   S. Kalia, B.S. Kaith, I. Kaur, Pretreatments of natural fibres and their application as reinforcing material in polymer composites-a review. Polym. Eng. Sci. 49 (2009) 1253–1272. https://doi.org/10.1002/pi.1386

[66]   N.M. Stark, R.E. Rowlands, Effects of wood fibre characteristics on mechanical properties of wood/polypropylene composites, Wood Fibre Sci. 35 (2003) 167–174.

[67]   N.M. Stark, Effect of species and particle size on properties of wood-flour-filled polypropylene composites. Functional fillers for thermoplastics & thermosets, San diego, California; 1997.

[68]   N. Sombatsompop, K. Chaochanchaikul, C. Phromchirasuk, S. Thongsang, Effect of wood sawdust content on rheological and structural changes, and thermo-mechanical properties of PVC/sawdust composites, Polym. Int. 52 (2003) 1847–1855. https://doi.org/10.1002/pi.1386

[69]   N.E. Marcovich, M.M. Reboredo, M.I Aranguren, Dependence of the mechanical properties of woodflour–polymer composites on the moisture content, J. Appl.

Polym. Sci. 68 (1997) 2069–2076. https://doi.org/10.1002/(SICI)1097-4628(19980627)68:13<2069::AID-APP2>3.0.CO;2-A

[70]   N.E. Marcovich, M.I Aranguren, M.M. Reboredo, Modified woodflour as thermoset fillers Part I. Effect of the chemical modification and percentage of filler on the mechanical properties, Polymer 42 (2000) 815–825. https://doi.org/10.1016/S0032-3861(00)00286-X

[71]   P.A.d. Arcaya, A. Retegi, A. Arbelaiz, K.M. Kenny, I. Mondragon, Mechanical properties of natural fibres/polyamides composites, Polym. Compos. 20 (2009) 257–264. https://doi.org/10.1002/pc.20558

[72]   Y. Mamunya, M. Zanoaga, V. Myshak, F. Tanasa, E. Lebedev, C. Grigoras, V. Semynog, Structure and properties of polymer–wood composites based on an aliphatic copolyamide and secondary polyethylenes, J. Appl. Polym. Sci. 101 (2005) 1700–1710. https://doi.org/10.1002/app.23328

[73]   F. Yao, Q. Wu, Y. Lei, W. Guo, Y. Xu Y, Thermal decomposition kinetics of natural fibres: activation energy with dynamic thermogravimetric analysis, Polym. Degrad. Stab.93 (2007) 90–98. https://doi.org/10.1016/j.polymdegradstab.2007.10.012

[74]   N.E. Marcovich, M.M. Reboredo, M.I. Aranguren, Modified woodflour as thermoset fillers II. Thermal degradation of woodflours and composites, Thermochim Acta 372 (2000) 45–57. https://doi.org/10.1016/S0040-6031(01)00425-7

[75]   H. Bouafif, A. Koubaa, P. Perre, A. Cloutier, B. Riedl B, Wood particle/high-density polyethylene composites: thermal sensitivity and nucleating ability of wood particles, J. Appl. Polym. Sci. 113 (2009) 593–600. https://doi.org/10.1002/app.30129

[76]   P.A. Santos, M.A.S. Spinace, K.K.G. Fermoselli, M.A.D. Paoli, Polyamide-6/vegetal fibre composite prepared by extrusion and injection molding,Composites: Part A 38 (2007) 2404–2411. https://doi.org/10.1016/j.compositesa.2007.08.011

[77]   F.M.B. Coutinho, T.H.S. Costa, D.L. Carvalho, Polypropylene-wood fibre composites: Effect of treatment and mixing conditions on mechanical properties, J. Appl. Sci. 65 (1997) 1227–1235. https://doi.org/10.1002/(SICI)1097-4628(19970808)65:6<1227::AID-APP18>3.0.CO;2-Q

[78]   S.E. Selke, I. Wichman, Wood fibre/polyolefin composites, Composites: Part A 35
       (2004) 321–326. https://doi.org/10.1016/j.compositesa.2003.09.010

[79]   H. Peltola, E. Paakkonen, P. Jetsu, S. Heinemann, Wood based PLA and PP
       composites: Effect of fibre type and matrix polymer on fibre morphology,
       dispersion and composite properties, Composites: Part A 61 (2014) 13–22.
       https://doi.org/10.1016/j.compositesa.2014.02.002

[80]   K.O. Niska, M. Sain, Wood-Polymer Composites, Elsevier; Amsterdam, 2008.
       https://doi.org/10.1533/9781845694579

[81]   A.K. Bledzki, J. Izbicka, J. Gassan, Kunststoffe-Umwelt-Recycling, Stettin:
       Poland; 1995, 27–29.

[82]   A.A Klyosov, Wood-Plastic Composites, Wiley: NJ, 2007.
       https://doi.org/10.1002/9780470165935

[83]   A.K. Bledzki, O. Faruk, Creep and impact properties of wood fibre–polypropylene
       composites: influence of temperature and moisture content, Compos. Sci. Technol.
       64 (2004) 693–700. https://doi.org/10.1016/S0266-3538(03)00291-4

[84]   A.K. Bledzki, A.A. Mamun, J. Volk, Barley husk and coconut shell reinforced
       polypropylene composites: the effect of fibre physical, chemical and surface
       properties, Compos. Sci. Technol. 70 (2010) 840-846.
       https://doi.org/10.1016/j.compscitech.2010.01.022

[85]   O. Nabinejad, D. Sujan, M.E. Rahman, I.J. Davies, Effect of oil palm shell powder
       on the mechanical performance and thermal stability of polyester composites,
       Mater. Des. 65 (2015) 823-830. https://doi.org/10.1016/j.matdes.2014.09.080

[86]   R.L. Quirino, J. Woodford, R.C. Larock, Soybean and linseed oil-based
       composites reinforced with wood flour and wood fibres, J. Appl. Polym. Sci. 124
       (2012) 1520-1528. https://doi.org/10.1002/app.35161

[87]   H.P.S.A. Khalil, M.A. Tehrani, Y. Davoudpour, A.H. Bhat, M. Jawaid, A. Hassan,
       Natural fibre reinforced poly(vinyl chloride) composites: a review, J. Reinf. Plast.
       Compos. 32(5) (2013) 330–356. https://doi.org/10.1177/0731684412458553

[88]   J. Gassan, A.K. Bledzki, Alkali treatment of jute fibres: Relationship between
       structure and mechanical properties, J. Appl. Polym. Sci. 71(1991) 623–629.
       https://doi.org/10.1002/(SICI)1097-4628(19990124)71:4<623::AID-
       APP14>3.0.CO;2-K

[89] J.J. Balatinecz, Byung-Dae Park,The effects of temperature and moisture exposure on the properties of wood-fibre thermoplastic composites, J. Thermoplast. Comp. Mat. 10 (1997) 476–487. https://doi.org/10.1177/089270579701000504

[90] F.H.M.M. Costa, J.R.M. D'Almeida, Effect of water absorption on the mechanical properties of sisal and jute fibre composites, Polym.-Plast. Technol. Eng. 38 (1999) 1081–1094. https://doi.org/10.1080/03602559909351632

[91] M. Miki, N. Takakura, K. Kanayama, K. Yamaguchi, T. Iizuka, Effects of forming conditions on compaction characteristics of wood powders, Trans. Jpn. Soc. Mech. Eng. C 69 (2003)502–508. https://doi.org/10.1299/kikaic.69.502

[92] M. Miki, N. Takakura, K. Kanayama, K. Yamaguchi, T. Iizuka, Effects of forming conditions on flow characteristics of wood powders. Trans. Jpn. Soc. Mech. Eng. C 69 (2003) 766–772. https://doi.org/10.1299/kikaic.69.766

[93] M. Miki, N. Takakura, T. Iizuka, K. Yamaguchi, K. Kanayama, Possibility and problems in injection moulding of wood powders, Trans. Jpn. Soc. Mech. Eng. C 70 (2004) 2966–2972. https://doi.org/10.1299/kikaic.70.2966

[94] H. Kinoshita, K. Kaizu, K. Koga, H. Tokunaga, K. Ikeda, In: Proceeding of the Japan Society of Mechanical Engineers M & M 2007; 2007.

[95] M.D.H. Beg, K.L. Pickering, Reprocessing of wood fibre reinforced polypropylene composites. Part I, Effect on physical and mechanical properties, Composites: Part A 39 (2008) 1091-1100. https://doi.org/10.1016/j.compositesa.2008.04.013

[96] G.E. Myres, I.S. Cahaydi, C.A. Coberly, D.S. Ermer, Wood flour/polypropylene composites: influence of maleated polypropylene and process and composition variables on mechanical properties, Int. J. Polym. Mater. 15 (1991) 21–44. https://doi.org/10.1080/00914039108031519

[97] C. Eckert. Opportunities for natural fibres in plastic composites. In: Proceedings of progress in wood fibre-plastic composites conference, Toronto; May 25–26, 2000.

[98] D. Maldas, V. Kokta, Composite molded products based on recycled polypropylene and wood flour. J. Thermoplast. Compos. Mater. 8 (1995) 420-434. https://doi.org/10.1177/089270579500800405

[99] A. Kaymakci, N. Ayrilmis, Surface roughness and wettability of polypropylene composites filled with fast-growing biomass: Paulownia elongata wood, J. Compos. Mater. 48 (2014) 951–957. https://doi.org/10.1177/0021998313480199

[100] A.D. Cavdar, F. Mengeloglu, K. Karakus, Effect of boric acid and borax on mechanical, fire and thermal properties of wood flour filled high density polyethylene composites, Measurement, http://dx.doi.org/10.1016/j.measurement.2014.09.078.

[101] T. Umemura, Y. Arao, S. Nakamura, Y. Tomita, T. Tanaka, Synergy effects of wood flour and fire retardants in flammability of wood-plastic composites, Energy Proc. 56 (2014) 48–56. https://doi.org/10.1016/j.egypro.2014.07.130

[102] B. Nornberg, E. Borchardt, G.A. Luinstra, J. Fromm, Wood plastic composites from poly(propylene carbonate) and poplar wood flour–Mechanical, thermal and morphological properties, Eur. Polym. J. 51 (2014) 167–176. https://doi.org/10.1016/j.eurpolymj.2013.11.008

[103] G. Grubbstrom, K. Oksman, Influence of wood flour moisture content on the degree of silane-crosslinking and its relationship to structure–property relations of wood–thermoplastic composites, Compos. Sci. Technol. 69 (2009) 1045–1050. https://doi.org/10.1016/j.compscitech.2009.01.021

[104] W.V. Srubar III, S. Pilla, Z. C. Wright, C. A. Ryan, J.P. Greene, C.W. Frank, S.L. Billington, Mechanisms and impact of fibre–matrix compatibilization techniques on the material characterization of PHBV/oak wood flour engineered biobased composites, Compos. Sci. Technol. 72 (2012) 708–715. https://doi.org/10.1016/j.compscitech.2012.01.021

[105] M.S. Sreekala,J. George, M.G. Kumaran, S. Thomas, The mechanical performance of hybrid phenol–formaldehyde-based composites reinforced with glass and oil palm fibres, Compos. Sci. Technol. 62 (2002) 339–353. https://doi.org/10.1016/S0266-3538(01)00219-6

[106] H.D. Rozman, A. Hazlan, A. Abubakar, Preliminary study on mechanical and dimensional stability of rice husk–glass fibre hybrid polyester composites, Polym-Plast. Technol. Eng. 43 (2004) 1129–1140. https://doi.org/10.1081/PPT-200030059

[107] H.D. Rozman, G.S. Tay, R.N. Kumar, A. Abusamah, H. Ismail, Z.A.M. Ishak, Polypropylene–oil palm empty fruit bunch–glass fibre hybrid composites: a preliminary study on the flexural and tensile properties, Eur. Polym. J. 37 (2001) 1283–1291. https://doi.org/10.1016/S0014-3057(00)00243-3

[108] H. Jiang, P. Kamdem, B. Bezubic, P. Ruede, Mechanical properties of poly(vinyl chloride)/wood flour/glass fibre hybrid composites, J. Vinyl Addit. Technol. 9 (2003) 138–145. https://doi.org/10.1002/vnl.10075

[109] S. Mishra, A.K. Mohanty, L.T. Drzal, M. Misra, S. Parija, S.K. Nayak, S. S. Tripathy, Studies on mechanical performance of biofibre/glass reinforced polyester hybrid composites, Compos. Sci. Technol. 63 (2003) 1377–1385. https://doi.org/10.1016/S0266-3538(03)00084-8

[110] G.M. Rizvi, H. Semeralul, Glass fibre reinforced wood/plastic composites, J. Vinyl Addit. Technol.14 (2008) 39–42. https://doi.org/10.1002/vnl.20135

[111] A. Arbelaiz, B. Fernandez, G. Cantero, R. Llano-Ponte, A. Valea, I. Mondragon, Mechanical properties of flax fibre/polypropylene composites. Influence of fibre/matrix modification and glass fibre hybridization, Composites Part A 36 (2005) 1637–1644. https://doi.org/10.1016/j.compositesa.2005.03.021

[112] Y. Zhang, S. Zhang, P. Choi, Effects of wood fibre content and coupling agent content on tensile properties of wood fibre polyethylene composites, Holz. Roh. Werkst. 66 (2008) 267-274. https://doi.org/10.1007/s00107-008-0246-4

[113] Y. Habibi, W.K.E. Zawawy, M.M. Ibrahim, A. Dufresne, Processing and characterization of reinforced polyethylene composites made with lignocellulosic fibres from Egyptian agro-industrial residues, Compos. Sci. Technol. 68 (7–8) (2008) 1877-1885. https://doi.org/10.1016/j.compscitech.2008.01.008

[114] F. Sliwa, N.E. Bounia, F. Charrier, G. Marin, F. Malet, Mechanical and interfacial properties of wood and bio-based thermoplastic composite, Compos. Sci. Technol. 72 (2012) 1733–1740. https://doi.org/10.1016/j.compscitech.2012.07.002

[115] N.E. Marcovich, A.N. Ostrovsky, M.I. Aranguren, M.M. Reboredo, Resin–sisal and wood flour composites made from unsaturated polyester thermosets, Compos. Interf. 16 (2009) 639–657. https://doi.org/10.1163/092764409X12477430713668

[116] M. Valente, F. Sarasini, F. Marra, J. Tirillo, G. Pulci, Hybrid recycled glass fibre/wood flour thermoplastic composites: Manufacturing and mechanical characterization. Composites: Part A 42 (2011) 649–657. https://doi.org/10.1016/j.compositesa.2011.02.004

[117] G. Cantero, A. Arbelaiz, F. Mugika, A. valea, I. Mondragon. Mechanical behavior of wood/polypropylene composites: Effects of fibre treatments and ageing processes, J. Reinf. Plast. Compos. 22(1) (2003) 37-50. https://doi.org/10.1177/0731684403022001495

[118] D.D. Stokke, Fundamental aspects of wood as a component of thermoplastic composites, J. Vinyl Addit. Technol. 9 (2003) 96–104. https://doi.org/10.1002/vnl.10069

[119] O. Nabinejad, D. Sujan, M.E. Rahman, I.J. Davies, Effect of filler load on the curing behavior and mechanical and thermal performance of wood flour filled thermoset composites, J. Clean. Prod. 164 (2017) 1145-1156. https://doi.org/10.1016/j.jclepro.2017.07.036

[120] H. Jeske, A. Schirp, F. Cornelius. Development of a thermogravimetric analysis (TGA) method for quantitative analysis of wood flour and polypropylene in wood plastic composites (WPC). Thermochimica Acta 543 (2012) 165– 171. https://doi.org/10.1016/j.tca.2012.05.016

[121] L. Soccalingame, A. Bourmaud, D. Perrin, J-C. Benezet, A. Bergeret, Reprocessing of wood flour reinforced polypropylene composites: impact of particle size and coupling agent on composite and particle properties, Polym. Degrad. Stab. (2015). https://doi.org/10.1016/j.polymdegradstab.2015.01.020

[122] M. Bengtsson, K. Oksman. The use of silane technology in crosslinking polyethylene/wood flour composites. Composites Part A 37 (2006) 752–765. https://doi.org/10.1016/j.compositesa.2005.06.014

[123] M. Bengtsson, K. Oksman, Silane crosslinked wood plastic composites: processing and properties, Compos. Sci. Technol. 66 (2006) 2177–2186. https://doi.org/10.1016/j.compscitech.2005.12.009

[124] M. Bengtsson, K. Oksman, N.M. Stark, Profile extrusion and mechanical properties of crosslinked wood–thermoplastic composites, Polym. Compos. 2006:184–194. https://doi.org/10.1002/pc.20177

[125] M. Sain, S.H. Park, F. Suhara, S. Law, Flame retardant and mechanical properties of natural fibre-PP composites containing magnesium hydroxide, Polym. Degrad. Stab. 83 (2004) 363-367. https://doi.org/10.1016/S0141-3910(03)00280-5

[126] M. Garcia, J. Hidalgo, J. Garmendia, J. Garcia-Jaca, Wood-plastics composites with better fire retardancy and durability performance. Compos Part A 40 (2009) 1772-1776. https://doi.org/10.1016/j.compositesa.2009.08.010

[127] N.M. Stark, R. White, S. Mueller, T. Osswald, Evaluation of various fire retardants for use in wood flour-plyethylene composites, Polym. Degrad. Stab. 95 (2010) 1903-1910. https://doi.org/10.1016/j.polymdegradstab.2010.04.014

[128] L. Danyadi, T. Janecska, Z. Szaboc, G. Nagy, J. Moczo, B. Pukanszky, Wood flour filled PP composites: Compatibilization and adhesion. Compos. Sci. Technol. 67 (2007) 2838–2846. https://doi.org/10.1016/j.compscitech.2007.01.024

[129] A. Jacob, WPC industry focuses on performance and cost. Reinf. Plast. 50 (2006) 32–33. https://doi.org/10.1016/S0034-3617(06)71010-4

[130] A.K. Bledzki, O. Faruk, V.E. Sperber, Cars from bio-fibres, Macromol. Mater. Eng.291 (2006) 449–457. https://doi.org/10.1002/mame.200600113

[131] B. Pukanszky, B. Turcsanyi, F. Tudos, Effect of interfacial interaction on the tensile yield stress of polymer composites. In: Ishida H, editor. Interfaces in polymer ceramic and metal matrix composites. New York: Elsevier; 1988. 467–477.

[132] B. Pukanszky, Influence of interface interaction on the ultimate tensile properties of polymer composites, Composites 21(1990) 255–262. https://doi.org/10.1016/0010-4361(90)90240-W

[133] B. Pukanszky, Particulate filled polypropylene: structure and properties. In: Karger-Kocsis J, editor. Polypropylene: structure blends and composites, vol. 3. London: Chapman & Hall; 1995. 1–70. https://doi.org/10.1007/978-94-011-0523-1_1

Thermoset Composite
Materials Research Foundations **38** (2018)

Materials Research Forum LLC
doi: http://dx.doi.org/10.21741/9781945291876

Chapter 3

Experimental and Analysis of Jute Fabric with Silk Fabric Reinforced Polymer Composites

M.R. Sanjay[1,*], K.N. Bharath[2], R. Vijay[3], D. Lenin Singaravelu[3], A. Vinod[4], M. Jawaid[5], Anish Khan[6]

[1]Department of Mechanical Engineering, Ramaiah Institute of Technology, Bengaluru, Karnataka, India

[2]Department of Mechanical Engineering, G M Institute of Technology, Davangere, India

[3]Department of Production Engineering, National Institute of Technology, Tiruchirappalli-620015, Tamil Nadu, India

[4]Department of Mechanical Engineering, Sri Lakshmi Ammaal Engineering College, Tiruvanchery, Selaiyur, Chennai-600126, Tamil Nadu, India

[5]Department of Biocomposite Technology, Institute of Tropical Forestry and Forest Products, Universiti Putra Malaysia, UPMSerdang, Selangor, Malaysia

[6]Center of Excellence for Advanced Materials Research, Chemistry Department, Faculty of Science, King Abdulaziz University, Jeddah, Saudi Arabia

mcemrs@gmail.com*

Abstract

In the present work, mechanical and physical properties of the fabrics extracted from jute are determined and compared with silk fabric. The specimen preparation and testing are carried out as per ASTM standards. Jute and silk based composite materials are fabricated using the hand lay-up technique as per stacking sequence. Further, these jute and silk fabrics were chemically treated and the effect of this treatment on fabric strength is studied. All the composite laminates were prepared with the proportion of 60-40%. Tests like tensile test, bending test, impact test, water absorption test of jute and silk fabric reinforced epoxy based composite laminates were conducted and evaluated. The behaviour of the composite under different tests was analyzed with the help of performance curves. Water absorption test plots for different periods indicated that jute and silk based epoxy composites offer better resistance. SEM analysis shows the fracture surface of the jute and silk fabric–matrix interface. On the other hand, ANOVA was used in one way to find the significant difference between the best mechanical composite specimens. The ANSYS analysis was used to gain knowledge of the stress distribution during the tensile and flexural tests of the composites. The results obtained from various

Thermoset Composite

Materials Research Forum LLC

Materials Research Foundations **38** (2018)

doi: http://dx.doi.org/10.21741/9781945291876

tests show that these composites can potentially replace automobile parts like bumpers, doors, and chemical containers.

Keywords

Jute, Silk, Mechanical Properties, Water Absorption, SEM, ANOVA, ANSYS, Epoxy

Contents

3.1 Introduction

Concerns about the preservation of natural sources and recycling have led to renewed interest in biomaterials with the focus on renewable raw materials. As a result, new types of composites based on plant fibers have been developed in recent years. Natural fiber reinforced composites offer a good mechanical performance and eco-friendliness. The application of natural fiber based composites is increasing rapidly. This is especially related to certain problems concerning the use of synthetic fiber reinforced composites [1-3]. Natural fibers are cheap, renewable, completely or partially recyclable, and biodegradable. Plants such as jute, flax, cotton, hemp, ramie, bamboo, etc., as well as wood, used from time immemorial as a source of lignocellulosic fibers, are more and more often applied as the reinforcement of composites. Such natural fiber containing composites are more environmentally friendly, and are used in transportation such as automobiles, railway coaches, aerospace, military applications, building and construction industries (ceiling, panelling, and partition laminates), packaging, consumer products, etc. Natural fibers include those made from plants, animal and mineral sources. Naturally occurring fibers can be classified according to their origin [4-6].

Thermoset Composite

Materials Research Foundations 38 (2018)

Materials Research Forum LLC

doi: http://dx.doi.org/10.21741/9781945291876

Mohanty et al. [7] determined the influence of different surface modifications of jute on the performance of the bio composites. More than a 40% improvement in the tensile strength occurred as a result of reinforcement with alkali treated jute. Jute fabric content also affected the bio composite performance and about 30% by weight of jute showed optimum properties of the bio composites. Mansur and Aziz [8] studied bamboo-mesh reinforced cement composites, and found that these reinforcing materials could enhance the strength and toughness of the cement matrix, and increase its tensile, flexural, and impact strengths significantly. On the other hand, jute fabric-reinforced polyester composites were tested for the evaluation of mechanical properties and compared with wood composite, and it was found that the jute fabric composite has better strengths than wood composites. Luo and Netravali [9] discussed the tensile and flexural properties of the green composites with different pineapple fiber content and compared with the virgin resin. Sisal fiber is fairly coarse and inflexible. It has good strength, durability, ability to stretch, affinity for certain dyestuffs and resistance to deterioration in seawater. Sisal ropes and twines are widely used for marine, agricultural, shipping, and general industrial use. Raghu et al. [10] discussed the abrupt improvement in the silk fiber epoxy laminate. Composites of untreated and alkali treated silk-sisal unsaturated polyester-based hybrid composites were prepared by using the hand lay-up technique. The chemical resistance of the treated and untreated silk/sisal hybrid composites to various acids, alkalis, and solvents was studied. Kweon et al. [11] introduced an idea of silk fibroin/chitosan blends as potential biomedical composites as the mechanical properties of silk fibroin are greatly enhanced with increasing chitosan content. Lee et al. [12] examined the effect of silk/poly (butylene's succinate) (PBS) bio-composites. They found that the mechanical properties including tensile strength, fracture toughness and impact resistance, and thermal stability of bio-composites would be greatly affected by their manufacturing processes. Moreover, a good adhesion between the silk fiber and PBS matrix was found through the observation and analysis by SEM. Annamaria et al. [13] discovered in the studies that environmentally-friendly biodegradable polymers can be produced by blending silk sericin with other resins. Identified that polyurethane foams incorporating sericin are said to have excellent moisture-absorbing and desorbing properties.

3.2 Materials and methods

Composite mats were prepared for treated conditions. The chosen composite includes materials such as jute fabric, silk fabric and matrix material (Lapox L-12).

Jute is a bast fiber obtained from inner bast tissues of the plant stem. The fibers are bound together by gummy materials which keep the fiber bundles cemented with non-fibrous tissues of jute bark. These encircling soft tissues must be softened, dissolved and washed away so that the fiber can be obtained from the stem. This is done by steeping the stems in water and is known as retting. The optimum water temperature for retting is 808F. Micro-organisms (mainly bacillus bacteria) decompose the gums and soften the tissues in 5 to 30 days depending upon temperatures and the type of water used. It has been found that the presence of higher amounts of calcium and magnesium tends to increase the tenacity of fiber [14]. In the present research, hessian yarns which are made from long jute fibers are used. The yarn is obtained by a spinning process which depends upon the class of goods being made; however, there are features common to all systems. Jute must be softened and lubricated with batch oil so that the fiber may be processed without excessive fiber breakage and waste. The messy nature of the reeds must be split-up and fibers separated as far as possible. Fibers are drawn evenly into slivers or loose untwisted strands and then drawn out to the desired yarn thickness. At the spinning frame the material is given its final drafting down to the required weight and the fibers are twisted together to form yarn, this is then wound on to bobbins. Twisting is done by flyers rotating at speeds of 3550–4000 rpm.

Silk is a protein fabric like wool. This gives it many of its characteristics. Out of the numerous species of silk moths, scientists have enumerated about 70 silk moths, which are of some economic value. Although the bulk of the worlds silk supply comes from the domesticated silk moth, the other varieties of silk (include most of the Thai silk fabrics) are known as wild silk, as they are grown in remote forest trees in natural conditions. Silk is the strongest of the nature fabrics, offers luxuriousness, resiliency, good drapability and beauty. This means that silk is very strong in terms of tensile strength. It can withstand a lot of pulling type pressure without breaking. However, silk will not stand up to the heavy wear that other fabrics will. It has only fair abrasion resistance, tends to water-spot easily, and is expensive [15].

3.3 Preparation of composites

The Jute, Silk and Lapox epoxy resin composites used in this work were made with the hand lay-up techniques. A mould made of mild steel square rod of dimension of 200x200 mm^2 was used to prepare the laminate. In order to get the required dimension of 200x200x10 mm^3 of laminate, two frames of the same dimension were placed adjacently on the mould. First, one pours the calculated amount of Lapox L-12 resin over the surface where the working was carryed out. Then squeegee and spread the resin over the surface. Then place the reinforcement jute and silk fabric in place at the proper orientation called for in the plans. Be very careful not to distort the mats. Then, using a squeegee begin to press gently from the Centre of the mat making sure that the squeegee moves in the same directions as the fabrics of the mat. Keep the fabrics straight and press the fabric into the resin while working the resin up through the jute and silk mat. The lay-up assembly was pressed in a press. The excess resin was allowed to squeeze out. The laminate was cured at ambient conditions for a period of about 24 hours. The laminate so prepared has a size of 200x200x10 mm^3.

The jute and silk fabrics are aggregated very tightly. As the jute and silk is mixed with dust and dirt it is very difficult to clean it up. The polymers are very adhesive in nature. So it sticks to the container in which it is taken, so a little extra amount of it is taken very precisely. Whereas the hardener is a very active item and it should be handled very carefully. A little more amount of hardener can completely spoil the casting. In the process we experienced that even two extra drops of hardener made the reaction exothermic. Another problem that we faced during the process of casting is the high volume of jute & silk. In the preparations of the composites presented in Table 1, 50% of the fabric is not that difficult but with 60% it becomes very difficult to fit inside the mould. It has to be laminate several times and in the process many polymer hardener mixture stick together. The weights has to be put very accurately because if we put a little extra weight the polymer leaks out through the side of the laminate.

Table 1 Stacking sequence and Percentage volume of composites.

Specimen code	Stacking sequence	% Of matrix volume	% Of reinforcement volume (Jute fabric)	% Of reinforcement volume (Silk fabric)
A	SSSSSJSJSJSJSJSJSJSJSJSJSJSJSJSJSJSJSJSSSSS	40	30	30
B	JSJSJSJSJSJSJSSJSSSSSSSJSSJSSJSJSJSJSJSJSJ	40	30	30
C	JJJJJJSSSSSSSSSSSSSSSJJJJJSSSSSSSSSSSSSSSSJJJJJJ	40	30	30
D	JJJJJJJJJSSSSSSSSSSSSSSSSSSSSSSSSSSSSSSJJJJJJJJJ	40	30	30

3.4 Experimentation

According to ASTM D3039 square specimen of 180 mm length, 30 mm side and thickness 10 mm as that of prepared laminate is considered for the tensile test. The specimen was loaded in a universal testing machine until the failure of the specimen occurs. The three points bending flexural test provides values for the modulus of elasticity in bending, bending stress, bending strain and the bending stress-strain response of the material. The main advantage of a three point bending test is the ease of the specimen preparation and testing. However, this method has also some disadvantages: the results of the testing method are sensitive to the specimen and loading geometry and strain rate. According to ASTM D 790 the composite specimens were prepared for bending test. Each test specimen of 30 mm width, length 180 mm and thickness 10 mm were prepared. The span (centre to centre distance between roller supports) for each specimen is 100 mm. The specimen is loaded at the centre of the span through a loading cell. The test is carried out until the specimen completely fails. Charpy impact test is performed on commercially available machines in which a pendulum hammer is released from a standard height to contact a beam specimen (either notched or unnotched) with a specified kinetic energy. A horizontal simply supported beam specimen is used in the Charpy test. The energy absorbed in breaking the specimen, usually indicated by the position of a pointer on a calibrated dial attached to the testing machine, is equal to the difference between the energy of the pendulum hammer at the instant of impact and the energy remaining in the pendulum hammer after breaking the specimen. Most natural fabrics absorbed more water compared to synthetic fabrics. Water is predominantly absorbed at the fabric interface and matrix. The effect of this absorbed water is to degrade the properties such as tensile strength. The specimens were prepared from a 12 mm thickness laminate with size 50 mm wide and 50 mm long. After preparing the test specimens, they are soaked in normal water. The variable parameters like mass, length, width and thickness are note down every day (Fig. 1).

Figure 1 Specimens dipped in water for moisture absorption test.

Thermoset Composite   Materials Research Forum LLC
Materials Research Foundations **38** (2018)   doi: http://dx.doi.org/10.21741/9781945291876

3.5   Results and discussions on experimentation

The tensile experiments were conducted on each jute-silk fabric (60-40) specimen and the results showed that the ultimate stress is around 25–35 MPa with 2.5-2.7% elongation at break. Fig. 4 shows different sequences of jute-silk fabric failure at different loads. And Fig. 5 shows the stress–strain diagram for the jute-silk fabric is identical and the nature of fracture exhibit like a ductile. Fig. 6 shows the maximum bending strength for composite laminate of composition A is 17.6 KN and that of B, C, and D composition composite laminates having 15.5, 15.8 & 14.9 respectively. Hence the specimen A have greater strength than B, C & D. Fig. 7 shows fracture surface of composite specimens after tensile load. The maximum load at which breaking takes place for A, B, C and D is given by 1.002, 1.0026, 1.004 and 1.015 KJ respectively. It is observed that sequence laminate D absorbs more energy than other sequence plates. Fig. 8 shows the water absorption curves for jute-silk fabric composite laminates. The percentage of moisture content increases with increase in number of days. Later the moisture absorption becomes almost constant after 20 days which is an indication of saturation. Maximum moisture absorption is 15% for jute-silk fabric composite.

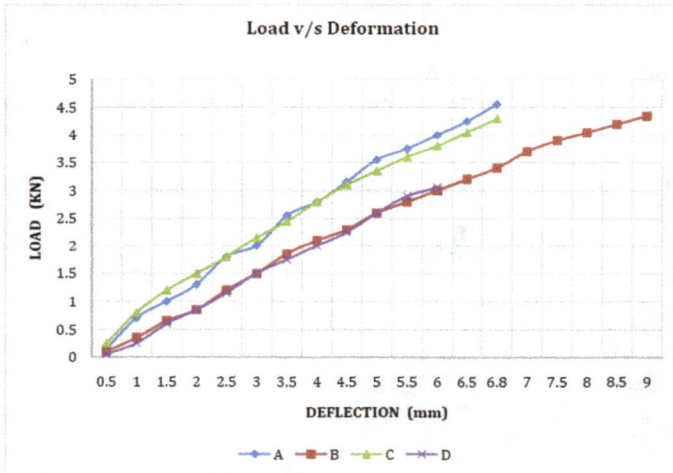

Figure 4 Load v/s deflection graph.

Thermoset Composite   Materials Research Forum LLC
Materials Research Foundations **38** (2018)   doi: http://dx.doi.org/10.21741/9781945291876

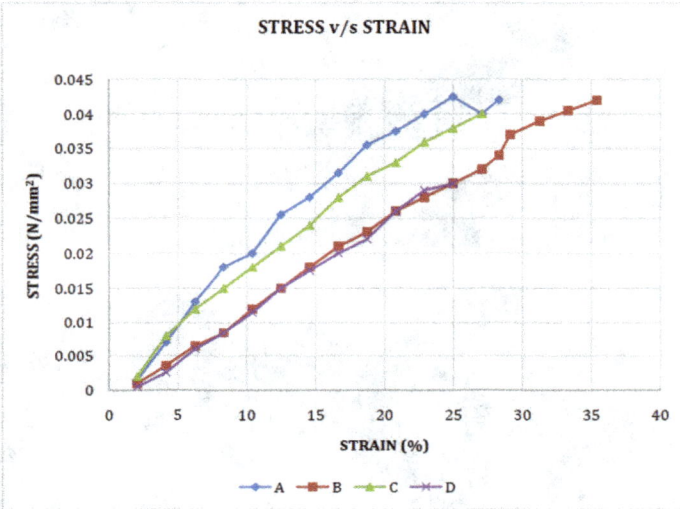

Figure 5 Stress v/s Strain graph.

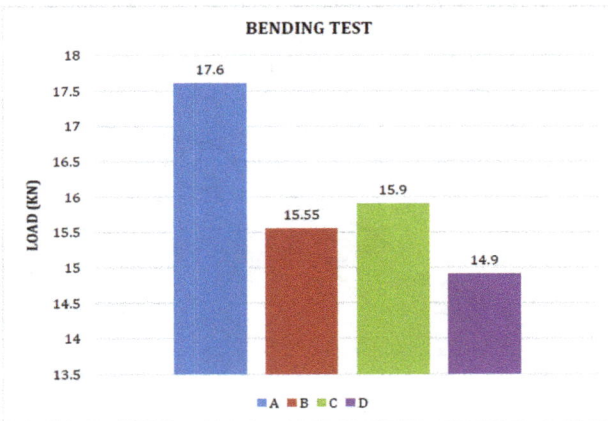

Figure 6 Bending strength of the specimens.

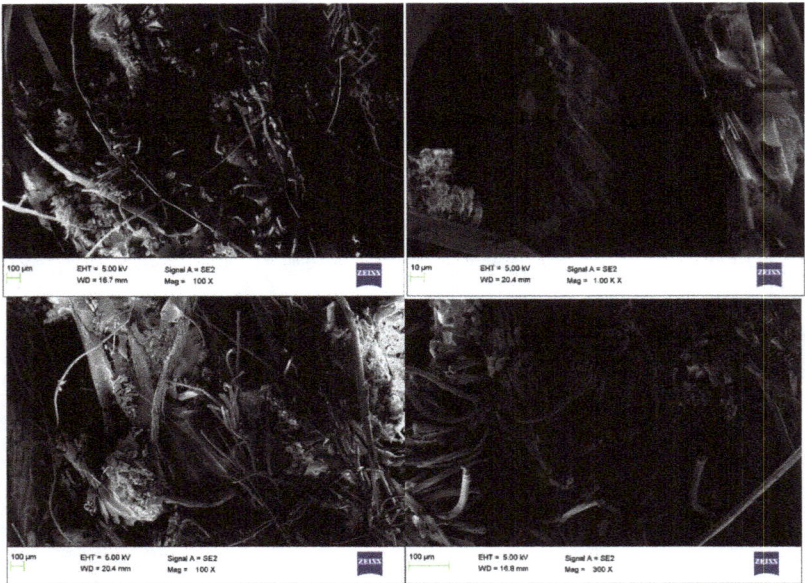

Figure 7 Fractured surface of jute-silk fabric composites after tensile test.

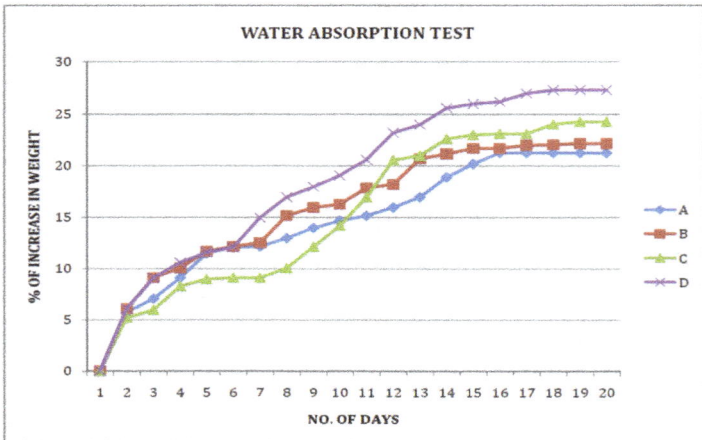

Figure 8 Water absorption results of composites.

3.6 Analysis

The tensile tests for the manufactured composites are carried out in a universal testing machine (UTM) according to ASTM D 638 standards. Three specimens of each composites were tested for consistency of the result. The manufactured four composites are tested for the tension test and the results are shown in the Fig. 9. From the test results it is clear that composite B shows better tensile strength of 0.0425 Mpa with a corresponding strain percentage of 36 when compared to the other three composite. From figure, it is clear that composite B has the same tensile strength as the composite A, enabling 1.06 times superior strength than composite C and 1.4 times greater than composite D. This phenomenon is greatly influenced by the stacking sequence of jute and silk which is clearly shown in the Table 2. The stacking sequence of the composite A starts with silk for the first 5 layers followed by 31 alternative layers of jute and silk, ends with 5 layers of silk. Composite B has 41 alternate layers of jute and silk, composite C has 6 layers of jute at the top followed by 12 layers of silk, 5 layers of jute and again 12 layers of silk, ends with 6 layers of jute. Composite D has 9 layers of jute at the top and bottom, between which are 25 layers of silk stacked. This stacking variation alters the mechanical properties of each specimen [16]. From scholarly articles it is clearly known that the tensile strength of jute is 133 Mpa and silk fiber is 250 Mpa [17]. Which means tensile strength of jute fiber is 1.87 times inferior to the silk fiber. But in comparison jute fibers are stiffer than the silk fibers. When tensile load is attributed to the composite C&D there is internal crack propagation generation from the jute fibers in composite C, prolongs to the matrix and the silk fibers, which induces faster failure. While when tensile load is applied to composite B due to the alternate sequence of the jute and silk, the tear developed from is stopped by silk. This consequence is attributed to the stiffness ratio of the silk fiber. And it posses same stress value as composite A. In composite B, as it posses alternate sequence of jute and silk leads to a superior tensile stress and strain when compared to the other composites, this phenomenon was caused by the interlocking of the matric with the jute fibers and silk in an alternate manner, and silk in the outer layer posses higher tensile strength compared to jute encourages a better mechanical property.

The three point bending test or flexural test is carried out in a Universal testing machine (UTM) according to ASTM D 3039 standards. Flexural test is a combination of tensile and compression. In detail the upper part of the specimen is subjected to compression load and bottom part of the specimen is subjected to tensile load. The test results are shown in Fig. 10. From the test results it is clear than the composite B possess superior flexural strength when compared to the other composites. This phenomenon is caused to the stacking sequence as shown in the Table 2. From the Table, composite B posses a flexural strength of 0.062 which is 1.05 times greater than composite A, 1.16 times greater than composite C and 1.24 times greater than composite D. The stacking sequence of Composite B explains that, the outer silk layer of the composite elongates when it is subjected to tension, adding it has a better tensile strength and the bottom layer of the composite compresses when it is subjected to compression load. The whole process defines the flexural load. As a fact, it restricts the crack propagation to move to further layers. Adding value, by the alternating layers of jute and silk. The occurrence of this phenomenon enhances the flexural strength of the composite B, thus making it superior than the other composites. While in other composites A, C, D the stacking sequence is started by jute fiber followed by silk that initiates a crack propagation when subjected to the flexural load and prolongs to the silk fiber leaning to an immediate flexural failure.

Table 2 ANOVA results for composites.

ANOVA FOR TENSILE INDIVIDUAL SPECIMENS						
COMPOSITE A						
Source of Variation	*SS*	*df*	*MS*	*F*	*P-value*	*F crit*
Between Groups	0.00	2	0.00	0.07348705	0.92926737	3.21994229
Within Groups	0.0090362	42	0.00021514			
Total	0.0090678	44				
COMPOSITE B						
Source of Variation	*SS*	*df*	*MS*	*F*	*P-value*	*F crit*
Between Groups	3.4105E-05	2	1.70526E-05	0.08724009	0.91658593	3.16824596
Within Groups	0.0105552	54	0.00019546			
Total	0.0105893	56				
COMPOSITE C						
Source of Variation	*SS*	*df*	*MS*	*F*	*P-value*	*F crit*
Between Groups	2.4142E-05	2	1.20714E-05	0.06796459	0.93440394	3.23809615
Within Groups	0.0069269	39	0.00017764			
Total	0.0069510	41				
COMPOSITE D						

Thermoset Composite
Materials Research Foundations **38** (2018)

Materials Research Forum LLC
doi: http://dx.doi.org/10.21741/9781945291876

Source of Variation	SS	df	MS	F	P-value	F crit
Between Groups	2.2153E-07	2	1.10769E-07	0.00098540	0.99901510	3.25944630
Within Groups	0.0040467	36	0.00011241			
Total	0.0040469	38				

Statistical analysis is carried out for the tensile specimens of the manufactured composites. Three specimens of each composite are subjected to test for the check of consistency. Analysis of variance or ANOVA is carried out for each composites to find the variance of the specimens within the composites and are shown in Table 2. From the Table, it is clear that the calculated F-Test of the manufactured composite is less than the statistical table value F_{crit}. This phenomenon denotes the H_0 hypothesis, states that there is no significant difference between the manufactured composites. The P-Value of the F-Test is 0.05, so that there is 95% of satisfactory significant level within the composites.

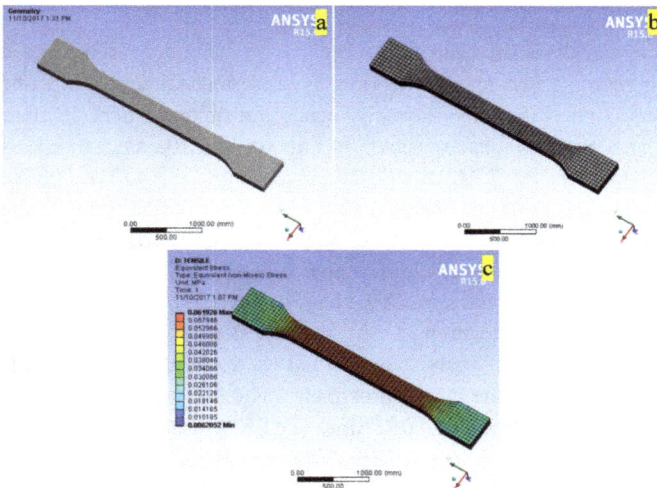

Figure 9 Tensile analysis for composite B.

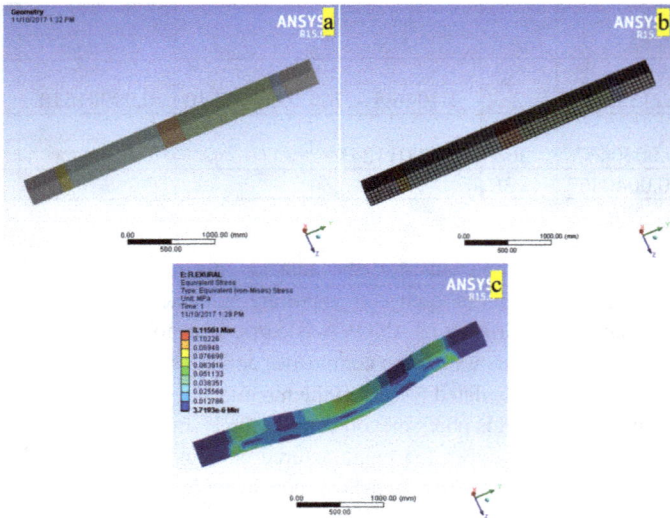

Figure 10 Flexural analysis for composite B.

A finite element analysis has been carried out for the manufactured composite to predict the stress distribution, and to validate with the experimental result. It is impossible to study the deflection curve experimentaly, hence the finite element simulation is been carried out. The tensile results are shown in Fig. 10. Fig. 10a are the three dimensional models created using CREO PARAMETRIC according to the ASTM D 638 and ASTM D 790 standards. The 3D models are then imported into ANSYS R 15 for predicting the maximum tensile and flexural load. Fig. 10 & 11 show the mesh model of the tensile and flexural specimens. Where the tensile specimen is meshed into 2920 elements and 15041 nodes, and the flexural specimen has 2840 elements and 15527 nodes.. From Fig. 10c it is clearly found that the tensile strength of the composite B is approximately 0.053 Mpa which is approximately equal to the experimental value of 0.042 Mpa. From Fig. 11 c the flexural strength of composite B is approximated as 0.076 Mpa which is approximately equal to the experimental result of 0.062 Mpa. The FEM values are obtained exactly from the contour chart, where the failure occurred experimentally.

Conclusion

As jute and silk fabrics are used as material for the textile industry they are cheaply available and it shows the perfect utilization of nature product. Jute and silk fabrics are bio-degradable and highly crystalline with well aligned structure. So it has been known that they also have higher tensile strength than glass fiber or synthetic organic fabrics, good elasticity, excellent resilience and hence, do not induce a serious environmental problem like in glass fibers. Glass fibers have serious drawbacks like non-renewable, non-recyclable, non-biodegradable and high energy consumption in manufacturing process. Therefore, jute and silk fabrics are a perfect replacement for glass fibers composites. Jute and silk fabric composites exhibit average values for the tensile strength, flexural strength and impact strength. ANSYS stress analysis showed better stress distribution in the composites and highest stress transfer among all the other composites in both tensile and flexural nature. ANOVA showed negligible significant differences thereby proving the consistent results.

References

[1] M. R. Sanjay, P. Madhu, M. Jawaid, S. Pradeep, P. Senthamaraikannan, S. Senthil, Characterization and Properties of Natural Fiber Polymer Composites: A Comprehensive Review, J. Clean. Produc. 172 (2018) 566-581. https://doi.org/10.1016/j.jclepro.2017.10.101

[2] M. R Sanjay, B. Yogesha, Studies on Natural/Glass Fiber Reinforced Polymer Hybrid Composites: An Evolution, Mater. Tod. Proc. 4 (2017) 2739–2747. https://doi.org/10.1016/j.matpr.2017.02.151

[3] O. Fruk, A. K. Bledzki, H. Fink, M. Sain, Biocomposites reinforced with natural fibers: 2000–2010. Prog. Poly. Sci. 37(2012) 1552–1596. https://doi.org/10.1016/j.progpolymsci.2012.04.003

[4] M. R. Sanjay, G. R. Arpitha, L. Laxmana Naik, K. Gopalakrisha, B. Yogesha, Applications of Natural Fibers and Its Composites: An Overview. Natur. Resou. 7 (2016) 108–114. https://doi.org/10.4236/nr.2016.73011

[5] D. Saravana Bavan, G. C. Mohan Kumar, Potential use of natural fiber composite materials in India, J. Reinf. Plastic. Compos. 29 (2010) 3600–3613. https://doi.org/10.1177/0731684410381151

[6] M. R Sanjay, B. Yogesha, Studies on Mechanical Properties of Jute/E-Glass Fiber Reinforced Epoxy Hybrid Composites, J. Min. Mater. Charact. Eng. 4(2016) 15–25. https://doi.org/10.4236/jmmce.2016.41002

[7] A. K. Mohanty, M. A. Khan, G. Hinrichsen, Influence of chemical surface modification on the properties of biodegradable jute fabrics-polyester amide composites, Compos A: App. Sci. Manuf. 31 (2000) 143- 150. https://doi.org/10.1016/S1359-835X(99)00057-3

[8]   M. A. Mansur, M. A. Aziz, Study of Bamboo-Mesh Reinforced Cement Composites, Int. Cement Compos. Lightweight Concrete, 5 (1983)165–171. https://doi.org/10.1016/0262-5075(83)90003-9

[9]   S. Luo, A. N. Netravali, Interfacial and mechanical properties of environment-friendly 'green' composites made from pineapple fibers and polyhydyoxybutyrate-co-valerate) resin, J. Mater. Scie. 34 (1999) 3709-3719. https://doi.org/10.1023/A:1004659507231

[10]   K. Raghu, P. NoorunnisaKhanam, S. Venkata Naidu, Chemical resistance studies of silk/sisal fiber reinforced unsaturated polyester-based hybrid composites, J. Reinf. Plastic. Compos. 29 (2010) 343–345. https://doi.org/10.1177/0731684408097770

[11]   H. Kweon, Hyun Chul Ha, In Chul Um, Young Hwan Park, Physical properties of silk fibroin/chitosan blend films, J. App. Polym. Scie. 80 (2001) 928-934. https://doi.org/10.1002/app.1172

[12]   S. M. Lee, D. Cho, W. H. Park, S. G. Lee, S. O. Han, T. DrzalL, Novel silk/poly (butylenes succinate) biocomposites: the effect of short fiber content on their mechanical and thermal properties, Compos. Scie. Tech. 65 (2005) 647-657. https://doi.org/10.1016/j.compscitech.2004.09.023

[13]   S. Annamaria, R. Maria, M. Tullia, S. Silvio, C. Orio. The microbial degradation of silk: a laboratory investigation, Inter. Biodeterior. Biodegrad. 42 (1998) 203-211. https://doi.org/10.1016/S0964-8305(98)00050-X

[14]   A. N. Shah, S. C. Lakkad, Mechanical properties of jute reinforced plastics, Fib. Scie. Techn. 15 (1981) 41–46. https://doi.org/10.1016/0015-0568(81)90030-0

[15]   A. U. Ude, A. K. Ariffin, C. H. Azhari, An Experimental Investigation on the Response of Woven Natural Silk Fiber/Epoxy Sandwich Composite Panels Under Low Velocity Impact, Fib. Poly. 14 (2013) 127–132.

[16]   S. P. Priya, S. K. Rai, Studies on the mechanical performance of PMMA toughened epoxy–silk and PC toughened epoxy–silk fabric composites. J. Reinf. Plastic. Compos. 25(2006) 33–41. https://doi.org/10.1177/0731684406055453

[17]   K.S. Ahmed, S. Vijayarangan, Tensile, flexural and interlaminar shear properties of woven jute and jute-glass fabric reinforced polyester composites. J. Mater. Process. Techn. 207(2008) 330–335. https://doi.org/10.1016/j.jmatprotec.2008.06.038

Thermoset Composite
Materials Research Foundations **38** (2018)

Materials Research Forum LLC
doi: http://dx.doi.org/10.21741/9781945291876

Chapter 4

Biosourced Thermosets for Lignocellulosic Composites

A. Pizzi*

LERMAB, University of Lorraine, 27 rue Philippe Seguin, 88000 Epinal, France

Dept. of Physics, King Abdulaziz University, Jeddah, Saudi Arabia

Abstract

Recent developments and trends in the field thermoset binders for thermoset lignocellulosic composites are reviewed. The more recent developments in tannin thermoset binders without the use of aldehyde-yielding compounds and Lignin binders are discussed. The combination of these for natural environment-friendly matrices for non-woven fibre mats is also briefly reviewed. Topical developments in protein-based binders, such as soy protein thermosets are addressed. New trends in carbohydrate thermosets as modifiers of existing composite binders, by forming furanic compounds then used as building blocks are addressed as well. Epoxidized unsaturated vegetable oil thermoset binders for lignocellulosic composites are described and an example of cashew nut shell oil modified by ozonolysis to yield thermoset resins by self-condensation is explored.

Keywords

Biobased Thermosets, Biobased Adhesives, Wood Composites, Fibre Composites, Tannins, Lignins, Proteins, Carbohydrates, Unsaturated Oils, Cashew Nut Liquid, Cardanol

Contents

4.1 Introduction

Biobased binders for lignocellulosic composites, such as wood or fibers composite panels, from renewable raw materials have now been a topic of considerable interest for many years. This interest, already present since the 1940s, became more intense with the world's first oil crisis in the early 1970s and subsided again as the cost of oil decreased. At the beginning of the 21st century this interest is becoming intense again for a number of reasons. The foreseen future scarcity of petrochemicals still appears to be reasonably far into the future. It is a contributing factor but, at this stage, it is not the main motivating force. The main impulse of today's renewed interest in bio-based thermoset resins for composites is the acute sensitivity of the general public to anything that has to do with the environment and its protection. It is not even this concern *per se* that motivates such an interest. There are rather very strict, almost crippling, government regulations which are just starting to be put into place to allay the environmental concerns of the public. All these factors play together and reinforce each other in contributing to the increasing interest in biobased thermosets for lignocellulosic materials.

First of all, it is necessary to define what is meant by bio-based thermoset resins, or binders from renewable, natural, non-oil-derived raw materials. This is necessary because in its broadest meaning the term might be considered to include urea-formaldehyde resins, urea being a non-oil derived raw material. This of course is not the case. The term "bio-based thermoset resin" has come to be used in a very well specified and narrow sense to only include those materials of natural, non-mineral, origin which can be used as such or after small modifications to reproduce the behaviour and performance of synthetic resins. Thus, only a limited number of materials can be currently included, at a stretch, in the narrowest sense of this definition. These are tannins, lignin, carbohydrates, unsaturated oils, proteins, and dissolved wood. The bio-based thermoset binders approach does not mean, however, to go back to the technology of natural product resins as they existed up to the 1920s and 1930s before they were supplanted by synthetic thermoset resins. The bio-based adhesives of which we are talking about here are derived from natural materials, but using or requiring novel technologies, formulations and methods. All these materials and approaches have been and will be further modified in the future in light of present day modern chemical knowledge.

Of the classes of bio-based thermosets mentioned above, in the case of tannins and lignins their interest has been directed primarily at substituting synthetic phenol-formaldehyde (PF) resins, due to the phenolic nature of these two classes of compounds. In some, but not all, of these cases some formaldehyde is still used, and in the case of lignin some other additives. It is then necessary to distinguish between biobased thermoset composite binders in which a limited amount of synthetic additives are still used, and bio-based thermoset composite binders where no synthetic additives are used.

This chapter deals with reviewing the newer technologies that can be implemented in thermoset bio-binders and composites, and not the already used, industrial technologies, that as good as they can be, are already described in depth in other reviews [1,2]

Wood products are the largest volume lignocellulosic biocomposites, and wood adhesives constitute more than 65% by volume of all the adhesives used in the world. It is such large volume that renders hard the substitution of existing synthetic adhesives with biosourced products, products possibly obtained from wood itself. For example, to substitute urea-formaldehyde (UF) binders, the worldwide consumption of which is of the order of 11 million tons resin solids per year is perfectly possible but not really feasible in the short term. It is for this reason that this review will address first this aspect.

4.2 Urea, also a natural material for wood adhesives

Urea is also a natural raw material. It is also obtained industrially in enormous quantities by catalytic reaction of oxygen and nitrogen of the air on glowing coals or other glowing carbon material, even charcoal or wood. The material to substitute, although even this can be of natural origin, is then formaldehyde, now classified toxic and oncogenic. It pays, seeing the volumes involved, to start concentrating to develop urea-based lignocellulosic binders using aldehydes that are not toxic, nor volatile, but still maintaining the commercially-sought after clear or white appearance of UF resins as such adhesives can be classified as natural too, urea being in the greater proportion. While many approaches can be taken to develop partially biosourced urea-based thermoset binders for lignocellulosic composites in this manner, recently a first important success in the bonding of plywood with this approach has been achieved [3)]. Resins based on urea-glyoxal for textiles are well known [4)] but these are low condensation resins not adaptable for wood. Hybrid resins urea-formaldehyde-another aldehyde have been the initial target of several researcher, with good results for plywood [5-7]. The problem of

Thermoset Composite

Materials Research Forum LLC

Materials Research Foundations 38 (2018)

doi: http://dx.doi.org/10.21741/9781945291876

these is that formaldehyde, although in much lower proportion, is still there. An old technology based on urea-furfural resins [8] is not a good substitute for urea-formaldehyde not only for the lower reactivity but mainly for the dark colour of the resin imparted by the condensation of furfural. Thus, the first truly urea binder for plywood without any formaldehyde that can be classified as mainly biosourced has only recently been developed opening a new chapter on natural environment friendly adhesives [3].

The nonvolatile and nontoxic aldehyde glyoxal (G) was used to substitute formaldehyde to react with urea (U) to synthesize a urea-glyoxal (UG) resin under weak acid conditions (pH = 4-5). The strength of the bonded plywoods was tested, and the curing process of the UG resin was studied by dynamic mechanical analysis (DMA). The results showed that the bonded plywood with dry shear strength of 0.98 MPa could be directly used as interior decoration and furniture material without formaldehyde emission in dry conditions [3]. The results of DMA analysis indicate that the cured system has best mechanical properties at 138.4 °C to 182.4 °C.

It is likely that such type of approach to urea-based resins can be expanded in future using other aldehydes, but it can constitute a major line of investigation in the short to medium term as regarding to biosourced adhesives for thermoset lignocellulosic composites. A further step forward in this direction was the use of ionic liquids as hardeners and additives to decrease the energy of activation of hardening of such urea-glyoxal binders rendering possible lignocellulosic composites such as particleboard needing much faster hot-pressing times. The results indicated that ionic liquids can be used as an efficient catalyst for UG resin. The composites prepared with ionic liquids had higher mechanical strength and dimensional stability compared to those made from urea-glyoxal resins containing traditional hardeners

Melamine can also be classified as biosourced as it is prepared by trimerisation of biosourced urea. Melamine-formaldehyde resins are a very important binder for thermoset lignocellulosic composites, presenting resistance to chemicals and water. To eliminate formaldehyde in melamine–formaldehyde (MF) resins again the nonvolatile and nontoxic aldehyde glyoxal (G) was reacted with melamine to prepare novel melamine-glyoxal (MG) resins [9]. As for urea-glyoxal binders the energy of activation of cross-linking and hardening of the MG resins was found to be higher than for MF resins. This rendered possible some applications such as MG resins impregnated paper surface overlays but showed such resins severe limitations as binders for thermoset lignocellulosic composites. They simply could not be used as binders for composites as their hardening started to occur at 180°C, too high a temperature for the core on present day lignocellulosic composites. For melamine too ionic liquids were used for unblocking this bottleneck for their use as thermoset binders. Melamine-glyoxal-glutaraldehyde adhesive resins for bonding wood panels were prepared by a single step procedure, namely reacting melamine with glyoxal and simultaneously with a much smaller proportion of glutaraldehyde (MGG' resin) [10]. No formaldehyde was used. The inherent slow hardening of this resin was overcome by the addition of N-methyl-2-pyrrolidone hydrogen sulphate ionic liquid as adhesive hardener in the glue-mix. The plywood strength results obtained were comparable with those obtained with melamine-formaldehyde resins pressed under the same conditions. Instrumental chemical analysis allowed the identification of the main oligomer species obtained and of the different types of linkages formed, as well as to indicate the multifaceted role of the ionic liquid. The role of ionic liquids was identified.

1. IL appears to catalyse the reaction of aldehydes, and aldehydes pre-reacted with urea or melamine, to yield aldol condensation. It allows aldol condensation even of aldehydes that have been pre-reacted with melamine or even with wood lignin. Reaction with lignin opens the possibility of reaction with the lignin in the wood substrate, and thus of some covalent bonding between an aldehyde based binder and the lignocellulosic substrate. While to advance such a hypothesis is possibly premature, it might also contribute to explain the good bonds obtained with aldehyde adhesives catalysed by ionic liquids.

2. IL catalyses the hardening of melamine-aldehyde resins to decrease hardening temperature, energy of activation of the hardening/condensation reaction and thus equally to improve their performance at equal temperature.

3. Moreover, IL affects the wood substrate by demethylation of lignin [10] rendering the substrate even more receptive to any type of adhesion.

However, it cannot be excluded that also other effects are induced by the presence of ILs.

4.3   Tannin thermoset binders for wood adhesives

The word tannin has been used loosely to define two different classes of chemical compounds of mainly phenolic nature: hydrolysable tannins and condensed tannins. The former are mixtures of simple phenols such as pyrogallol and ellagic acid and of esters of a sugar, mainly glucose, with gallic and digallic acids [2]. Their lack of macromolecular structure in their natural state, the low level of phenol substitution they allow, their low nucleophilicity, limited worldwide production, and higher price have somewhat decreased their chemical and economical interest. Some references to their utilisation, even industrial utilisation do however exist and are noteworthy:

In 1973 as a consequence of the first oil crysis, Norsechem, a malaysian adhesives factory belonging to a norwegian paint group, producing phenol-formaldehyde resins for South East Asia plywood manufacturers, was forced to implement a technological change that was maintained for three years in industrial production [11]. This technology consisted in substituting 33% by weight of phenol with chestnut tannin extract, an hydrolysable tannin, in the phenol-formaldehyde resin during its preparation. Up to 50% substitution in the laboratory was claimed but no glue-mix or other information on this was given [11]. The motive of this was purely economical, as phenol prices had skyrocketed and the cost of chestnut tannin extract was at that time much lower. In the only written reference that exists to this development [11], although results and basic material proportions are reported, resin formulations are not disclosed. A more recent research work has taken up again this approach using more modern phenol-formaldehyde formulations. Phenol-formaldehyde-chestnut tannin adhesives satysfying the relevant standards in which phenol:tannin weight ratios 30:70 were obtained and successfully used [12].

Condensed tannins, on the other hand, constituting more than 90% of the total world production of commercial tannins (220,000 tons per year), are both chemically and economically more interesting for the preparation of adhesives and resins. Condensed tannins and their flavonoid precursors are known for their wide distribution in nature and particularly for their substantial concentration in wood and bark of various trees. These include various *Acacia* (wattle or mimosa bark extract), *Schinopsis* (quebracho wood extract), *Tsuga* (hemlock bark extract), and *Rhus* (sumach extract) species, from which commercial tannin extracts are manufactured, and various *Pinus* bark extract species.

Condensed tannin extracts, consists of flavonoid units that have undergone varying degrees of condensation. They are invariably associated with their immediatte precursors (flavan-3-ols, flavan-3,4-diols), other flavonoid analogs [1,2], carbohydrates and traces of amino- and imino-acids [13,14]. Monoflavonoids and nitrogen-containing acids are present in concentrations which are too low to influence the chemical and physical charcteristics of the extract as a whole. However, the simple carbohydrates (hexoses, pentoses and disaccharides) and complex glucuronates (hydrocolloid gums) as well as oligomers derived from hydrolysed hemicelluloses are often present in sufficient quantity. Equally carbohydrate chains of various length [15] are also sometime linked to flavonoid unit in the tannin.

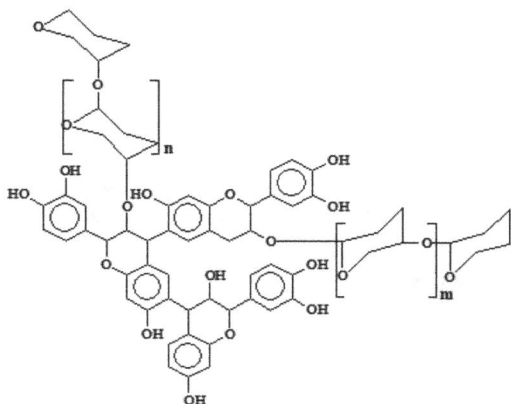

All these materials are often present in sufficient quantities to decrease and/or increase viscosity, and excessive variation in their percentages alters the physical properties of the natural tannin extract independently of the contribution of the degree of condensation of the tannin.

The repeating unit of tannin oligomers are based on different patterns of A and B-rings. To mainly profisetinidin and prorobinetinidin type tannins such as quebracho wood and mimosa bark tannins, are linked as counterpart generally the more common procyanidin type tannins such as pine bark and other tannins.

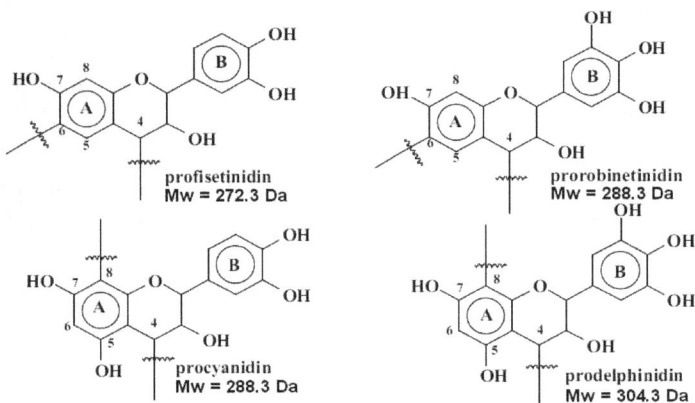

Reactivity and orientation of electrophilic substitutions of flavonoids. The relative accessibility and reactivity of flavonoid units is of interest for their use in resins and adhesives. The C8 site on the A-ring is the first one to react, for example with an aldehyde, and is the site, when free, of highest reactivity [1,6]. The C6 site on the A-ring is also very reactive but less than the C8 site as this latter presents lower steric hindrance too [1,6]. The reactions involve in general only these two sites on the A-ring. The B-ring is particularly unreactive. A low degree of substitution at the 6' site of the B-ring can occur. In general at higher pHs such as pH 10 the B-ring start to react too contributing to cross-linking as well [1,5,6,16,17].

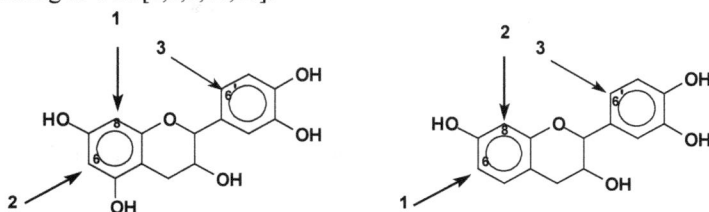

Thus, for catechins and phlorogucinol A-ring type flavonoids the reactivity sequence of the sites is C8 > C6 > C6' when these are free. For robinetinidin and fisetinidin, thus for resorcinol A-ring type flavonoids the reactivity sequence is modified to C6 > C8 > C6'. Due to the greater accessibility and lower possibility of steric hindrance of the C6 site [1,5,6,18]. The curve of gel time of flavonoid tannins with aldehydes has always the shape of a bell curve. The longer gel time is at around pH 4 and fastest gel times at lower pHs and higher pHs. The curve reaches an almost asymptotic plateau of very high reactivity and short gel time at around pH 10 and higher and a fast reactivity too at pHs lower than 1-2 [1,5,6]. The shape of this curve is always the same, but the gel time value is different for different tannins, being slower for mimosa and quebracho, and much faster for procyanidin-type tannins (such as pine) [1,5,6,16,17].

Finally, the reactions of tannins with formaldehyde and other aldehydes, that has led to their extensive application as wood adhesives, will not be treated here as extensively reviewed in depth already [1,5,6].

4.4 New technologies for industrial tannin adhesives

Extensive, up-to-date and in depth reviews of the technology of tannin adhesives based on the classical technology of tannin-formaldehyde resins already exist [1,13] and are beyond the scope of this review, as these technologies are commercial, now for several years, and used in a number of countries. It is sufficient to state here that tannin-

Materials Research Forum LLC

doi: http://dx.doi.org/10.21741/9781945291876

formaldehyde adhesives of very low emission (E0), fast pressing times and using unmodified tannin extracts are well-known, are used commercially and their technology is commercially and perfectly mastered [1,13]. The technologies of interest here are the new ones based either on no addition of aldehydes, or on the use of hardeners which are non-emitting or manifestly non toxic.

The quest to decrease or completely eliminate formaldehyde emissions from wood panels bonded with adhesives, although not really necessary in tannin adhesives due to their very low emission (as most phenolic adhesives), has nonetheless promoted some research to further improve formaldehyde emissions. This has centered into two lines of investigation: (i) the use of hardeners not emitting at all simply because either no aldehyde has been added to the tannin, or because the aldehyde cannot be liberated from the system, and (ii) tannins autocondensation. Methylolated nitroparaffins and in particular the simpler and least expensive exponent of their class, namely trishydroxymethyl nitromethane [19,20], belong to the first class. They function well as hardeners of a variety of tannin-based adhesives while affording considerable side advantages to the adhesive and to the bonded wood joint. In panel products such as particleboard, medium density fiberboard and plywood, the joint performance which is obtained is of the exterior/marine grade type, while a very advantageous and very considerable lengthening in glue-mix pot-life is obtained. Furthermore, the use of this hardener is coupled with such a marked reduction in formaldehyde emission from the bonded wood panel to reduce emission exclusively to the formaldehyde emitted by heating just the wood (and slightly less, thus functioning as a mild depressant of emissions from the wood itself). Furthermore, trishydroxymethyl nitromethane can be mixed in any proportion with traditional formaldehyde-based hardeners for tannin adhesives, its proportional substitution of such hardeners inducing a proportionally marked decrease in the formaldehyde emissions of the wood panel without affecting the exterior/marine grade performance of the panel. Medium density fibreboard (MDF) industrial plant trials confirmed all the properties reported above and the trial conditions and results are reported [20,21]. Equally, formulations in which tannins are hardened by furfuryl alcohol just as a glue-mix hardener or even are prereacted in a reactor with furfuryl alcohol have been used and extensive tempered hardboard plant trials carried out [9]. A cheaper but as equally effective alternative to these approaches are the use of acetone-formaldehyde resins as hardeners or even better is the use of hexamine as a tannin hardener.

4.5 Tannin-Hexamethylenetetramine (Hexamine) adhesives and adhesives with alternative aldehydes

Under many wood adhesive application conditions, contrary to what was thought for many years in the past, hexamine used as a hardener of a fast reacting species is not at all a formaldehyde-yielding compound, yielding extremely low formaldehyde emissions in bonded wood joints [22]. ^{13}C NMR evidence has confirmed [23-25] that the main decomposition (and recomposition) mechanism of hexamine under such conditions is not directly to formaldehyde. It rather proceeds through reactive intermediates, hence mainly through the formation of reactive imines and iminoaminomethylene bases **(Fig. 4.1)**.

^{13}C NMR evidence has also confirmed [22-25] that in presence of chemical species with very reactive nucleophilic sites, such as melamine, resorcinol and condensed flavonoid tannins, hexamine does not decompose to formaldehyde and ammonia. Instead, the very reactive but unstable intermediate fragments react with the tannin, melamine, etc. to form aminomethylene bridges before any chances to yield formaldehyde. These are also stable up to 5 hours at temperatures as high as 120°C. The intermediate fragments of the decomposition of hexamine pass first through the formation of imines followed by their decomposition to imino-methylene bases. The latter present only one positive charge as the second methylene group is stabilized by an imine-type bond [22-25] **(Fig. 4.1)**.

Any species with a strong real or nominal negative charge under alkaline conditions, be it a tannin, resorcinol or another highly reactive phenol, be it melamine or another highly reactive amine or amide, or be it an organic or inorganic anion is capable of reacting with the intermediate species formed by decomposition (or recomposition) of hexamine far more readily than formaldehyde [22-25]. This explains the capability of wood adhesive formulations based on hexamine to give bonded panels of extremely low formaldehyde emissions. If no highly reactive species with strong real or nominal negative charge is present then decomposition of hexamine proceeds rapidly to formaldehyde formation as reported in the previous literature [26].

On this basis, the use of hexamine as a hardener of a tannin, hence a tannin-hexamine adhesive is a very environmentally-friendly proposition. Formaldehyde emissions in a great chamber have been proven to be so low to be limited exclusively to what is generated by the wood itself, hence truly E0 panels. The panels obtained with tannin-hexamine adhesives, according to under which conditions they are manufactured, can satisfy both interior and exterior grade standard specification requirements [27]. Steam injection presses recently have shown to be better suited to give better results for exterior grade boards using tannin-hexamine adhesives [28]. Typical laboratory particle board and full industrial scale plant trials yielded panel strength results as indicated in Table 4.1 [29].

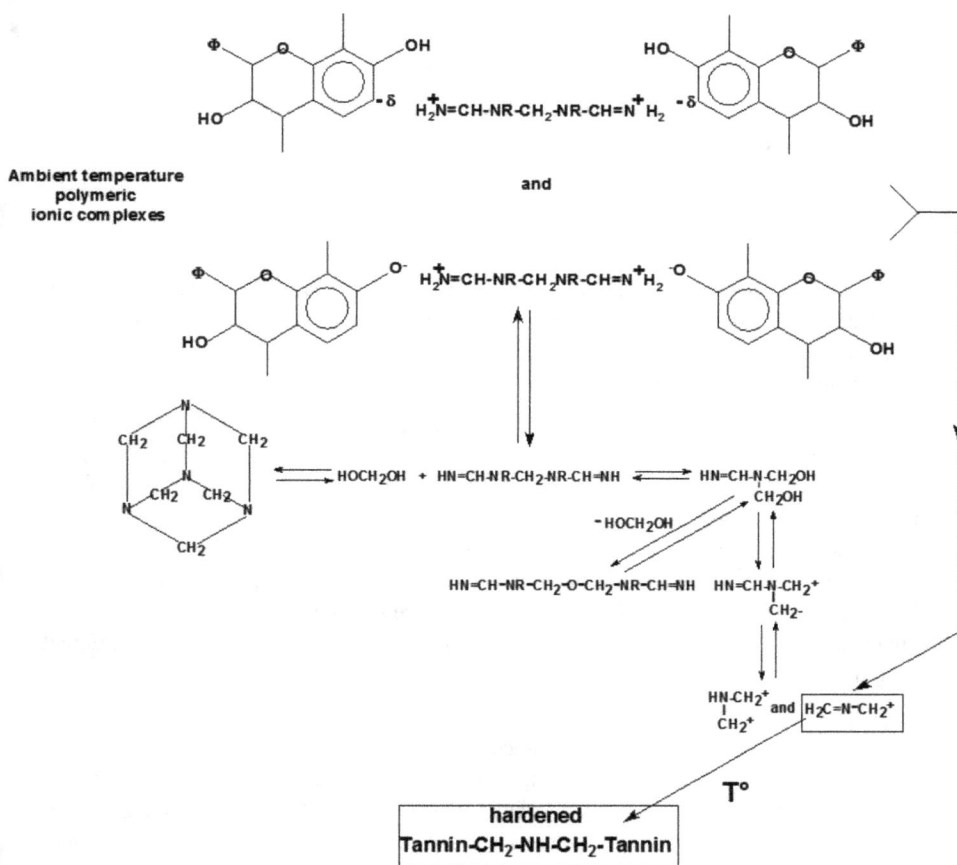

Figure 4.1 Schematic representation of the decomposition of hexamine to iminoamino methylene basis in presence of a reactive species such as tannin to form (i) ionic polymeric complexes at ambient temperature and (ii) stable benzylamine covalently bridged network in hardening, at higher temperature, without producing or releasing any formaldehyde.

Table 4.1. Typical results of laboratory and industrial particleboard bonded with mimosa tannin extract hardened with 6.5% hexamine [15].

	IB dry (MPa)	IB 2h boil tested, redried (MPa)	density (kg/m3)
Laboratory	0.92	0.27	711
Industrial	0.58	-	680

Comparable results as those reported in Table 4.1 are obtained with pine tannins or other procyanidins hardened with hexamine [30]. In the same reference, catalysis of the reaction in the presence of small amounts of accelerators such as a zinc salt allow even better results or faster press times.

As regards to alternative aldehydes two different aldehydes have been used successfully to obtain lignocellulosic thermoset composites, namely furfural, and even better furfuryl alcohol, and vanillin and its derivates. As regards vanillin an adhesive based on the reaction of a very fast reacting procyanidin-type condensed tannin, namely purified pine bark tannin, and vanillin, a food-grade non-toxic slow-reacting aldehydes derived from lignin was shown to satisfy well the relevant standards for bonding wood particleboard. Vanillin and a dialdehyde derivative of vanillin were the aldehydes used. The oligomers obtained and their distribution has also been determined by a number of different analytical techniques [31].

Several aldehydes other than formaldehyde were tested also with tannin based thermosets [32,33]. Network formation, cure characteristics and bonding performance of tannin-based thermoset resins were investigated in order to establish structure-property relationships between the stage B and stage C. Tannin-aldehyde and base-catalyzed auto-condensed tannin resins were synthesized and characterized for molecular weight distribution, cure kinetics and cure chemistry. Resins prepared with highly reactive aldehydes, such as formaldehyde or glyoxal, exhibited a significant extent of hetero-condensation reactions, fast cure kinetics, a high storage modulus and good solvent stability of the stage C-resin [32,33]. In contrast, thermoset resins prepared with bulky aldehydes of low reactivity, such as furfural [13] and citral [33], were dominated by auto-condensation reactions, and exhibited slower cure kinetics, a lower storage modulus and solvent-stability of the stage C-resin, alike those neat auto-condensed tannin resins.

However, all resin systems fulfilled the standard requirements for thermoset composites bonding for interior applications. Additionally, storage modulus increase during cure was found to be a good predictor of the stiffness of the wood-bonded assembly, useful to discriminate between the auto-condensation (see following paragraph) and hetero-condensation cure chemistries.

A case apart is the coreaction of tannin with furfuryl alcohol without aldehydes. This reaction mainly used for the preparation of thermoset foams [34,35] it is also usable for lignocellulosic thermoset composites, but attention must be paid to bond these under alkaline pH [36], or if acid setting is used [36,37] then a self-neutralizing system needs to be used to avoid acid degradation of the lignocellulosic substrate.

4.6   Hardening by tannins autocondensation

The auto-condensation reactions characteristic of polyflavonoid tannins have only recently been used to prepare adhesive polycondensates hardening in absence of aldehydes [38]. This auto-condensation reaction is based on the opening under either alkaline or acidic conditions of the O1-C2 bond of the flavonoid repeating unit and the subsequent condensation of the reactive center formed at C2 with the free C6 or C8 sites of a flavonoid unit on another tannin chain [38-42]. Although this reaction may lead to considerable increases in viscosity, gelling does not generally occur. However gelling takes place when the reaction occurs (i) in presence of small amounts of dissolved silica (silicic acid or silicates) catalyst and some other catalysts [38-42], and (ii) on a lignocellulosic surface [38].

As in the case of other formaldehyde-based resins, the interaction energies of tannins with cellulose obtained by molecular mechanics calculations [43] tend to confirm the effect of surface catalysis induced by cellulose also on the curing and hardening reaction of tannin adhesives. The considerable energies of interactions obtained can effectively explain weakening of the heterocyclic ether bond leading to accelerated and easier opening of the pyran ring in a flavonoid unit, as well as the facility with which hardening by auto-condensation can occur. In the case of the more reactive procyanidins and prodelphinidin type tannins, such as pine tannin, cellulose catalysis is more than enough to cause hardening and to produce boards of strength satisfying the relevant standards for interior grade panels [38]. The internal bond (IB) strengths of interior particleboard prepared using different types of commercial tannins are shown **in Fig. 4.2.**

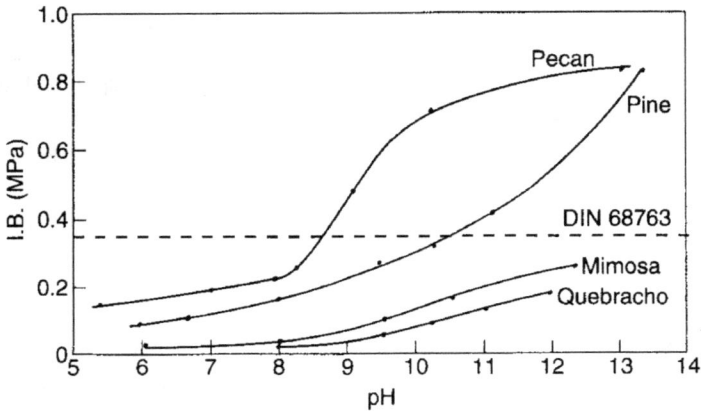

Figure 4.2 Dry internal bond (IB) strength as a function of tannin solution pH of laboratory particleboards prepared with for different commercial tannin extracts without any aldehyde hardener, using the lignocellulosic substrate-induced tannin auto-condensation reaction.

The slower reacting tannins can yield an upgraded IB strength of the board when mixed with small amounts of faster reacting tannins such as pine bark or pecan nut tannins. In the case of the less reactive tannins, however, such as mimosa and quebracho, the presence of a dissolved silica or silicate catalyst of some type is the best manner to achieve panel strength as required by the relevant standards [1]. Auto-condensation reactions have been shown to contribute considerably to the dry strength of wood panels bonded with tannins, but to be relatively inconsequential in contributing to the bonded panels exterior grade properties which are rather determined by polycondensation reactions with aldehydes [43-45]. Combination of tannin auto-condensation and reactions with aldehydes, and combination of radicals with ionic reactions, have been used to decrease both the proportion of aldehyde hardener as well as to decrease considerably the already low formaldehyde emissions yielded by the use of exterior tannin adhesives [43-45]. A variation on the same theme of wood adhesives by tannin auto-condensation is acid-catalysed oxidative condensation [46].

These technologies, tannin auto-condensation and hexamine hardening are already in industrial use and are under consideration further by number of industrial groups worldwide.

Thermoset Composite
Materials Research Foundations 38 (2018)

Materials Research Forum LLC
doi: http://dx.doi.org/10.21741/9781945291876

Tannin-Hexamine thermoset resins have been used, at a different level, for fibrous high strength composites in which a non-woven natural fibres mat, fibres such as linen, kenaf, hemp etc., is impregnated with up to 50%-60% of total;weight with tannin-hexamine resin [47]. Both higher density thin composites as well as lower density thicker composites were prepared. Two natural matrices types were used: (i) commercial mimosa flavonoid tannin extract with 5% hexamine added as hardener, and (ii) a mix of mimosa tannin+hexamine with glyoxalated organosolv lignin of low molecular weight, these two resins mixed 50/50 by solids content weight. The results obtained were good [47]. The composites made with the mix of tannin and lignin resins as a matrix remained thermoplastic after a first pressing. The flat sheets prepared after the first pressing were then thermoformed into desired shape.

4.7 Lignin adhesives

Much has been written about, and much research has been conducted in the use of lignins as binders for lignocellulosic composites. It can safely be said that this natural raw material has probably been the most intensely researched one as regards composite binders application. Lignins are phenolic materials, they are abundant and of low cost but they have lower reactivity towards formaldehyde, or other aldehydes, than even phenol. Lignin is a phenolic polymer that is one of the main polymeric constituents of wood.

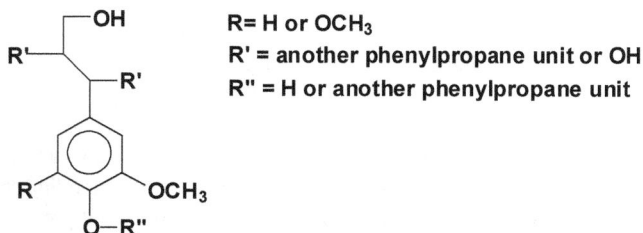

R= H or OCH$_3$
R' = another phenylpropane unit or OH
R" = H or another phenylpropane unit

It is generally produced in great quantities as waste from paper pulp mills. It is composed of repeating phenylpropane units. Extensive reviews on a number of proposed technologies of formulation and application do exist, and the reader is referred to these in earnest [1, 48-58]. This field is however remarkable for how small has been the industrial success in using these materials. In general, lignin and lignosulphonates have been mixed in smaller proportions to synthetic resins, such as PF resins [51-54], and even UF resins [55], to decrease their cost as composite binders. Their low reactivity and lower level of reactive sites, however, conjures that for any percentage of lignin added the cost advantage is abundantly lost in the lengthening of the composite manufacturing time this causes. The only step forward that has found industrial application in the last twenty years is to pre-react in a reactor lignin with formaldehyde to form methylolated lignin, thus to do part of the reaction with formaldehyde first, and then add this methylolated lignin to PF resins at the 20% to 30% level [51,53] and to-day up to the 50% level [59]. These resins have been used in some North American plywood mills [51]. Plywood is a lignocellulosic composite for which the pressing time is not a factor for determining the output rate of the factory and hence one can afford to use relatively long press times to obtain good results [51].

None of the many binder systems based on pure lignin resins, hence without synthetic resin addition, have succeeded commercially at an industrial level, at least not for long times. Some were tried industrially but for one reason or another, too long a pressing time, high corrosiveness for the equipment etc., they did not meet with commercial success. Still notable among these is the Nimz system based on the networking of lignin in presence of hydrogen peroxide [1,2,57,58]. Only one system is used successfully still to-day, but this only for high density hardboard, in several mills worldwide. This is the Shen system, based on the self coagulation and cross-linking of lignin by a strong mineral acid in the presence of some aluminium salt catalysts [57,58,60,61]. However, attempts to extend this system to the industrial manufacture of other lignocellulosic composites such as medium density fiberboard (MDF) are known to have failed.

Of interest in the MDF field is also the system of adding laccase enzyme-activated lignin to the fibers or activating the lignin in situ, in the lignocellulosic fibres also by enzyme treatment [62,63]. The results obtained however, yielded composites that did not satisfy the relevant standards, and this at very long high temperature manufacturing times. The researchers involved obviated this successfully by adding some 1% isocyanate (PMDI) to the board [62] and pressing at acceptably short press times, or by extending the pressing times to ridiculous lengths (100 s/mm board thickness while industrial press times are of the order of 3 to 7 s/mm board thickness) [64]. In the former case a synthetic adhesive had to be used, with the same result obtained by pressing untreated hardboard, a 100 year old process, hence just wasting expensive time and enzymes. The second case instead illustrates even more clearly where the problem lies and what breakthrough is necessary: enzyme mobilizing lignin works, but not fast enough. The breakthrough necessary is a new, strong catalyst of the enzymatic action capable of allowing hot-pressing times of industrial significance. This has not been found, or even considered, as yet.

A promising new technology based on lignin use for thermoset lignocellulosic composite binders is relatively recent and uses again pre-methylolated lignin in presence of small amounts of a synthetic PF resin, and polymeric 4'4'-diphenyl methane diisocyanate

(PMDI) [65,66]. The proportion of pre-methylolated lignin used is of 65% of the total adhesive, the balance being made-up of 10-15% PF resin and of 20-25% PMDI. This thermoset resin presses at very fast pressing times, well within the fastest range used to-day industrially, contains a high proportion of lignin, and yields exterior grade lignocellulosic composites [45,46]. PF resin and methylolated lignin cure accelerators such as triacetin, other esters, or others, can also be used [65,66] notwithstanding that just the presence of the PMDI already gives a considerable acceleration to the curing rate (Table 4.2). The system is based on cross-linking caused by the simultaneous formation of methylene bridges and of urethane bridges, overcoming with the latter the need for higher cross-linking density that has been one of the problems which has stopped lignin utilisation in the past.

Table 4.2. Results of laboratory particleboard bonded with methylolated kraft lignin-based adhesives co-reacted at the glue-mix level with PMDI and a synthetic PF resin. The proportions of the three materials are by weight.

	Press time (s/mm)	IB dry (MPa)	IB 2h boil tested redried (MPa)	density (kg/m^3)
MDI/PF/Kraft Lignin (methylolated) 22/26/52 by weight + ester accelerator	10	0.85	0.53	687
MDI/PF/Kraft Lignin (methylolated) 11/26/63 by weight + ester accelerator	20	0.68	0.53	696
MDI/PF/Kraft Lignin (methylolated) 11/26/63 by weight + ester accelerator	10	0.53	0.43	680

Recently, on the same principle, a thermoset binder for thermoset wood composites based on a 50/50 mix of tannin/hexamine with pre-glyoxalated lignin, glyoxal being a non-toxic and non-volatile albeit less reactive aldehyde, has been developed yielding interesting results being 100% bio-based [67]. Oxazolidines have also been used as lignin thermoset resin hardeners of low formaldehyde emission [68].

Other than the above there is not much new in this field, but just literature rehashing older systems all based on the substitution of some phenol in PF resins. In general these papers do not seem to be aware of the slow pressing time problem, and they do not address it, perpetuating the myth of PF/lignin resins while repeating the same age-old errors. They lead new researchers in the field to believe they are doing something worthwhile with parameters that do not satisfy the requirements of press rate of the panel manufacturing industry.

All the above does not mean that there may not be new opportunities in lignin-based thermoset lignocellulosic composites. It simply means that further new, alternative technologies need to be developed before lignin can be used commercially and extensively for application to such composites.

4.8 Protein adhesives

Intense research induced by the sponsorship of the USA United Soybean Board has revived interest in protein adhesives, soya protein primarily, but also others. These technologies must not be confused with the age-old protein and bone glues used in carpentry, or blood used as an additive in plywood glue-mixes. Certainly, of the traditional technologies, one stands out head and shoulder above the others, namely casein adhesives. These are still produced, and still used industrially to very good effect in some special plywood and related products. They work as such, they are very environmentally friendly and their technology was completely mastered a long time ago. They are definitely strong candidates for future expansion, in their actual form or with some further technological improvement. New is some interesting work on the correlation between the bond strength obtained and the molecular architecture of the protein [69].

Soy protein hydrolysates- and soyflour-based adhesives instead are definitely new. Both addition to traditional synthetic wood adhesives, as well as their use as panels adhesives after partial hydrolysis and modifications have been reported, and with acceptable results it appears [70-73]. These products are used industrially/commercially but only one company in the USA, although the pressing times appear to be long in relation to what envisaged with synthetic adhesives in the panel industry today. Aldehyde-prereacted soy [74] or other proteins (such as wheat gluten for example [75]) resins mixed with synthetic adhesives have been reported with good results [74-76] especially where the reinforcing synthetic resin is 20%-30% MDI (polymeric 4,4' diphenylmethane diisocyanate) and some use of them can be envisaged in the future too.

4.9 Carbohydrate adhesives

Carbohydrates in the form of polysaccharides, gums, oligomers and monomeric sugars have been employed in adhesive formulations for many years. Carbohydrates can be used as wood panel adhesives in three main ways: (i) as modifiers of existing PF and UF adhesives, (ii) by forming degradation compounds which then can be used as adhesives building blocks, and (iii) directly as wood adhesives. The second route above leads to furanic resins. Furanic resins, notwithstanding that their basic building blocks, furfuraldehyde and furfuryl alcohol, are derived from the acid treatment of the carbohydrates in waste vegetable material, are considered today as purely synthetic resins [77]. This opinion might need to change. However, both compounds are relatively expensive, very dark-coloured and furanic resins have made their industrial mark in fields where their high cost is not a disadvantage. They can be used very successfully for panel adhesives, they are used very successfully in other fields (as foundry core binders) but the relatively higher toxicity of furfuryl alcohol before it is reacted is a problem that will have to be taken into consideration if these resins are to be considered for wood products.

The use of carbohydrates directly dissolved in strong alkali as wood panel adhesives is not a new concept, but is an interesting and a topical one today. This technology has been extensively reported [78]. All sorts of agricultural cellulosic materials have been successfully adapted to this technology and the technology and its application has been extensively reported in the past [78].

Research on the first route has centered particularly on the substitution of carbohydrates for parts of PF resins. It has been reported that at laboratory level up to 50% to 55% of phenol in a PF resin can be substituted with a variety of carbohydrates, from glucose to polymeric, tree-derived hemicelluloses [79-83]. Apparently reducing sugars could not be used directly as they are degraded to saccharinic acids under the acid conditions required in the formulation of the resin. Reducing sugars can be used to successfully modify PF resins if they are reduced to the corresponding alditols or conveted to glycosides. Some carbohydrates appeared to be incorporated into the resin network mainly through ether bridges [79]. Generally the resin is prepared by co-reacting phenol, the carbohydrate in high proportion, a lower amount of urea and formaldehyde. Extensive and rather successful industrial trials of these resins have also been reported [83].

Carbohydrate-based adhesives in which the formulation starts with the carbohydrate itself have also been reported, but the acid system used during formulation readily degrades the original carbohydrate to furan intermediates which then polymerize. An interesting concept that was advanced early on in carbohydrate adhesives research was the conversion of the carbohydrate to furanic products *in situ*, which then homopolymerize as well as react with the lignin in wood.

Several research groups [79,84,85] have recently described the use of liquefied products from cellulosic materials, literally liquefied wood, which showed good wood adhesive properties. Lignocellulosic and cellulosic materials were liquefied in presence of sulphuric acid under normal pressure using either phenol or ethylene glycol. The cellulosic component in wood was found to lose its pyranose ring structure when liquefied. The liquefied product contains phenolic groups when phenol is used for liquefaction. In the case of ethylene glycol liquefaction, glucosides were observed at the initial stage of liquefaction and levulinates after complete liquefaction.

4.10 Unsaturated oil adhesives

Saturated and unsaturated vegetable oils are now widely available as a bulk commodity for a variety of purposes and at very acceptable prices. All resin research to date has focused on oils that contain at least one double bond. These oils are predominantly a mixture of triglycerides, hence esters, with a small quantity of free fatty acid, the small proportion of free fatty acid being dictated both by the plant species and the extraction conditions. As the number of unsaturations increases so does the overall molecule reactivity and its potential for side reactions.

Until fairly recently only two examples could be found in the literature where seed oil derivatives were being employed as wood adhesives. Linseed oil, for example, has been used to prepare a resin that can be used as an adhesive or surface coating material [86,87]. The chemistry of this resin centers on an epoxidation of the oil double bonds followed by cross-linking with a cyclic polycarboxylic acid anhydride to build-up molecular weight. The reaction is started by the addition of a small amount of polycarboxilic acid.

When the epoxidized oil resin was evaluated as a wood adhesive in composite panels it could be tightly controlled through the appropriate selection of triglycerides and polycarboxylic anhydrides. This apparently enables a wide range of materials with quite different features to be manufactured. The use as wood adhesives is one among the many uses, the focus of the development being more on plastic materials. The literature states that this plastic is well suited for use as a formaldehyde free binder for wood fibres and wood particles including fibers and chips from cereal residues, such as straw and fibre mats.

The literature on this resin [86] claims that cross-linking can be varied through the addition of specialised catalysts and several samples were prepared at a range of temperatures (120°-180°C) that exhibited high water tolerance even at elevated temperatures, but no actual test data were included. Since the resin of reference [86], research in a number of other countries by imitators has produced very similar epoxidized oil resins. These are suitable for a number of applications, but the writer has tested one or two of them finding that for wood adhesives application these resins have two major defects: (i) their hot-pressing time is far too slow to be of any interest in wood panel products, with the exception perhaps of plywood (for which they have not been tried), and (ii) they are relatively expensive. Unless the slow hot-pressing problem is overcome, and at a reasonable price, these resins will remain at the stage of potential interest. There is no doubt that these resins can be of interest in other fields, but it is symptomatic that no industrial use for wood panel adhesives has been reported as yet, or is known to have occurred. Thus the use of unwoven natural fibre mats impregnated with epoxidized unsaturated oils and hardened by using maleic anhydride has potential for applications in which the resin forms a high proportion of the composite [87].

Bioresins based on soy bean and other oils have been developed also by other groups, mainly for replacement of polyester resins [88]. These liquid resins were obtained from plant and animal triglycerides by suitably functionalising the triglyceride with chemical groups (e.g., epoxy, carboxyl, hydroxyl, vinyl, amine, etc.) that render it polymerisable. The reference claims that excellent inexpensive composites were made using natural fibres such as hemp, straw, flax and wood in fibre, particle and flake form. That soy oil based resins have a strong affinity for natural fibres and form a good fibre-matrix interface as determined by scanning electron microscopy of fractured composites. The reference also stated that these resins can be viewed as candidate replacements for phenol-formaldehyde, urethane and other petroleum-based binders in particleboard, MDF, OSB and other panel types. However, no actual test data has been supplied, and no industrial use in wood panel adhesives has actually been reported as yet.

Cashew nut shell liquid, mainly composed of cardanol but containing also other compounds (**Fig. 4.3**), is an interesting candidate for wood based resins. Its dual nature, phenolic nuclei+unsaturated fatty acid chain, makes a potential natural raw material for the synthesis of water resistant resins and polymers. Cardanol resins are known from the past, but their use has not been very diffuse simply because the raw material itself was rather expensive. The price however appears to be more affordable now since the extensive cashew nut plantations in Mozambique are in production.

The phenol, often resorcinol, group and/or double bonds in the chain can be directly used to form hardened networks. Alternatively, more suitable functional groups such as aldehyde groups and others can be generated on the alkenyl chain. Generally, modifications of this kind take several reaction steps, rendering the process too expensive for commercial exploitation in wood adhesives.

However, the Biocomposite Center in Wales [89,90] has developed a system of ozonolysis [89,90] in industrial methylated spirit [90] through which an aldehyde function is generated on the alkenyl chain of cardanol. The first reaction step yields as major product a cardanol hydroperoxide that following reduction by glucose or by zinc/acetic acid yields a high proportion of cardanolaldehyde groups. These cross-link with the aromatic groups of cardanol itself, thus a self condensation of the system yielding hardened networks (**Fig. 4.4**).

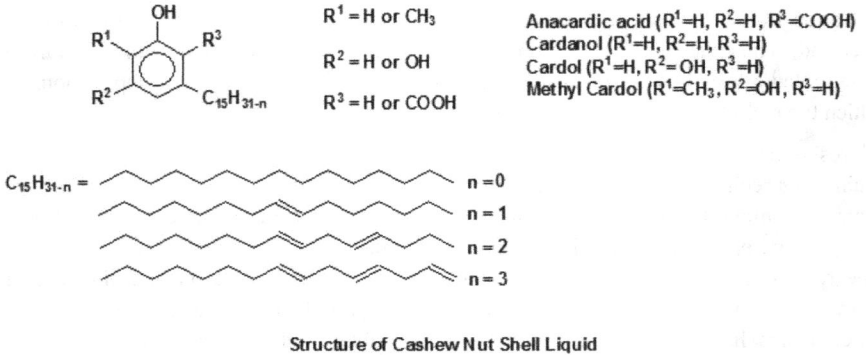

R^1 = H or CH_3

R^2 = H or OH

R^3 = H or COOH

Anacardic acid (R^1=H, R^2=H, R^3=COOH)
Cardanol (R^1=H, R^2=H, R^3=H)
Cardol (R^1=H, R^2=OH, R^3=H)
Methyl Cardol (R^1=CH_3, R^2=OH, R^3=H)

$C_{15}H_{31-n}$ =

n = 0
n = 1
n = 2
n = 3

Structure of Cashew Nut Shell Liquid

Figure 4.3   Chemical composition and type of compounds in cashew nut shell liquid (CNSL).

Table 4.3   Laboratory results for ozonolysis/reduction cashew nut shell liquid bonded particleboard. Single lap shear bonds cured at 180°C for 3 minutes and internal bond results of laboratory particleboard (percentage resin load of 10%) (press times, board density and other conditions were not reported) [89].

	Lap shear bond (MPa)	IB dry (MPa)	IB 2h boil (MPa)
Phenol-formaldehyde control	5.55	0.69	*
CNSL aldehyde (Zinc/acetic acid reduced)	6.77	1.05	0.58
CNSL aldehyde(glucose reduced)	6.02	*	*

* not reported

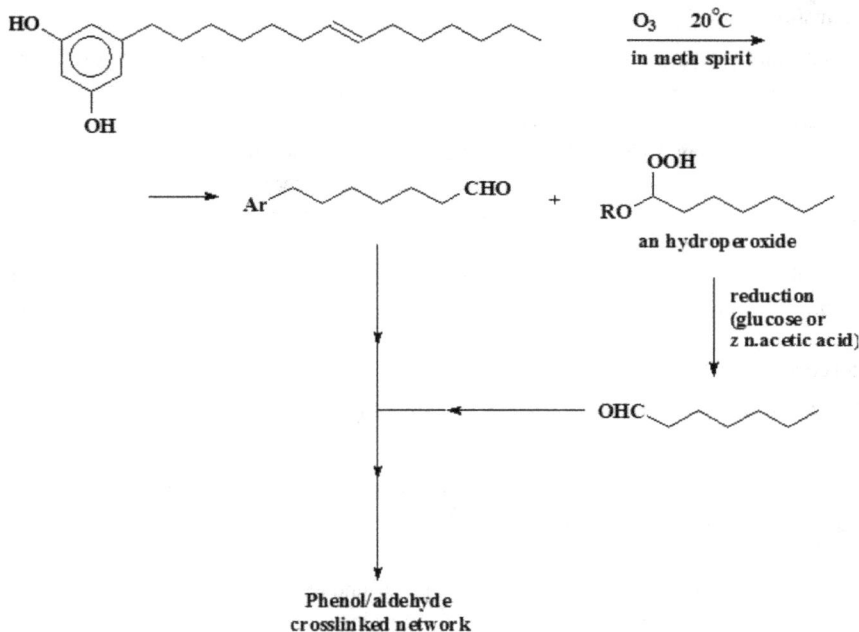

Figure 4.4 Ozonolysis of CNSL to produce cardanolaldehyde and a hydroperoxide, the latter further reduced to another aldehyde. The two aldehyde groups and the reactive sites on the aromatic ring react readily to form cross-linked network during hardening.

Exploratory laboratory particle board and lap shear bonding yielded the results shown in Table 4.3. The results reported are good.

Nonetheless neither the press times used nor other essential conditions that could help to evaluate the economical feasibility of these products were reported [89]. It remains to evaluate if the cost of the ozonolysis allows wood adhesives of a suitably low cost and, again, if the pressing times can match those of industrial resins. However, the resorcinolic structure of the cardanol phenol group would appear to indicate that the molecule should be able to achieve industrial pressing times.

Conclusions

In conclusion, bio-based thermoset binders for lignocellulosic thermoset composites are alive and well, and research on their development and application is definitely expanding. Very serious industrial interest exists in these products, always with an eye on their environmental acceptability and also on their economic and technical viability. Some clear industrial applications are already emerging. Almost certainly, further materials for bonding such composites will emerge in the future, as well as further improvements in the materials presented in this review will occur. There is equally no doubt that at first perhaps for niche applications, and later on for more widespread applications the use of these materials is likely to expand.

References

[1]   A. Pizzi, Advanced Wood Adhesives Technology, Marcel Dekker, New York, 1994

[2]   A. Pizzi, Natural Phenolic Adhesives 2: Lignin, in A.Pizzi, K.L.Mittal (Eds.), Handbook of Adhesive Technology, 2nd Edition, Marcel Dekker, New York, 2003, pp. 589-598.

[3]   S. Deng, G. Du, X. Li, .A. Pizzi, Performance and reaction mechanism of zero formaldehyde-emission urea-glyoxal (UG) resin, J.Taiwan Inst.Chem.Engineers, 45 (2014) 2029-2038. https://doi.org/10.1016/j.jtice.2014.02.007

[4]   H. Petersen, Process for the production of formaldehyde-free finishing agents for cellulosic textiles and the use of such agents, Textilveredlung 2 (1968) 51-62.

[5]   J. Zhang, H. Chen, A. Pizzi, Y. Li, Q. Gao, J. Li, Characterisation and application of urea-formaldehyde-furfural co-condensed resins as wood adhesives, BioResources 9 (2014) 6267-6276. https://doi.org/10.15376/biores.9.4.6267-6276

[6]   Y.F. Zhang, X.R. Zeng, B.Y. Ren, Synthesis and structural characterization of urea-isobutyraldehyde-formaldehyde resins, J.Coatings Technol.Res. 6 (2009) 337-344. https://doi.org/10.1007/s11998-008-9126-4

[7]   S. Deng, A. Pizzi, G. Du, Jizhi Zhang, Jun Zhang, Synthesis and chemical structure of a novel glyoxal-urea-formaldehyde (GUF) co-condensed resins with different MMU/G molar ratios by ^{13}CNMR and MALDI-TOF-MS, J.Appl.Polym.Sci. (2014). https://doi.org/10.1002/app.41009

[8]   E.E. Novotny and W.W. Johnson, U.S. Patent 1,827,824 (1931).

[9]   S. Deng, A. Pizzi, G. Du, M.C. Lagel, L. Delmotte, S. Abdalla, Synthesis, structure characterization and application of melamine-glyoxal adhesive resins, Eur.J.Wood Prod. 76 (2018) 283-296. https://doi.org/10.1007/s00107-017-1184-9

[10]   X. Xi, A. Pizzi, S. Amirou, Melamine-Glyoxal-Glutaraldehyde wood panel adhesives without formaldehyde, Polymers, 10 (2018) 22: 1-18.

[11]   E. Kulvik, Chestnut wood tannin extract in plywood adhesives, Adhesives Age 19 (1976) 19-21.

[12]   S. Spina, X. Zhou, C. Segovia, A. Pizzi, M. Romagnoli, S. Giovando, H. Pasch, K. Rode, L .Delmotte, Phenolic resin adhesives based on chestnut hydrolysable tannins, J.Adh.Sci.Technol. 27 (2013) 2103-2111. https://doi.org/10.1080/01694243.2012.697673

[13]   A. Pizzi, Tannin based adhesives, in A. Pizzi (Ed.), Wood Adhesives Chemistry and Technology, Vol. I, Marcel Dekker, New York, 1983, pp. 177-246.

[14]   A. Pizzi, Natural Phenolic Adhesives 1: Tannin, in A.Pizzi, K.L.Mittal (Eds.), Handbook of Adhesive Technology, 2nd Edition, Marcel Dekker, New York, 2003, pp.573-588. https://doi.org/10.1201/9780203912225.ch27

[15]   S. Drovou, A. Pizzi, C. Lacoste, J. Zhang, S. Abdalla, F.M. Al-Marzouki, Flavonoid tannins linked to long carbohydrate chains – MALDI ToF analysis of the tannin extract of the african locust bean. Ind.Crops Prod. 67 (2015) 25-32. https://doi.org/10.1016/j.indcrop.2015.01.004

[16]   W.E. Hillis, G. Urbach, The reaction of (+) catechin with formaldehyde, J.Chem.Technol.Biotechnol. 9(1959), 474-482

[17]   W.E. Hillis, G. Urbach, Reaction of polyphenols with formaldehyde, J.Chem.Technol.Biotechnol. 9(1959), 665-673.

[18]   D.G. Roux, D. Ferreira, H.K.L. Hundt, E. Malan, Structure, stereochemistry, and reactivity of natural condensed tannins as basis for their extended industrial application. J.Appl.Polym.Sci., Appl.Polym.Symp. 28 (1975) 335-353.

[19]   A. Trosa, A. Pizzi, A no-aldehyde emission hardener for tannin-based wood adhesives, Holz Roh Werkst., 59 (2001), 266-271. https://doi.org/10.1007/s001070100200

[20]   A. Trosa, Développement et application industrielle de résines thermodurcissables à base de produits naturels de déchet et leur produits de copolymérisation avec des résines synthétiques pour application aux panneaux composites de bois, Ph.D. thesis, University Henri Poincaré – Nancy 1, Nancy, France (1999)

[21]   A. Trosa, A. Pizzi, Industrial hardboard and other panels binder from tannin/furfuryl alcohol in absence of formaldehyde, Holz Roh Werkst. 56 (1998), 213-214. https://doi.org/10.1007/s001070050301

[22]   A. Pizzi, Hardening mechanism of tannin adhesives with hexamine, Holz Roh Werkst. 52 (1994), 229. https://doi.org/10.1007/BF02619098

[23]   F. Pichelin, C. Kamoun, A. Pizzi, Hexamine hardener behaviour – effects on wood glueing, tannin and other wood adhesives, Holz Roh Werkst. 57(1999), 305-317. https://doi.org/10.1007/s001070050349

[24]   C. Kamoun, A. Pizzi, Mechanism of hexamine as a non-aldehyde polycondensation hardener, Holzforschung Holzverwertung, 52(2000), 16-19.

[25]   C. Kamoun, A. Pizzi, M. Zanetti, Upgrading of MUF resins by buffering additives – Part 1: hexamine sulphate effect and its limits, J.Appl.Polym.Sci. 90(2003), 203-214. https://doi.org/10.1002/app.12634

[26]   J.F. Walker, Formaldehyde, Am.Chem.Soc.Monogr.Ser. 159, 1964.

[27]   A. Pizzi, W. Roll, and B. Dombo, European patent EP-B 0 648 807 (1994); German patent DE 44 06 825 A1 (1995); USA patent 5,532,330 (1996)

[28]   F. Pichelin, SWOOD unpublished results, 2004

[29]   A. Pizzi, Chemistry and technology of cold-and thermosetting tannin-based exterior wood adhesives, Ph.D Thesis , University of the Orange Free State, Bloemfontein, South Africa, 1978.

[30]   A. Pizzi, J. Valenzuela, C. Westermeyer, Low-formaldehyde emission, fast pressing, pine and pecan tannin adhesives for exterior particleboard, Holz Roh Werkst., 52 (1994), 311-315. https://doi.org/10.1007/BF02621421

[31]   F.J. Santiago-Medina, G. Foyer, A. Pizzi, S. Calliol, L. Delmotte, Lignin-derived non-toxic aldehydes for ecofriendly tannin adhesives for wood panels, Int.J.Adhesion Adhesives, 70 (2016), 239-248. https://doi.org/10.1016/j.ijadhadh.2016.07.002

[32]   A. Ballerini, A. Despres, A. Pizzi, Non-toxic, zero-emission tannin-glyoxal adhesives for wood panels, Holz Roh Werkst., 63 (2005), 477-478. https://doi.org/10.1007/s00107-005-0048-x

[33]   R. Böhm, M. Hauptmann, A. Pizzi, C. Friederich, M.-P. Laborie, The chemical, kinetic and mechanical characterization of tannin-based adhesives with different

crosslinking systems, Int.J.Adhesion Adhesives, 68 (2016), 1-8.
https://doi.org/10.1016/j.ijadhadh.2016.01.006

[34] N.Meikleham, A. Pizzi, Acid and alkali-setting tannin-based rigid foams,
      J.Appl.Polym.Sci. 53 (1994), 1547-1556.
      https://doi.org/10.1002/app.1994.070531117

[35] C. Lacoste, A. Pizzi, M.C. Basso, M.-P. Laborie, A. Celzard, Pinus pinaster
      tannin/furanic foams: Part 1, Formulation, Ind.Crops Prod. 52 (2014) 450-456.
      https://doi.org/10.1016/j.indcrop.2013.10.044

[36] U.H.B. Abdullah, A. Pizzi, Tannin-furfuryl alcohol wood panel adhesives without
      formaldehyde, Eur.J.Wood Prod. 71 (2013), 131-132.
      https://doi.org/10.1007/s00107-012-0629-4

[37] P. Luckeneder, J. Gavino, R. Kuchernig, A. Petutschnigg, G. Tondi, Sustainable
      phenolic fractions as basis for furfuryl alcohol-based co-polymers and their use as
      wood adhesives, Polymers, 8 (2016), 396, 1-15.

[38] N. Meikleham, A. Pizzi, A. Stephanou, Induced accelerated autocondensation of
      polyflavonoid tannins for phenolic polycondensates, Part 1: 13C NMR, 29Si
      NMR, X-ray and polarimetry studies and mechanism, J.Appl.Polym.Sci. 54
      (1994), 1827-1845. https://doi.org/10.1002/app.1994.070541206

[39] A. Pizzi, A. Stephanou, Comparative and differential behaviour of pine vs. pecan
      nut tannin adhesives for particleboard, Holzforschung Holzverwertung, 45 (1993),
      30-33.

[40] A. Pizzi, N. Meikleham, Induced accelerated autocondensation of polyflavonoid
      tannins for phenolic polycondensates - Part III: CP-MAS 13C NMR of different
      tannins and models, J.Appl.Polym.Sci. 55 (1995), 1265-1269.
      https://doi.org/10.1002/app.1995.070550812

[41] A. Pizzi, N. Meikleham, A. Stephanou, Induced accelerated autocondensation of
      polyflavonoid tannins for phenolic polycondensates - Part II: cellulose effect and
      application, J.Appl.Polym.Sci. 55 (1995), 929 - 933.
      https://doi.org/10.1002/app.1995.070550611

[42] A. Pizzi, N. Meikleham, B. Dombo, W. Roll, Autocondensation-based, zero-
      emission, tannin adhesives for particleboard, Holz Roh Werkst. 53 (1995), 201-
      204. https://doi.org/10.1007/BF02716424

[43] R. Garcia, A. Pizzi, A. Merlin, Ionic polycondensation effects on the radical
      autocondensation of polyflavonoid tannins-An ESR study, J.Appl.Polym.Sci., 65

(1997), 2623-2632. https://doi.org/10.1002/(SICI)1097-4628(19970926)65:13<2623::AID-APP4>3.0.CO;2-D

[44] R. Garcia, A. Pizzi, Polycondensation and autocondensation networks in polyflavonoid tannins, Part 1: final networks, J.Appl.Polym.Sci. 70 (1998), 1083-1091. https://doi.org/10.1002/(SICI)1097-4628(19981107)70:6<1083::AID-APP5>3.0.CO;2-K

[45] R. Garcia, A. Pizzi, Polycondensation and autocondensation networks in polyflavonoid tannins, Part 2: polycondensation vs. autocondensation, J.Appl.Polym.Sci. 70 (1998), 1093 - 1110. https://doi.org/10.1002/(SICI)1097-4628(19981107)70:6<1093::AID-APP6>3.0.CO;2-J

[46] H. Yamaguchi, Japan patent 2004-143385 (2004)

[47] A. Pizzi, R. Kueny, F. Lecoanet, B. Masseteau, D. Carpentier, A. Krebs, F. Loiseau, S. Molina, M. Ragoubi, High resin content natural matrix-natural fibre biocomposites, Ind.Crops Prod., 30 (2009), 235-240. https://doi.org/10.1016/j.indcrop.2009.03.013

[48] P. Blanchet, A. Cloutier, B. Riedl, Particleboard made from hammer milled black spruce bark residues, Wood Sci.Technol. 34 (2000), 11-19. https://doi.org/10.1007/s002260050003

[49] F. Lopez-Suevos, B. Riedl, Effects of Pinus pinaster bark extracts content on the cure properties of tannin-modified adhesives and on bonding of exterior grade MDF, J.Adhesion Sci.Technol. 17 (2003), 1507-1522. https://doi.org/10.1163/156856103769207374

[50] S. Kim, H.-J. Kim, Curing behavior and viscoelastic properties of pine and wattle tannin-based adhesives studied by dynamic mechanical thermal analysis and FT-IR-ATR spectroscopy, J.Adhesion Sci.Technol. 17 (2003), 1369-1384. https://doi.org/10.1163/156856103769172797

[51] L.R. Calvé, Fast cure and pre-cure resistant cross-linked phenol-formaldehyde adhesives and methods of making same. Can. Pat. 2042476 (1999)

[52] K. Shimatani, Y. Sono, T. Sasaya, Preparation of moderate-temperature setting adhesives from softwood kraft lignin. Part 2. Effect of some factors on strength properties and characteristics of lignin-based adhesives, Holzforschung 48 (1994), 337-342. https://doi.org/10.1515/hfsg.1994.48.4.337

[53] D. Gardner, T. Sellers Jr., Formulation of a lignin-based plywood adhesive from steam-exploded mixed hardwood lignin, Forest Prod.J. 36 (1986), 61-67.

[54]   W.H. Newman, W.G. Glasser, Engineering plastics from lignin-XII. Synthesis and performance of lignin adhesives with isocyanate and melamine, Holzforschung, 39 (1985), 345-353. https://doi.org/10.1515/hfsg.1985.39.6.345

[55]   V.I.Azarov, N.N.Koverniskii, G.V.Zaitseva, *Izvestjia Vysshikh Uchnykh Zavedenii, Lesnai Zhurnal*, 5, 81-83 (1985), in A. Pizzi, Types, processing and properties of bioadhesives for wood and fibers, in Advances in Biorefineries, Chapter 23, Woodhead publishing, 2014, pp. 736-770.

[56]   L. Viikari, A. Hase, P. Quintus-Leina, K. Kataja, S. Tuominen and L. Gadda, European Patent EP 95030 A1 (1999)

[57]   H.H. Nimz, Lignin-based adhesives, in A.Pizzi (Ed.), Wood Adhesives Chemistry and Technology, Vol. 1, Marcel Dekker, New York, 1983, pp. 247-288.

[58]   H.H. Nimz, G. Hitze, The application of spent sulfite liquor as an adhesive for particleboards, Cell.Chem.Technol. 14 (1980), 371-382.

[59]   S. Valkonen, C. Hübsch, Lignin based binders: an industrial reality, latest developments, Wood Adhesives 2017, Forest Products Society, Atlanta, October 2017.

[60]   K.C. Shen, Spent sulphite liquor binder for exterior waferboard, Forest Prod.J., 24 (1974), 38-44.

[61]   K.C. Shen, Spent sulphite liquor binder for exterior waferboard., Part 2, Forest Prod.J., 27 (1977), 32-38.

[62]   A. Kharazipour, A. Haars, M. Shekholeslami, A. Hüttermann, Enzymgebundene holzwerkstoffe auf der basis von lignin und phenoloxidasen, Adhäsion, 35 (1991), 30-36.

[63]   A. Kharazipour, C. Mai, A. Hüttermann, Polyphenols for compounded materials, Polym.Degrad.Stabil. 59 (1998), 237-243. https://doi.org/10.1016/S0141-3910(97)00157-2

[64]   C. Felby, L.S. Pedersen, B.R. Nielsen, Enhanced auto adhesion of wood fibers using phenol oxidases, Holzforschung, 51 (1997), 281-286. https://doi.org/10.1515/hfsg.1997.51.3.281

[65]   A. Pizzi, A. Stephanou, Rapid curing lignins-based exterior wood adhesives, Part 1: diisocyanates reaction mechanisms and application to panel products, Holzforschung, 47 (1993), 439-445. https://doi.org/10.1515/hfsg.1993.47.5.439

[66]   A. Pizzi, A. Stephanou, Rapid curing lignins-based exterior wood adhesives, Part 2: Acceleration mechanisms and application to panel products, Holzforschung, 47 (1993), 501-506. https://doi.org/10.1515/hfsg.1993.47.6.501

[67]   P. Navarrete, H.R. Mansouri, A. Pizzi, S. Tapin-Lingua, B .Benjelloun-Mlayah, S. Rigolet, Synthetic-resin-free wood panel adhesives from low molecular mass lignin and tannin, J.Adhesion Sci.Technol. 24 (2010), 1597-1610. https://doi.org/10.1163/016942410X500972

[68]   N. El-Mansouri, Q. Yuan, F. Huang, Characterization of alkaline lignins for use in phenol-formaldehyde and epoxy resins, BioResources, 6 (2011), 2647-2662.

[69]   M.C .Lagel, A. Pizzi, A. Redl, Phenol-wheat protein-formaldehyde adhesives for wood-based panels, ProLigno. 10 (2014), 3-17; ,Eur.J.Wood Prods. 73 (2015), 439-448. https://doi.org/10.1007/s00107-015-0904-2

[70]   Z. Zhong, X.S. Sun, D. Wang, J.A. Ratto, Wet strength and water resistance of modified soy protein adhesives and effects of drying treatment, J.Polym.Environm. 11(2003), 137-144. https://doi.org/10.1023/A:1026048213787

[71]   Y. Liu, K. Li, Chemical modification of soy protein for wood adhesives, Macromol.Rapid Comm. 23(2002) 739-742. https://doi.org/10.1002/1521-3927(20020901)23:13<739::AID-MARC739>3.0.CO;2-0

[72]   X. Sun, K. Bion, Shear strength and water resistance of modified soy protein adhesives, J.American Oil Chemists Soc. 76 (1999) 977-980. https://doi.org/10.1007/s11746-999-0115-2

[73]   N.S. Hettiarachy, U. Kalapotly, D.J. Myers, Alkali-modified soy protein with improved adhesive and hydrophobic properties, J.American Oil Chemists Soc. 72 (1995), 1461-1464. https://doi.org/10.1007/BF02577838

[74]   G.A. Amaral-Labat, A. Pizzi, A.R. Goncalves, A. Celzard, S. Rigolet, Environment-friendly soy flour-based resins without formaldehyde, J.Appl.Polym.Sci., 108 (2008), 624-632. https://doi.org/10.1002/app.27692

[75]   H. Lei, A. Pizzi, P. Navarrete, S. Rigolet, A. Redl, A. Wagner, Gluten protein adhesives for wood panels, J.Adhesion Sci.Technol. 24 (2010), 1583-1596. https://doi.org/10.1163/016942410X500963

[76]   J.M. Wescott, C.R. Frihart, L. Lorenz, Durable soy-based adhesives, Proceedings Wood Adhesives 2005, Forest Products Society, Madison, Wisconsin (2006)

[77] M.N. Belgacem, A. Gandini, Furan-based adhesives, in A.Pizzi and K.L.Mittal (Eds), Handbook of Adhesive Technology, 2nd edition, Chapter 30, Marcel Dekker, New York, 2004, pp 615-634.

[78] C.-M. Chen, State of the art report: adhesives from renewable resources, Holzforschung Holzverwertung, 48 (1996), 58-60.

[79] A.H. Conner, B.H. River, L.F. Lorenz, Carbohydrates-modified PF resins for wood panels, J.Wood Chem.Technol. 6 (1986), 591-596. https://doi.org/10.1080/02773818608085246

[80] A.H. Conner, L.F. Lorenz, B.H. River, Carbohydrate-modified PF resins formulated at neutral conditions, ACS Symposium series, 385 (Adhesives from Renewable Resources) pp 355-369 (1989)

[81] A. Trosa, European Patent EP 924280 (1999)

[82] K.C. Shen, Patent WO 9837148 (1998); European patent EP 102778 (1997)

[83] A. Trosa, A. Pizzi, Industrial hardboard and other panels binder from waste lignocellulosic liquors/phenol-formaldehyde resins, Holz Roh Werkst. 56 (1998), 229-233. https://doi.org/10.1007/s001070050307

[84] M.H. Alma, M. Yoshioka, Y. Yao, N. Shiraishi, Preparation of sulfuric acid-catalyzed phenolated wood resin, Wood Sci.Technol. 32 (1998), 397-308. https://doi.org/10.1007/BF00702897

[85] M.H.Alma, M.Yoshioka, Y.Yao and N.Shiraishi, The preparation and flow properties of HC1 catalyzed phenolated wood and its blends with commercial novolak resin, Holzforschung, 50 (1996), 85-90. https://doi.org/10.1515/hfsg.1996.50.1.85

[86] R. Miller, U. Shonfeld, Company Literature, Preform Raumgliederungssysteme GmBH, Esbacher Weg 15, D-91555 Feuchtwangen, Germany, 2002

[87] A. Sauget, M.Sc.report, ENSTIB, University Henri Poincaré - Nancy 1, June 2011.

[88] R.P. Wool, Proceedings of the second European Panel Products Symposium, Bangor, Wales, 1998

[89] J. Tomkinson, Adhesives based on natural resources, Chapter 2.3 in M. Dunky, A. Pizzi, M. Van Leemput (Eds.), Wood Adhesion and Glued Products: Wood Adhesives, Chapter 2.3, European Commission, Directorate General for Research, Brussels, pp 46-65 (2002)

[90] P.S. Bailey, Ozonation in Organic Chemistry, Academic Press, New York, 1978.

Thermoset Composite   Materials Research Forum LLC
Materials Research Foundations **38** (2018)   doi: http://dx.doi.org/10.21741/9781945291876

Chapter 5

Hybrid Bast Fibre Strengthened Thermoset Composites

Sekhar Das[1*], Shantanu Basak[2], Pintu Pandit[3] and Amiya Kr. Singha[4]

[1]ICAR-Central Institute for Research on Cotton Technology, Mumbai- 400 019, India

[2]Institute of Chemical Technology, Department of Fibres & Textile Processing Technology, Matunga, Mumbai - 400 019, India

[3]Department of Jute and Fibre Technology - University of Calcutta, India

[4]ICAR-Central Sheep and Wool Research Institute, Avikanagar, Rajasthan-304501, India

*Sekhar.Das@icar.gov.in

Abstract

Flax, jute, hemp, kenaf, and ramie are well-known bast fibres which gain popularity nowadays due to its gaining popularity as renewable, environmentally friendly, and biodegradable raw material for textiles, industrial applications, pulp, and paper as well as for composite applications. Bast fibres have been serving as a traditional reinforcing material of human society since the prehistoric age. Bast fiber-reinforced thermoset composites gain popularity in various fields from household articles to automobiles. The prime advantages of bast fibers over manmade fibers have been their low cost, light weight, high specific strength, renewability, and biodegradability. Bast fibre has some drawbacks such as high moisture absorption, poor dimensional stability and poor wettability that makes it incompatible with hydrophobic resins. Some surface modification treatments are described in this chapter to overcome the limitation of bast fibres as reinforcement materials.

Keywords

Bast fibres, Composite, Surface Modification, Mechanical Property

Contents

5.1   Introduction

Bast fibre crops have been serving as a traditional basic need of human society since the prehistoric age. Archaeological evidence confirmed that Egyptian people knew the use of flax fibre 7000 years ago. The well-known bast fibres are flax, jute, hemp, Kenaf and ramie. These fibres are gaining popularity as renewable, environmentally friendly, and biodegradable raw material for textiles, industrial applications, pulp and paper, as well as for structural applications. Because of their excellent mechanical properties, flax and hemp have, historically, been used in applications such as cordage and textiles. Bast fibres are obtained from the outer layer of bast surrounding the plant stem. Their main purpose is to provide strength to the long plant stem. So bast fibres have a low extension at the break and high modulus compare to leaf or fruit fibres which makes it suitable for use as a reinforcing agent in polymer matrix composites. Among the all natural fibres, bast fibres are consider most popular reinforcing material in polymer matrix composites. Presently there is renewed attention in these fibres as potential reinforcement application in polymer matrix composites. Since the 1990s, natural cellulosic fibre based composites have gained popularity as realistic alternatives to glass fibre reinforced polymer composites in many applications. In general, the density of glass and bast fibres are around 2.6 g/cm^3 and 1.4 g/cm^3 and glass fibre is also costlier than bast fibres [1,2].

The several advantages of natural fibres as compared to man-made glass and carbon fibres are low cost, comparable specific tensile properties, low density, non-irritation to the skin, non-abrasive to the equipment, less health risk, reduced energy consumption, renewability, recyclability and biodegradability [3].

The mechanical characteristics for most of the bast fibres permit the substitution of synthetic, glass and carbon fibres in a wide range of industrial products. The weight of the natural fibres is about two thirds and the consumption of primary energy for their production is only one third that of glass fibres at a comparable strength. Therefore, natural fibres embedded in plastics will soon compete strongly with conventional reinforcing fibres. Remarkable strength, high stiffness, and dimensional stability of light constructional elements make it possible to essentially reduce the weight of aircraft, trains, trucks, and cars. Technical designers have long been talking about the "end of the metal age," and a silent revolution will take place in the construction of aircraft and vehicles during the next ten years. The industrial application of natural fibres requires making high-quality fibres continuously available in large quantities at competitive prices and independently of weather conditions and annual yields. Conventional processing technologies cannot meet the strict demands of modern industries. Consequently, new technologies have to be developed in order to successfully set up powerful process plants for natural bastfibre [4,5]. The processing needed for bast fibres depends on the quality of the input material (the straw), the type of end-product being produced, and the specific capabilities of the machinery on which it will be produced. This chapter shows aspects, trends, data, and the possibilities for such processing from the present point of view. In view of economic efficiency, all components of the bast fibre plants have to be processed and utilized [6,7]. The main emphasis of this chapter is on the utilization of fibres and the treatment of the accompanying shives/hurds, however, will not be discussed in detail.

5.2 Bast fibre

Bast fibres are normally extracted from the stem of the plants. These fibres are the conductive cells of phloem and provide strength to the stem of the plant. Some of the important bast fibres, available in the market are flax, hemp, ramie, jute, kenaf, etc. Recently, some of the bast fibres have been extracted also from the stem of wild plants like nettle, wisteria, linden, etc. Normally, bast fibre has high tensile strength and low extensibility and used for making ropes, yarn, paper and composite materials [1,2] [1,2].

5.2.1 Surface morphology and elemental composition analysis

Surface morphology of the flax fibre showed a rough structure with the presence of the small holes in its surface. It also looks shiny and some amount of the strong stiff nature are also present in the morphology of the flax fibre. In addition many marks, hitches, ditches are also observed on the flax fibre surface. Cross section of the concern fibre showed holes which have a large diameter and spongy behavior. SEM pictures of the hemp fibre showed a rough, irregular appearance (Fig 5.1B) with the presence of the lines in its surface. It may be due to the presence of the non-cellulosic encrusting substances on the fibre surface that give hemp fibre a rough and irregular appearance [2,3,8]. On the contrary, the gray jute fibre showed comparatively a smooth and regular surface appearance, and the cotton fibres presented even cleaner and smoother surfaces. Non-cellulosic material present on the hemp fibre surfaces may be responsible for their rough and irregular appearance. Cross-section based morphological behaviour of the jute and the other bast fibres are also represented in Fig. 5.1. It shows that the cross-section of the hemp fibre having sponge-like behaviour with the presence of tiny holes in its surfaces. On the other hand, cross-sectional view of the jute fibre showed the presence of very small pin like holes in its surface [8–11]. These holes present in the cross-section of the bastfibre reduce the bulk density of the fibre. Surface morphology of the aforementioned bastfibre are shown in the Fig. 5.1.

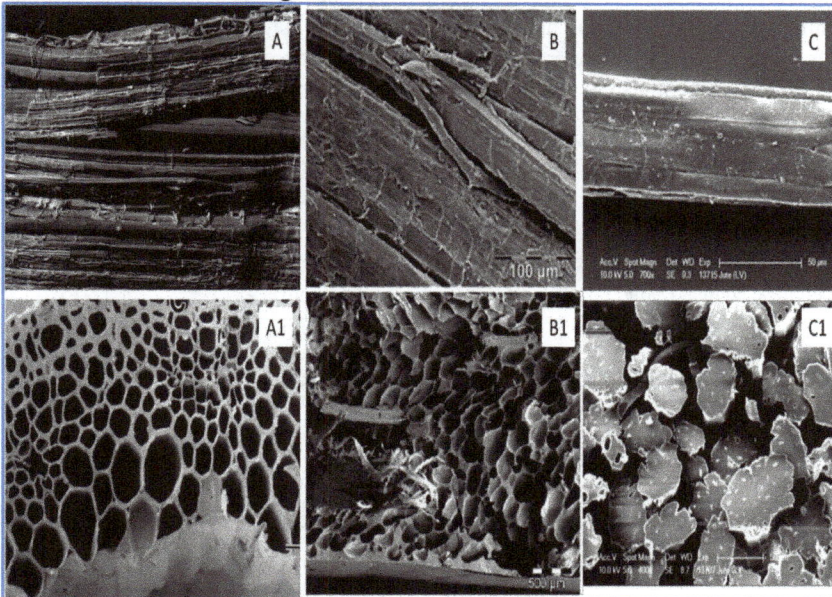

Fig 5.1 SEM surface and cross-sectional morphology of (A and A1) Flax, (B and B1) hemp, and (C and C1) Jute fibres [1–3,8].

5.2.2 Structural composition and the physical properties of the bast fibre

The chemical composition of a fibre affects its appearance, structure, properties, and processability. In this context, compositions and the properties of the extracted bast fibres have been reported (Table 5.1). It can be seen that the jute fibre contains 50-55% cellulose, whereas the flax and hemp fibre contain around 65-87% and 80% cellulose, respectively [7,11]. The hemicellulose content of the jute fibre is quite higher rather than the flax and the hemp fibre. It signifies the fact that the jute fibre has more amorphous portion and can absorb larger amounts of the dye and the other finishing chemicals at lower temperature compared to the flax and the hemp fibre. Moreover, hemicellulose content of the jute fibre also affected the moisture regain property of the fibre which is also more rather than the jute and cotton fibre [10]. Lignin, the binding cementing constituent that hold the individual cells together, are present in higher amount in the jute fibre (11.8-14.2%), followed by the hemp and flax fibre. As far as the crystallinity is concerned, jute fibre is less crystalline (56.9%) than the flax fibre (65-70%) and almost similar to lignocellulosic hemp fibres, which has crystallinity range of 50-55% [8,9]. This might be possibly due to the presence of more amorphous hemicellulose (40%) in it that is significantly higher than flax and hemp fibres. Due to its less crystalline region, jute fibre is more accessible to water, dyes and other finishing chemicals. Indeed flax and hemp also has a high moisture regain of 13%, which is more than the jute fibre.

Table 5.1 Physical and chemical properties of the different bast fibre.

Composition and properties	Flax	Hemp	Jute
Cellulose content (%)	65-87	Under 80%	59-72
Hemicellulose content (%)	nil	Nil	12-15
Lignin (%)	small	2-3	11.8-14.2
Non-cellulosic wall materials (%)	1	2-3	3-4
Crystallinity index	65-70	66.9	50-55
Moisture regain %	12	13	13.8
Length (mm)	Up to 900 mm	914- 4572	150-360
Diameter (μm)	11-22	11-22	15-25
Strength (g/ tex)	30-60	23-50	35-59

5.2.3 Composition and the properties of the different bast fibre

Jute fibre is a multi-cellular lignocellulosic fibre with the initial modulus of 1150 g/tex and the tenacity is in the rage of 27-44g/tex. It is composed of cellulose (50-52%), hemicellulose (25-27%) and lignin (10-12%). Cellulose part of the jute fibre is composed of a linear chain of the gluco pyranose unit connected by 1-4β glucosidic linkage. It is crystalline in nature. Spiral angle in the secondary cell wall of the cellulose is around 6 to 8°. On the other hand, the hemi-cellulose part is composed of xylan, pentosane, etc., and it is amorphous in nature. Because of the amorphous nature of the hemicellulose, it helps to absorb moisture from the atmosphere (8.5% moisture regain) and also assist to improve the uptake of the dye and other finishing chemicals from the dye bath. However, hemi-cellulose part of the jute fibre has a tendency to degrade in the presence of alkali and high temperature [12]. On the contrary, the lignin part is branched structure and amorphous in nature. Mainly it is present in between the plant cells and helps to improve the rigidity of the plant. It consists of three different alcohols named sinapyl, cumaryl and the colephenyl alcohol. Indeed, total lignin is a polyphenolic high molecular weight structure. The molecular weight of the lignin varies with the maturity of the plant cell wall. Lignin material, present in the jute fabric helps to absorb the harmful UV rays of the sun (due to its polyphenolic nature) and also delivered hygienic antimicrobial coating on the fibre surface. It also helps to attract both the anionic and the cationic dye from the dye bath. It may be due to the presence of the phenolic –OH group of the lignin which can easily attached to the positively charged basic dye molecule. Lignin is slightly reddish to yellowish in color due to the presence of the anthraquinone based chromophore present into its structure. This natural color has been damaged easily when it is bleached in the presence of alkaline hydrogen peroxide solution or by chlorine-based hutex bleaching process [13].

Flax fibre also contains lignin but in lesser amount compared to the jute fibre and the initial modulous of the concern fibre is even more than jute. Concerning the chemical properties, jute fibre is attacked by the acid as most of the glucosidic linkage present in the cellulose chain has been broken. On the other hand, alkali degrades the amorphous portion of the jute fibre like hemi-cellulose etc., Alkali degrades the xylan and the pentosan part of the hemicellulose and help to hydrolyze the jute fibre, depending on the concentration level of the alkali.

Flax fibre is another example of the bast fibre and mostly used by the composite industries, clothing and the sewing or surgical thread making units. Apart from it, the concern fibre also has been used for making upholstery fabrics, canvas, table wear and curtains of the room for its excellent aesthetic properties. Normally the flax fibre (light greyish in colour) has been extracted from the stem of the flax plant and by nature it is

Thermoset Composite   Materials Research Forum LLC
Materials Research Foundations **38** (2018)   doi: http://dx.doi.org/10.21741/9781945291876

soft, lustrous, flexible and strongest fibre. It contains 65-68% cellulose with small amount of the lignin present in its structure. Spiral angle of the secondary cell wall of the cellulose of the flax fibre is in the range of 6-8°. It is one of the major reason of the high initial modulous and the strength of the flax fibre compared to the jute and the cotton fibre. The tenacity range of the flax fibre is around 6.5 to 8 gm/ den while the length of the fibre is in the range of 18-30 inches. Specific gravity of the flax fibre is around 1.54 g/cc. Most important physical property of the flax fibre is its heat resistance behavior. It may be due to the presence of a large number of hydrogen bonds and the long molecular chains in its structure. By prolonged heating, flax fibre forms black char mass which is the characteristics of the cellulosic based polymer structure. Concerning the chemical properties, flax fibre is easily attacked by the acid rather than the alkali due to its cellulosic nature. As the basic unit of the flax fibre is cellulosic in nature it can be easily dyed by the vat dye, direct dye, etc. [1].

Hemp fibre is also similar like flax fibre. However, the amount of the lignin present in the hemp fibre is slightly more than the extent of the lignin present in the flax fibre. Concerning the swelling behaviour, flax fibre swells easily and firstly compared to the hemp fibre due to the presence of tubes in its structure which contracts in a serpentine fashion. Another major difference between the two fibre is the dyeability. Due to the presence of more amount of the lignin, hemp fibre can be dyed by the cationic basic dye molecule. However, flax fibre can not be dyed by the cationic dyes. Like flax fibre, Hemp fibre is also very much suitable for making composite materials [3,8].

Ramie fibre is an example of one of the strongest bast fibre and like cotton fibre, its strength has been increased in wet condition. It has some special capability such as shape holding, wrinkle resistance and to introduce a silky lusture to the finished material. In addition, the major advantage of the ramie fibre is that it is not affected by any kind of bacteria, fungi or microorganism kind of material. However, ramie fibre lacks in resiliency, elasticity property and it has very low abrasion resistance [1,13–15].

5.3   Advantage and limitation of bast fibre as reinforcing material

Natural cellulosic fibres have several advantages over synthetic fibres as reinforcing the material. The main advantages of natural fibres are low cost, low density, comparable specific strength, lower energy requirement, lower pollutant emissions, wide availability, acoustic properties, eco-friendly renew-able natural resource, etc. [12,16–20].

In bast fibres from a plant source (flax, hemp, ramie, jute, etc.) found that the micro fibrils are well aligned with respect to the fibre axis. Wood cells have an additional structural parameter with important influence on the mechanical properties, the so-called micro fibril angle (MFA). The MFA is the angle of that helix with respect to the longitudinal cell axis. The smaller the MFA (i. e., the steeper the helix) the higher is the longitudinal Young's modulus of the wood cell. In woody plant tissue, the embedding matrix is reinforced by lignin. Several other limitations such as high moisture absorption, poor dimensional stability, incapability with hydrophobic resins, poor wettability, ageing rapidly at moderately high temperature, etc. hinders its use in high tech applications. Some of the chemical, physical and surface treatments may also be able to eliminate these problems which are described in the literature [13,18,20,21].

5.4   Surface modification of bast fibres

Bast fibres such as jute, flax, ramie, etc. are the most used cellulosic fibre for composite application. The three basic constituent components of bast fibres are cellulose, hemicelluloses, and lignin. Cellulose is the major structural component of bastfibres [22]. Cellulose macromolecule is a semi-crystalline polysaccharide made up of anhydro-d-glucose, which contains three alcohol hydroxyls (-OH). Hemi-cellulosic component is branched, fully amorphous and containing many hydroxyl and acetyl groups in their molecule. Lignins are highly complex, amorphous, mainly aromatic, polymers of phenyl propane units but have the least water sorption than the cellulose and hemicelluloses. These hydroxyls groups are able to form hydrogen bonds with hydroxyl groups of water molecule present in the air. Therefore, all the bast fibres are hydrophilic in nature [7]. The main limitation of cellulosic bast fibre is their hydrophilic nature, which reduces their compatibility with hydrophobic polymer matrices [3,6,23].

Bast fibre contains a trace amount of waxy substance on their surface which affects the fibre matrix bonding and surface wetting property. The presence of free water molecule in fibre reduces the adhesive characteristics with most binder resins. The water molecule performs as a separating agent in the fiber-matrix interface. An optimum fiber-matrix interface bond is important to achieve high mechanical properties of composites [24,25].

5.5   Methods for surface modification of natural fibres

A number of surface modification methods have been used on lignocellulosic fibres in the past in order to remove the organic compounds attached to the hydrophilic surfaces of plant cell walls. Physical and chemical methods can be used to modify the fibre surface for a better interface.

5.3.1 Physical methods

In this method, the chemical composition of the fibres does not change. The physical treatments change the fibre surface properties and structure which influence the better mechanical bonding to polymers. Various types of physical surface modification methods are applied such as stretching, calandering [7,26], the production of blended yarns, thermo treatment [27,28], and corona treatment. UV treatment and plasma treatment, etc. Corona treatment is one of the most widely used treatments for surface modification [29]. Corona discharge treatment improves fibre to matrix adhesion and wetting properties of the fibre. These treatments introduce polar groups on the fibre surface due to oxidation and other chemical changes in the surface region. Surface oxidation is the main chemical mechanism for corona treatment. The increase of adhesion properties due to electrets formation, elimination of weak boundary layers, increased surface roughness due to pitting, and the introduction of polar groups due to oxidation and other chemical changes in the surface region of fibre [30,31].

The same type of effects can be achieved by cold plasma treatment. Plasma can be defined as an ionized gas containing a mixture of ions, electrons, neutral and excited molecules and photons. Cold plasma is created by a high frequency electric current passes in the presence of a feeding gas into relatively low temperature [32,33]. A variety of surface modification can be achieved by plasma treatment using different types of gases medium. Plasma treatment introduces surface crosslinking, increase or decrease of the surface energy and the introduction of reactive free groups on the fibre surface. The intensity of plasma treatment depends on the treatment duration, magnitude of power, distance from the plasma nozzle to the sample, etc. Plasma and Corona discharge treatment are successfully used to modify bast fibre surfaces and improve mechanical properties of bast fibre based composites [34,35].

5.5.2 Chemical methods

Several chemical methods are employed to make compatible the bast fibres with hydrophobic matrixes. Alkaline treatment is one of the most important chemical treatments of bast fibres when used to reinforce thermo sets resins.

5.5.2.1 Alkali treatment

Alkaline treatment removes a certain amount of hemicelluloses, lignin, wax, and oils covering the external surface of the fibre cell wall [36,37]. After hemicelluloses removal by this process, the inter fibrillar region will be less dense and less rigid which makes the fibrils more able of rearranging themselves along the direction of tensile deformation. It is observed that the fibrils are the more arranged in shape as a result of alkali treatment.

Also, alkali treatment leads to decrease in the orientation angle of the fibril and an increasing degree of orientation. This treatment leads to increase in crystallinity index of due to the removal of the hemicelluloses and lignin, which cause better packing of the cellulose chains. This treatment disrupts the hydrogen bonding in the fibre network structure, thereby increasing surface roughness. The tensile strength and modulus of fibre is expected to increase with increasing crystallinity index and degree of molecular orientation [38,39]. Thus, the increasing fibre surface roughness and removal of wax and oils may be responsible for better fibre/matrix adhesion by giving rise to additional sites of mechanical interlocking [20].

5.5.2.2 Graft copolymerization

A useful method of surface chemical modification of natural bast fibres is graft copolymerization. This method makes the natural fibres compatible with several resin systems. Redox initiation has been used to graft vinyl monomers on to natural cellulosic fibres. The use of acid during the grafting reaction deteriorates the properties of natural cellulosic fibres [40,41]. Among all the redox initiating systems, $CuSO_4$-$NaIO_4$ is considered to be a less harmful system; no acid is used during the graft copolymerization reaction. It is observed that grafted cellulosic fibre has better fibre/matrix adhesion and improve mechanical properties of bast fibre based composites [42,43] .

5.5.2.3 Acetylation

Acetylation of bast fibres is a popular esterification method to introduce plasticization to cellulosic fibres. Acetylation is a reaction mechanism, introducing an acetyl functional group (CH_3COO-) into an organic compound. Cellulosic fibre modification with acetic anhydride substitutes the polymer hydroxyl groups of the fibre with acetyl groups. Acetylation significantly reduces the hygroscopic nature of fibre and the modified fibre becomes hydrophobic [44,45]. Modified fibre attracts less moisture and increases the dimensional stability of composites. Cellulosic fibres become more compatible with hydrophobic resins [46].

5.5.2.4 Treatment with isocyanate

A chemical containing functional group $-N=C=O$ termed as isocyanate compound. The isocyanate functional group is highly reactive with the hydroxyl groups of cellulose and lignin present in cellulosic fibres. This coupling agent treatment reduces the hydrophilic nature of the cellulosic fibre and creates better compatibility with the binder resin in the composites [47,48].

Thermoset Composite

Materials Research Forum LLC

Materials Research Foundations 38 (2018)

doi: http://dx.doi.org/10.21741/9781945291876

5.5.2.5 Other chemical treatments

Silane is a chemical compound with chemical formula SiH_4 which can be used as a coupling agent for cellulosic fibres. These coupling agents can able to reduce the number of hydroxyl groups present in the cellulosic fibre [37]. Organic peroxides decompose easily to form free radicals of the form RO; RO then reacts with the hydrogen group of the matrix and cellulose fibres. Generally, benzoyl peroxide and dicumyl peroxide are used in natural fibre surface modifications purpose [47]. Permanganate treatment is used for bast fibre surface modification purpose. This treatment works as the formation of MnO_3^- ion which initiate graft copolymerization with cellulose molecules. Mostly potassium permanganate solution is used for permanganate treatments of cellulosic fibres [47].

Conclusions

Bast fibres (flax, hemp, ramie, jute, kenaf etc.) has been extracted from the stem of the wild plants like nettle, wisteria, linden, etc. with high tensile strength and low extensibility mostly used for making ropes, yarn, paper and composite materials. Surface morphology of the flax fibre showed rough structure with the presence of small holes in its surface where as cross section of jute fibers showed the presence of very small pin like holes in the surface resulted reduce the bulk density of the fibre. It was found that the chemical composition of a fibre affects its appearance, structure, properties and processability which play a crucial role for composite application. The three basic constituent components of bast fibres are cellulose, hemicelluloses and lignin. Cellulose is the major structural component of bast fibres. The main limitation of cellulosic contained in bast fibre showed the hydrophilic nature which reduces their compatibility with hydrophobic polymer matrices. Lots of research have been going on in this direction either by modification of fibres or modify the resins by physical and chemical methods to make compatible the bast fibres with hydrophobic polymer matrixes. When used to reinforce thermosets resins it was found that alkaline treatment is one of the most important chemical treatments of bast fibres.

References

[1]   www. https://en.wikipedia.org/wiki/Bast_fibre.

[2]   A. V Kiruthika, A review on physico-mechanical properties of bast fibre reinforced polymer composites, J. Build. Eng. 9 (2017) 91–99. https://doi.org/10.1016/j.jobe.2016.12.003

[3]   J. Wiener, V. Kovačič, P. Dejlová, Differences between flax and hemp, AUTEX Res. J. 3 (2003) 58–63.

[4]   P. V Joseph, K. Joseph, S. Thomas, Short sisal fiber reinforced polypropylene composites: the role of interface modification on ultimate properties, Compos. Interfaces. 9 (2002) 171–205. https://doi.org/10.1163/156855402760116094

[5]   L. Chen, B. Wang, J. Chen, X. Ruan, Y. Yang, Characterization of dimethyl sulfoxide-treated wool and enhancement of reactive wool dyeing in non-aqueous medium, Text. Res. J. 86 (2016) 533–542. https://doi.org/10.1177/0040517515591784

[6]   A.K. Mohanty, M. Misra, L.T. Drzal, Surface modifications of natural fibers and performance of the resulting biocomposites: an overview, Compos. Interfaces. 8 (2001) 313–343. https://doi.org/10.1163/156855401753255422

[7]   A.K. Bledzki, S. Reihmane, J. Gassan, Properties and modification methods for vegetable fibers for natural fiber composites, J. Appl. Polym. Sci. 59 (1996) 1329–1336. https://doi.org/10.1002/(SICI)1097-4628(19960222)59:8<1329::AID-APP17>3.0.CO;2-0

[8]   N. Maneerat, N. Tangsuphoom, A. Nitithamyong, Effect of extraction condition on properties of pectin from banana peels and its function as fat replacer in salad cream, J. Food Sci. Technol. 54 (2017) 386–397. https://doi.org/10.1007/s13197-016-2475-6

[9]   Q.F. Sun, Y. Lu, Y.Z. Xia, D.J. Yang, J. Li, Y.X. Liu, Flame retardancy of wood treated by TiO2/ZnO coating, Surf. Eng. 28 (2012) 555–559. https://doi.org/10.1179/1743294412Y.0000000027

[10]   N. Chand, S.A.R. Hashmi, Mechanical properties of sisal fibre at elevated temperatures, J. Mater. Sci. 28 (1993) 6724–6728. https://doi.org/10.1007/BF00356422

[11]   A. Shukla, S. Basak, S.W. Ali, R. Chattopadhyay, Development of fire retardant sisal yarn, Cellulose. 24 (2017) 423–434. https://doi.org/10.1007/s10570-016-1115-7

[12]   S. Das, M. Bhowmick, S.K. Chattopadhyay, S. Basak, Application of biomimicry in textiles, Curr. Sci. 109 (2015) 893–901. https://doi.org/10.18520/cs/v109/i5/893-901

[13]   S. Das, S. Basak, M. Bhowmick, S.K. Chattopadhyay, M.G. Ambare, Waste paper as a cheap source of natural fibre to reinforce polyester resin in production of bio-

composites, J. Polym. Eng. 36 (2016) 441–447. https://doi.org/10.1515/polyeng-2015-0263

[14] S. Basak, K.K. Samanta, S.K. Chattopadhyay, P. Pandit, S. Maiti, Green fire retardant finishing and combined dyeing of proteinous wool fabric, Color. Technol. 132 (2016). https://doi.org/10.1111/cote.12200

[15] M.D. Teli, P. Pandit, Novel method of ecofriendly single bath dyeing and functional finishing of wool protein with coconut shell extract biomolecules, ACS Sustain. Chem. Eng. (2017). https://doi.org/10.1021/acssuschemeng.7b02078

[16] M.D. Teli, P. Pandit, Development of thermally stable and hygienic colored cotton fabric made by treatment with natural coconut shell extract, J. Ind. Text. (2017) 1528083717725113.

[17] M.D. Teli, P. Pandit, S. Basak, Coconut shell extract imparting multifunction properties to ligno- cellulosic material, (n.d.) 1–30. https://doi.org/10.1177/1528083716686937

[18] S. Das, Mechanical and water swelling properties of waste paper reinforced unsaturated polyester composites, Constr. Build. Mater. 138 (2017) 469–478. https://doi.org/10.1016/j.conbuildmat.2017.02.041

[19] L. Ammayappan, S. Das, R. Guruprasad, D.P. Ray, P.K. Ganguly, Effect of lac treatment on mechanical properties of jute fabric/polyester resin based biocomposite, (2016).

[20] S. Das, N. Shanmugam, A. Kumar, S. Jose, Potential of biomimicry in the field of textile technology, Bioinspired, Biomim. Nanobiomaterials. 6 (2017) 224–235. https://doi.org/10.1680/jbibn.16.00048

[21] Das S. Mechanical properties of waste paper/jute fabric reinforced polyester resin matrix hybrid composites. Carbohydr Polym 2017;172:60–7. https://doi.org/10.1016/j.carbpol.2017.05.036

[22] J. Van der Geer, J.A.J. Hanraads, R.A. Lupton, Clean Energy Project Analysis: Retscreen® Engineering & Cases Textbook, Small Hydro Project Analysis Chapter. J Sci Commun 163 (2000) 51–59.

[23] M.M. Kabir, H. Wang, K.T. Lau, F. Cardona, Chemical treatments on plant-based natural fibre reinforced polymer composites: An overview, Compos. Part B Eng. 43 (2012) 2883–2892. https://doi.org/10.1016/j.compositesb.2012.04.053

[24] S. Debnath, C.W. Nguong, S.N.B. Lee, A review on natural fibre reinforced polymer composites, World Acad. Sci. Eng. Technol. (2013) 1123–1130.

[25] L.T. Drzal, M. Madhukar, Fibre-matrix adhesion and its relationship to composite mechanical properties, J. Mater. Sci. 28 (1993) 569–610. https://doi.org/10.1007/BF01151234

[26] A.K. Bledzki, S. Reihmane, J. Gassan, Properties and modification methods for vegetable fibres for natural fibre composites, J Appl Polym Sci. 59 (1996) 1329–1336. https://doi.org/10.1002/(SICI)1097-4628(19960222)59:8<1329::AID-APP17>3.0.CO;2-0

[27] M.M. Hassan, M.H. Wagner, Surface Modification of Natural Fibers for Reinforced Polymer Composites, Prog. Adhes. Adhes. (2017) 1–44.

[28] P.K. Ray, A.C. Chakravarty, S.B. Bandyopadhaya, Fine structure and mechanical properties of jute differently dried after retting, J. Appl. Polym. Sci. 20 (1976) 1765–1767. https://doi.org/10.1002/app.1976.070200705

[29] D. Zhang, L.C. Wadsworth, Corona treatment of polyolefin films—a review, Adv. Polym. Technol. 18 (1999) 171–180. https://doi.org/10.1002/(SICI)1098-2329(199922)18:2<171::AID-ADV6>3.0.CO;2-8

[30] J. Gassan, V.S. Gutowski, Effects of corona discharge and UV treatment on the properties of jute-fibre epoxy composites, Compos. Sci. Technol. 60 (2000) 2857–2863. https://doi.org/10.1016/S0266-3538(00)00168-8

[31] M.N. Belgacem, P. Bataille, S. Sapieha, Effect of corona modification on the mechanical properties of polypropylene/cellulose composites, J. Appl. Polym. Sci. 53 (1994) 379–385. https://doi.org/10.1002/app.1994.070530401

[32] M.J. Shenton, M.C. Lovell-Hoare, G.C. Stevens, Adhesion enhancement of polymer surfaces by atmospheric plasma treatment, J. Phys. D. Appl. Phys. 34 (2001) 2754. https://doi.org/10.1088/0022-3727/34/18/307

[33] X. Yuan, K. Jayaraman, D. Bhattacharyya, Effects of plasma treatment in enhancing the performance of woodfibre-polypropylene composites, Compos. Part A Appl. Sci. Manuf. 35 (2004) 1363–1374. https://doi.org/10.1016/j.compositesa.2004.06.023

[34] X. Yuan, K. Jayaraman, D. Bhattacharyya, Mechanical properties of plasma-treated sisal fibre-reinforced polypropylene composites, J. Adhes. Sci. Technol. 18 (2004) 1027–1045. https://doi.org/10.1163/1568561041257478

[35] A. Baltazar-y-Jimenez, M. Bistritz, E. Schulz, A. Bismarck, Atmospheric air pressure plasma treatment of lignocellulosic fibres: Impact on mechanical

properties and adhesion to cellulose acetate butyrate, Compos. Sci. Technol. 68 (2008) 215–227. https://doi.org/10.1016/j.compscitech.2007.04.028

[36]   A. Valadez-Gonzalez, J.M. Cervantes-Uc, R. Olayo, P.J. Herrera-Franco, Effect of fiber surface treatment on the fiber–matrix bond strength of natural fiber reinforced composites, Compos. Part B Eng. 30 (1999) 309–320. https://doi.org/10.1016/S1359-8368(98)00054-7

[37]   A. Jähn, M.W. Schröder, M. Füting, K. Schenzel, W. Diepenbrock, Characterization of alkali treated flax fibres by means of FT Raman spectroscopy and environmental scanning electron microscopy, Spectrochim. Acta Part A Mol. Biomol. Spectrosc. 58 (2002) 2271–2279. https://doi.org/10.1016/S1386-1425(01)00697-7

[38]   D.S. Varma, M. Varma, I.K. Varma, Coir fibers: Part I: Effect of physical and chemical treatments on properties, Text. Res. J. 54 (1984) 827–832. https://doi.org/10.1177/004051758405401206

[39]   X. Li, L.G. Tabil, S. Panigrahi, Chemical treatments of natural fiber for use in natural fiber-reinforced composites: a review, J. Polym. Environ. 15 (2007) 25–33. https://doi.org/10.1007/s10924-006-0042-3

[40]   A.K. Mohanty, S. Parija, M. Misra, Ce (IV)-N-acetylglycine initiated graft copolymerization of acrylonitrile onto chemically modified pineapple leaf fibers, J. Appl. Polym. Sci. 60 (1996) 931–937. https://doi.org/10.1002/(SICI)1097-4628(19960516)60:7<931::AID-APP2>3.0.CO;2-N

[41]   P.C. Tripathy, M. Misra, S. Parija, S. Mishra, A.K. Mohanty, Studies of Cu (II)–IO4− initiated graft copolymerization of methyl methacrylate from defatted pineapple leaf fibres, Polym. Int. 48 (1999) 868–872. https://doi.org/10.1002/(SICI)1097-0126(199909)48:9<868::AID-PI230>3.0.CO;2-B

[42]   P. Ghosh, P.K. Ganguly, Jute fibre-reinforced polyester resin composites: effect of different types and degrees of chemical modification of jute on performance of the composites, Plast. Rubber Compos. Process. Appl. 20 (1993) 171–177.

[43]   A.K. Mohanty, M.A. Khan, S. Sahoo, G. Hinrichsen, Effect of chemical modification on the performance of biodegradable jute yarn-Biopol® composites, J. Mater. Sci. 35 (2000) 2589–2595. https://doi.org/10.1023/A:1004723330799

[44] C.A.S. Hill, H.P.S.A. Khalil, M.D. Hale, A study of the potential of acetylation to improve the properties of plant fibres, Ind. Crops Prod. 8 (1998) 53–63. https://doi.org/10.1016/S0926-6690(97)10012-7

[45] M.S. Sreekala, S. Thomas, Effect of fibre surface modification on water-sorption characteristics of oil palm fibres, Compos. Sci. Technol. 63 (2003) 861–869. https://doi.org/10.1016/S0266-3538(02)00270-1

[46] M.Z. Rong, M.Q. Zhang, Y. Liu, G.C. Yang, H.M. Zeng, The effect of fiber treatment on the mechanical properties of unidirectional sisal-reinforced epoxy composites, Compos. Sci. Technol. 61 (2001) 1437–1447. https://doi.org/10.1016/S0266-3538(01)00046-X

[47] A. Paul, K. Joseph, S. Thomas, Effect of surface treatments on the electrical properties of low-density polyethylene composites reinforced with short sisal fibers, Compos. Sci. Technol. 57 (1997) 67–79. https://doi.org/10.1016/S0266-3538(96)00109-1

[48] D. Maldas, B. V Kokta, C. Daneault, Influence of coupling agents and treatments on the mechanical properties of cellulose fiber–polystyrene composites, J. Appl. Polym. Sci. 37 (1989) 751–775. https://doi.org/10.1002/app.1989.070370313

Chapter 6

Nano-Carbon/Polymer Composites for Electromagnetic Shielding, Structural Mechanical and Field Emission Applications

Ashwini P. Alegaonkar [1], Prashant S. Alegaonkar [*, 2]

[1] Department of Chemistry, Savitribai Phule Pune University (formerly Pune University), Ganeshkhind, Pune 411 007, MS, India

[2] Department of Applied Physics, Defence Institute of Advanced Technology (DIAT), Girinagar, Pune 411 025, MS, India

Abstract

Carbon material, due to their stability, forms a number of polymer composites compounds; useful for several applications. Herein, we discussed properties of poly urathene composite incorporated with graphene like nano carbon (GNCs) utilized for X-band EMI shielding. GNCs showed good mechanical and superior thermal properties, at very low fraction, when added with epoxy. Subsequently, utility of carbon nanotubes/nylon fibre spun by electro-spinning is discussed followed by field emission analysis of nanotubes/polymer prepared by different dispersion routes.

Keywords

EMI Shielding, GNC Epoxy Composite, Field Emission of CNT-Ploymer, CNT-Nylon Composite

Contents

6.1   Introduction

Composite materials of carbon like CNTs, nanocarbons play a crucial role in various applications like electromagnetic shielding, field emission parameters, its composites with a polymer for use in many devices. Composites with epoxy, nylon 6, 6, polyurethane, poly(methyl methacrylate) (PMMA), polystyrene, polyaniline, acrylonitrile butadiene styrene and many other composites are reported and find applications in suitable fields. In the current chapter electromagnetic interference (EMI) shielding properties of graphene like nano carbon(GNCs) composite of poly urathene in X-band is discussed [1]. Further study of mechanical and thermal properties of GNCs composites is given [2]. MWNTs/nylon conductive composite nanofibers prepared by electro spinning and dispersibility of the thin multiwall carbon nanotubes (t-MWCNTs) in the composite [3], carbon nanotube composite are achieved by the chemical and mechanical dispersion routes to their field emission parameters [4].

6.2   Shielding parameters of GNCs/Polyurethane nanocomposites

For hybrid coatings, SE depends, specifically, on cooperative interactions between filler and host matrix. Dimensions, charge mobility, and structural robustness are, particularly, essential for fillers. For polymer host, important properties are conductivity, matrix density, design flexibility, and environmental compatibility. Due to their premium properties nano-carbons like carbon nanotubes (CNTs) and graphene in polymer as nanocomposites were reported to be an effective shielding constituents in the 8-12 GHz regime [5-9]. For multi walled carbon nanotubes incorporated in ethylene tertiary polymer, [10] PMMA, [11] and PS (polystyrene), [11] SE was reported, respectively, 28 (3.2), 40 (10), and 66 dB (20). Bracketed numbers indicate filler loading with maximum weight with shield thickness of 2−2.5 mm range. The values of SE were reported to be 20 (7) and 85 dB (0.5) with thickness ~ 2 mm, respectively, for PS [13] and carbon foam [14] incorporated with CNTs. For PANI, [15] PU, [16] and epoxy, [17] dispersed with single-walled nanotubes, the values of SEs were, respectively, found to be 30 (25), 17 (20), and 30 dB (15) with a variable range of shielding thickness from 0.5-1.0 cm. Moreover, graphene-based nano-composites shielding has been the focus of several recent investigations. [18] Graphene/PANI, [15] graphene/nitrile-rubber, [19] and graphene/epoxy [20] showed SE values, respectively, 35 (33), 57 (10), and 21 dB (8.8) with thickness ~ 5 mm of shield. Wherein, graphene incorporated in PVDF, [21] PMDS, [22,160] and PMMA [23] foam showed magnitude of SE, respectively, 20 (5), 20 (1.8), and 33.3 dB (0.8). Literature also exists on incorporating another type of nanocarbon in PANI, [24] PP, [25] and acrylonitrile [11] with reported SE 20-50 dB range at different shielding thicknesses.

In this work, a typical interface polarization mechanism is recognized between GNCs (1−25 wt %) and PU to mitigate shielding parameters for PU, in 8-12 GHz radar region. The scattering study is presented in light of interface bonding at GNCs/PU, GNCs dispersibility, and morphology/rheology of nano-composites, using vibration spectroscopy. The microwave measurements were performed on torroidal shaped samples to determine permittivity real and imaginary part, alternating current (ac) transport, skin depth, transmission coefficient, S_{21}, and shielding efficiency. The details of polarization mechanism are discussed. Broadly, the required PU thickness for about ~ 40% loss in S_{21} is more than a centimeter, whereas, almost 99.9% loss recorded for a millimeter thick PU, at 25 wt % loading of GNCs.

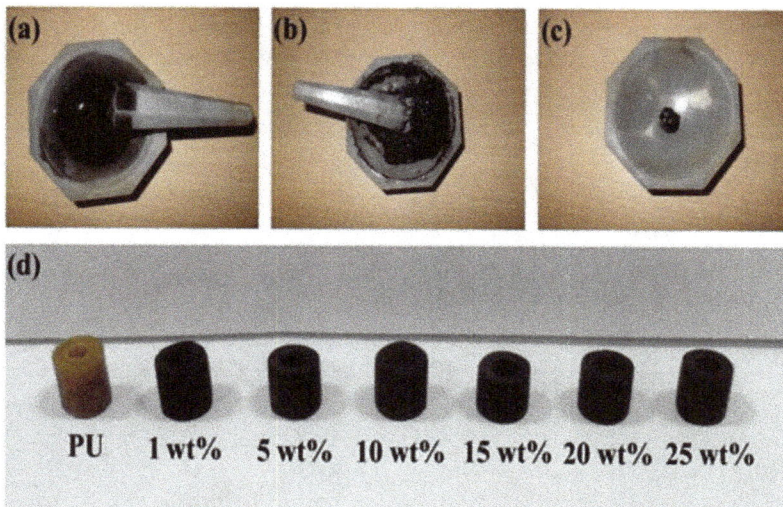

Fig 6.1: (Upper panel) a typical production batch (a) initial phase of the mixture (liquid state), (b) gel-type paste, and (c) thick paste, in a mortar and pestle (mixing and grinding ~3 h to obtain paste). (Lower panel, d) fabricated torroidal shaped specimen (dimensions: i.d., 3 mm; o.d., 7 mm; and height, 6 mm; postproduction edge polishing).

An injection moulding technique has been adopted for the preparation of nanocomposite samples using a cylindrical die. In brief, the methodology followed is sealing and simultaneous heat-pressing of the composite slurry at 120 °C, loaded $\sim$ 100 kg/cm^2, adiabatically, in a hydraulic press as shown in Fig 6.1.

6.2.2 Characterizations and measurements

(a) FTIR and Raman Spectroscopy

Chemical Bonding. Characteristic FTIR and respective Raman fingerprints recorded for PU, GNCs, and 1-25 wt % GNCs/PU nanocomposite samples are shown in plots a–d of Fig. 6.2. The graphs in Fig. 6.2 indicate band assignments for PU revealing characteristic inelastic vibration modes (in cm^{-1}) for urethane amide (C–N$^-$) @ 1184 and amide III @1271, δ(CH) amide III @1315 and δ(CH$_2$) ν_{sym} for N- C-O at 1441. Corresponding Rayleigh active amide band is broad and present @1196. Amide III and δ(CH) is completely merged in 1296. The emerging feature at 1515 is assigned to ν(Ar) of amide II ν(C−N) + δ(N−H), @ 1615 to ν(Ar), and @ 1657 to ν(Ar) amide I ν(C-O).

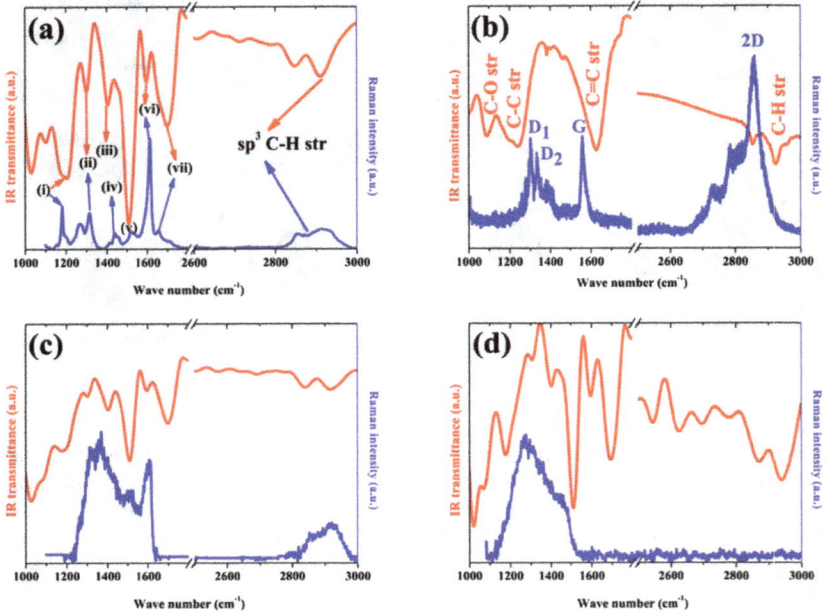

Fig. 6.2 Typical FTIR and Raman spectra (excitation wavelength: 785 nm) recorded for the systems. (a) PU, (b) GNCs, (c) 1 wt %, and (d) 25 wt % GNCs/PU nanocomposite samples. For PU the peak indexing is (i) Urethane amide (Raman active), C−N⁻ (IR), (ii) Urethane amide III (both Raman and IR), (iii) δ(CH) urethane amide III (IR), (iv) δ(CH) ν N C O (Raman), (v) ν (Ar)-urethane amide II ν(C−N) + δ(N−H) (IR), (vi) ν (Ar) (both IR and Raman), and (vii) ν (Ar) urethane amide I ν(C O) (both IR and Raman). The peak positions are indicated in the text.

Between 3000-2700, the modes appeared are attributed to C−H (sp³str) band, regardless of their nature to the rest of the PU macromolecule. Fig. 6.2 (b), is a typical band at 1088 cm^{-1} indicating epoxy presence (C−O str) group in virgin GNCs. At ∼ 2925, 1628, and 1300 are the bands allotted to C−H (sp³ str), C=C, and C−C stretching vibrations for GNCs, respectively [27,28]. The resonant modes associated with GNCs are assigned for

D-peak (in split form), G-band, and a broad 2D-peak; details in ref 123. Fig. 6.2(c-d), are recorded spectra for 1 and 25 wt % GNCs/PU nanocomposite systems; respectively. They displayed significant alterations to their parent counterparts. In plot (c), the emergence of a wide peak, at 1150-1400 interval, has merged all active inelastic modes of GNCs and PU. The band v(C−N)+ δ(N−H) is existent @ 1511 for 1wt % but terminated, thereafter, with successive GNCs loading. The modes related to v(Ar), amide backbone of PU, is invariant. The Raman active C− H str-mode is changed markedly, for PU, with 1wt % GNCs presence. Corresponding high intensity for GNCs compared with PU is indicative of hydrosorption site availability in GNCs. For the transmittance loss of IR bands, the characteristic trend observed is inconsistent with the change observed for Raman active modes in all samples, mostly. Broadly speaking, with GNCs incorporation in PU, any change in the band associated amide backbone, i.e. v(Ar), is peripheral. For these peaks, neither intensity nor vibration frequency is observed to be varied indicating single bonded backbone of polymer matrix mostly remains intact even though GNCs is added.

The summary of observations are: (a) for nonplanar, hydrogen bonded amide III δ(CH) peaks @1315 and 1500 cm^{-1} (-N−H-) there is decrease in Raman peak intensity, (b) doubly bonded oxygen (C=O@1700 cm^{-1}), δ(CH$_2$) v_{sym} N=C=O @1441 cm^{-1}, nitrogen (C=N- at 1250 cm^{-1}) is modified, and (c) sp^3 C−H str Raman active modes @3000 cm^{-1} are vanishing. Hydrogen bonding in the host matrix is modified heavily by GNCs with a change in doubly bonded moieties. Amongst all, the groups −N−H−, N=C=O, and −C=N− are nitrogen-based moieties having electron donating capacity; contributing to GNCs bonding via hydrosorption. Due to π-conjugation, in this, the double bonded sites are more reactive, and the effect is dominant at these sites, probably. The nature of hydrogen is, in general, tricky because they are neither electro-negative nor positive. The intermediate conclusion is that the nature of the interaction is lying at double bonded sites, dominantly, with hydrosorption in origin.

(b) Direct current conductivity:

The transport properties of PU and GNCs/PU composite samples were determined by a proto type two-probe method in the coaxial configuration using a standard Keithley 6487 picoammeter/voltage source equipped with the data acquisition software.

Recorded variations in dc conductivity (σ_{dc}) with respect to GNCs weight fraction (p) are shown in Fig 6.3. Below the weight percent threshold of ~ 5.0, the conductivity showed a dramatic decrease of ~ 7 orders of magnitude, indicative of the fact that above this threshold, the percolating network is formed in the PU matrix. Due to increase in the number of hydrosorpted conducting sites, the observed variations are seen.

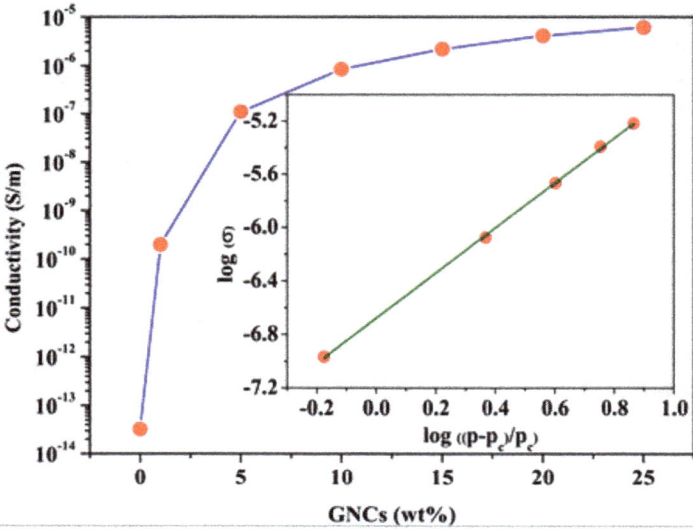

Fig 6.3 Variations in dc conductivity (σ_{dc}) (in logarithmic scale) as a function of weight fraction (p) of GNCs in the PU matrix. Measurements were performed using a standard two-probe technique, at room temperature. Inset: log–log profile for σ_{dc} vs log((p – p_c)/p_c). The straight line in the inset is fitted using least-squares methods for the obtained data using eq. 1 returning the best fit values p_c ~ 5.0 wt % and γ = 1.69 (correlation factor, 0.02).

The power law shown in the inset of Fig. 6.3 is broadly obeyed by the electrical conductivity [29]

$$\sigma \, \alpha \, (P - Pc)^\gamma \tag{1}$$

where, σ is conductivity of the composite, P, GNCs wt. fraction, P_c, percolation threshold, γ, critical exponent. Herein, we used weight fraction values of GNCs instead of volume fraction, as the density of GNCs is, approximately, estimated. The GNCs/PU nanocomposite conductivity agrees well with the percolation behavior for log(σ) with log((p–p_c)/p_c), predicted by Eq 1. An excellent fit is obtained for the data having a correlation factor of 0.02, with a profile as a straight line and p_c ~ 3.0 wt % and γ = 1.69. Relatively at the lower side, i.e. 3.0 wt % GNCs, the percolation threshold is seen, attributing to efficient GNCs dispersibility into the host matrix. In reported literature, theoretical values for γ in case of a three-dimensional (3D) percolating network are observed to be varied from 1.6 to 2.0 [30]. While for carbonaceous composites exponent values are, experimentally, reported to be varied from 0.7-3.1 [31-33]. In the subsequent section; analysis of X-band measurements has been presented which resembles discussions presented above.

(c) Morphological studies:

FESEM (Zeiss ΣIGMA) was used to study surface morphology of PU and GNCs/PU nanocomposite was investigated at beam voltage 5 kV. Using the cryo-fractured technique, the samples were prepared in which one PU sample was immersed, at a time, into a liquid nitrogen bath for about 5 min and the sample were allowed to reach near liquid nitrogen temperature. The sample was taken, subsequently, out and instantly broken. GNCs/PU nanocomposite samples were fractured in a similar fashion. Prior to FESEM imaging, the sample surface was subjected to gold coating; the process was carried out using a standard sputtering technique.

Surface morphology of the cryofracture samples (left panel) and magnified region (right panel) are shown in Fig. 6.4. As compared to the morphology of GNCs incorporated PU, the surface morphology of PU is found to be distinctly different. The uncorrugated zones in PU are observed as well as the microvoids that, inherently, exist in the host matrix. The synthesis route of PU has the origin of such microvoids which were formed by onset hot press curing of the samples. Gaseous species responsible out-diffusing during the curing could be forming these microvoids with random size distribution. They are mostly spherical in shape and a few were found to be elliptically shaped, coupled to each other and thought to be formed due to local fugacity and rheology of the polymer matrix. The homogeneous morphology of a typically magnified position of host matrix is seen in Fig. 6.4(e) and surrounded by the microvoids in Fig. 6.4(a). The morphological change associated with low weight % GNCs, such as 1, of the homogeneous portion of the matrix is seen in Fig. 6.4(b,f).

(d) Microwave Measurements:

Using a vector network analyzer (VNA, Agilent-E8364B) equipped with the coaxial transmission waveguide (HP broad frequency range coaxial 7 mm airline (85051-60007)) in the frequency range of 8.2–12.4 GHz, microwave measurements on PU and GNCs/PU samples were carried out. Schematic sketch of the distinctive electromagnetic shielding measurements setup is shown in Fig. 6.5. VNA was started for about 2 h, earlier to the measurements, for stabilizing the microwave source.

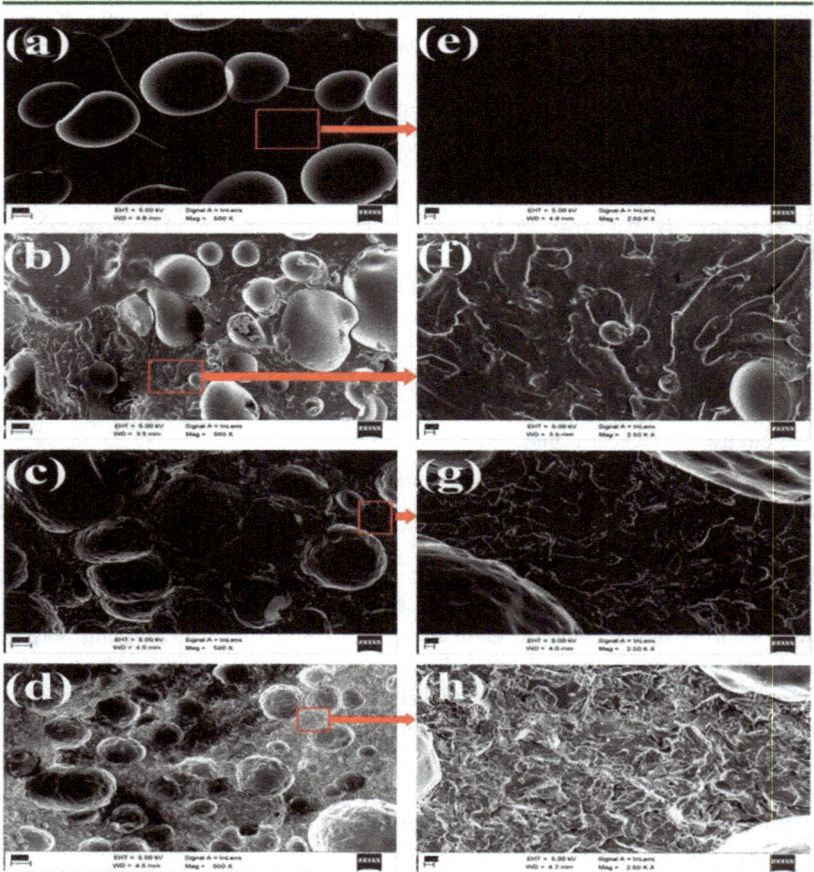

Fig. 6.4 Typical FESEM micrographs recorded for (a) PU, (b) 1 wt %, (c) 10 wt %, and (d) 25 wt % GNCs incorporated in PU. The rectangle coupled to the arrow indicates the respective magnified regions showed in images e−h.

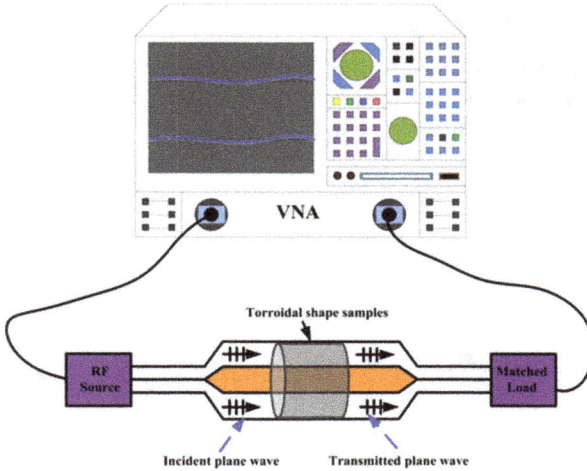

Fig. 6.5 Schematic representation of the electromagnetic scattering measurement setup comprised of the vector network analyzer, an RF source, matched load, and the sample under test in the coaxial transmission waveguide. The frequency range X-band, 8.2– 12.4 GHz.

Errors due to isolation, directivity, load match, and source match, etc., are minimized by full two-port calibration of the VNA performed on the test specimen along with the calibration, performed in both forward and reverse directions. From the measured scattering parameters, the complex permittivity ($\epsilon'- j\epsilon''$) and S_{21} parameters for composites were determined by standard Agilent software module 85071, based on the procedure given in the HP product note [34].

(e) Torroidal shape sample preparation:

The stipulations of the waveguide were the toroidal shaped samples with an inner diameter (i.d.) 3 mm and outer diameter (o.d.) 7 mm, for X-band measurements. A die assembly has been designed and developed, in order to fabricate such samples. The details of (a) design and fabrication of the die, (b) preparation of the nanocomposite paste, (c) the adiabatic hot-press technique, and (d) cutting and edge polishing protocols have been provided elsewhere.

6.2.3 Analysis of microwave parameters

Analysis of Scattering Parameters: Real and Imaginary Parts of Permittivity For the attenuation of an incident electromagnetic wave, three mechanisms are, basically, responsible; (a) reflection, (b) absorption, and (c) multiple internal reflection losses at the interface due to conductive fillers or porosity of the materials. While analysing permittivity response we determined the absorption properties of the test specimens in the X-band.

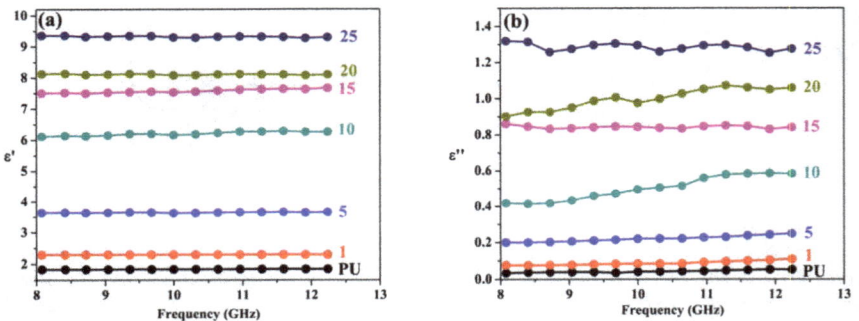

Fig. 6.6 Recorded complex microwave scattering data over the measured frequency regime: (a) real (ϵ') and (b) imaginary (ϵ'') parts of the permittivity spectra for PU and GNCs/PU nanocomposites. The numbers to the right-hand side in each profile indicate the weight percent of the GNCs.

For real (ϵ') and imaginary (ϵ'') counterparts of the permittivity function, plots (a) and (b) of Fig. 6.6 is the recorded frequency response spectra, respectively, for PU and GNCs/PU nanocomposites with variable GNCs wt. %. In both real and imaginary parts of the permittivity, a monotonic increase has been observed, over the frequency zone, with the increase in GNCs content.

For the real part of the permittivity, the response is nearly frequency independent. The flat response showed that PU, as well as nanocomposite, is a class of non-Debye solid [35][172]. Broadly, the frequency dependence of permittivity comes from the polarization mechanism via the *Clausis-Mossitte* relation. However, small percentage variations in the real part of the permittivity (i.e., ϵ'%) have been estimated for all samples. At actuals, differences are estimated over the full range frequency. The ϵ'% ~

1.03, for PU, whereas, for 5 wt %, ~ 1.11, resembling, closely, to that of PU. The ϵ'% is found to be saturated to ~16, with an increase in the loading of GNCs from 5 to 15 wt %. The percentage variation is negative, −3.17 thereafter, and −15.00, respectively, for 20 and 25wt % GNCs/PU nanocomposites. With rapid saturation in both directions, the flipping of magnitude of ϵ' is indicative of the flip in the operative dipole moment. Gradual trendy variations in frequency have been observed the measured for the imaginary part of permittivity. Small variation is observed at higher loadings, for, e.g., 10, 15, and 25wt % samples. Over the frequency regime, both real and imaginary parts of the permittivity are found to be varied, marginally. The comparison, on a relative scale, has been made across the categories of the samples by taking an average at log-normal scale (indicated in Fig. 6.6). At 25wt %, the value of the real part of permittivity is 9.31 ± 0.03. PU offers low (real) permittivity 1.84 ± 0.01, in contrast with an increase ~ 5x. The imaginary part of permittivity is 1.29 ± 0.03 (for 25wt %), which indicates an increase by a factor of ~30 with respect to the base value of PU (0.043 ± 0.006). Due to an increase in ac conductivity by enhancing active modes of charge transfer polarization via GNCs in PU, on a relative platform, increases in both the real and imaginary parts of the permittivity is observed.

Loss Tangent, Alternating Current Conductivity, and Skin Thickness: Using the relation, $\tan \delta_e = \epsilon''/\epsilon'$, the electrical losses inside the nanocomposite material under the test condition are quantified by calculated loss tangent ($\tan \delta_e$). In Table 6.1 the magnitudes of the average $\tan \delta_e$ (rad) measured for the systems are provided.

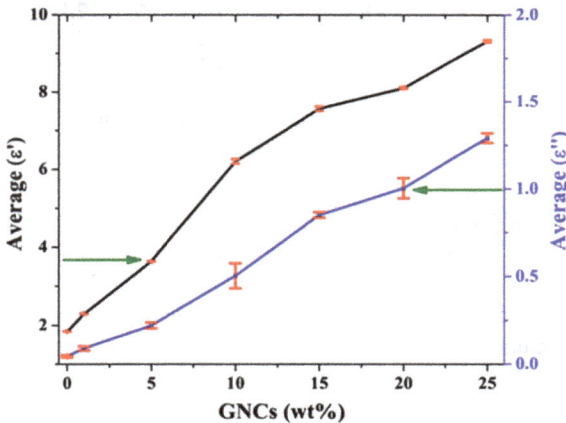

Fig. 6.7 Estimated magnitudes of real ϵ' and imaginary ϵ'' parts of the permittivity (average at logarithmic-normal scale) as a function of GNCs weight percent.

The loss component, for PU, is 0.024 ± 0.003. For the 25 wt % value, it is recorded to be 0.140 ± 0.003. The tan δ_e is increased by a factor of 6 times. Basically, tan δ_e is the quadrature part of the polarizability vector component in line with the incident electromagnetic field. Greater field losses are indicative of an increase in field arrest inside the nanocomposite material. So, after incorporating GNCs, in the host matrix, the field losses seem to be increased, marginally. Using, $\sigma_{ac} = 2\pi f \epsilon_0 \epsilon''$, relation the ϵ'' parameter the ac conductivity (σ_{ac}) of a dielectric material could be evaluated, where, σ_{ac} is measured in siemens per meter, ϵ_0 is the free space permittivity (8.854×10^{12} F/m), and f is the applied frequency in Hertz. For PU, the value of σ_{ac} is 0.248 ± 0.135, which is increased linearly by a factor of 30 times with sequential incorporation of GNCs until 25 wt %. As discussed, the observed increase is due to donor loaded nitrogen sites such as -N−H-, N=C=O.

Table 6.1 Magnitudes of Measured Loss Tangent, tan δ_e, and Alternating Current Conductivity, σ_{ac}, for PU and Different Weight Percent GNCs/PU Nanocomposites under Testing and C−N- attached to GNCs. The effect seems to be dominant at double bonded GNCs sites, due to π-conjugation with the host matrix.

samples		tan	δ, rad	σ_{ac}, S/m	
	P U	0.024	± 0.003	0.248	± 0.135
1	wt %	0.039	± 0.005	0.511	± 0.255
5	wt %	0.061	± 0.005	1.261	± 0.515
10	wt %	0.081	± 0.011	2.894	± 1.548
15	wt %	0.110	± 0.002	4.801	± 1.215
20	wt %	0.120	± 0.007	5.706	± 2.076
25	wt %	0.140	± 0.003	7.288	± 2.740

[a]The σ_{ac} is increased linearly by a factor of 30 with subsequent GNCs incorporation up to 25 wt %.

The incident EM field interacts with mobile charge carriers in the host medium to generate charge transfer displacement current quantified as the skin depth which is a thickness parameter (δ). This effect is responsible to couple incident field wiggles with charge carriers and a decisively set thickness of the shielding nanocomposite. The σ_{ac} is related to the magnitude of δ, given by $\delta = (1/(\pi f \mu_0 \mu_r \sigma_{ac})^{1/2})$, in which μ_0 is the free space permeability ($4\pi \times 10^{-7}$ H/m) and μ_r is the relative permeability for GNCs/PU nanocomposites, ≈ 1. Moreover, the skin thickness estimates the distance over which the field intensity decreases to 1/e of its original value (0.3678 mW) under the condition $\sigma_{ac} \gg 2\pi f \epsilon_0 \epsilon'$. In Fig. 6.7 the estimated values of skin depth in terms of thickness as a function of GNCs weight percent are shown, indicating, the magnitude of the skin thickness, δ, for PU ~ 10.44 ± 2.1 mm, whereas, 5.91 ± 1.09 mm is for the 3wt % GNC. The estimated values showed ~ 40% reduction in skin thickness of the sample. With the subsequent increase in GNCs weight percentage in PU (25wt % GNC) the thickness of shielding is reduced to 1.88 ± 0.25 mm. Reduction in the skin thickness is indicative of absorption of microwave power generating hindrance to free charge carrier propagation. The higher attenuation microwave power shows the ability to block charge carrier across the shielding thickness compounded with efficient coupling between field wiggles to the matrix. The amount of microwave absorption has been quantified from the scattering data, given in a subsequent section.

6.2.4 Efficient microwave absorbing properties:

The question to be asked: how much the level of an incident power (or power flux density) has decreased, after passing the specimen under test? The mode of measurement is typical transmission measurements (scalar S_{21} measurements) and obtained decibel value can provide the answer. In Table 6.2, the calculation of percentage values is presented to their power relationship. The penetrating power is reduced down to 1%, at ~ 20 dB shielding. To calculate the dB value, following equation is used to compute SE:

$$\frac{SE}{dB} = 10 \log \frac{P_T}{P_I} \qquad (2)$$

Where, incident power, P_I, and transmitted power, P_T.

The S_{21} is computed, as shown in Fig. 6.8.(a), having a flat response, consistent with the variations in permittivity. Since S_{11} and S_{12} data are not presented. Whereas, S_{21} alone could not be claimed for the total shielding effectiveness (SE_{TO}) and the uniformity of S_{21} is indicative of operative mechanism; as explained earlier surmises our results.

Table 6.2 Relationship between Shielding Effectiveness (SE) and Power Transmission (%).

SE, dB	power transmission, %	SE, dB	power transmission, %
0	100	12	6.25
1	81	13	5.00
2	62.80	14	4.00
3	50	15	3.13
4	40	16	2.50
5	31.60	17	2.00
6	25	18	1.56
7	20	19	1.20
8	16	20	1.00
9	12.50	25	0.316
10	10	30	0.1
11	7.90		

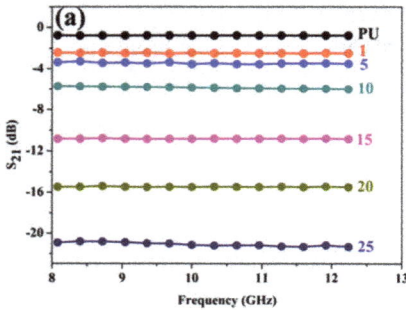

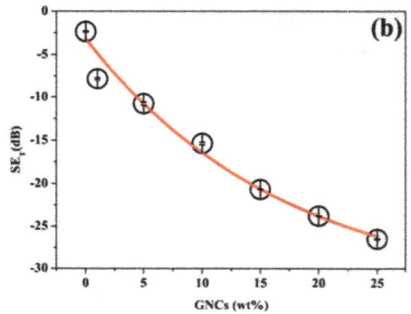

Fig. 6.8 (a) Variation in scalar S_{21} parameters measured in dB unit. The response is plotted over 8.2−12.4 GHz for PU and variable weight percent of GNCs in the composite. (b) Computed shielding effectiveness due to transmission loss, SE_T.

The plot of Fig. 6.8 (b), is the SE due to transmission (SET); measured using eq 9.3 for the estimated effective transmittance loss. For PU, the SE_T is highest and gradually decreased with filler wt %. The magnitude of SE_T, for PU, is −2.34 dB and found to be −10.76 dB on the percolation threshold and progressively, the value is enhanced to −26.45 dB @ 25wt % GNCs/PU composites, maintaining identical trend as that of S_{21}. So, we have plotted the average change. Showing, the magnitude of SE_T is round about 12 times lesser than 25wt %. Interestingly, percent variation in transmission loss has been estimated. The amount of transmission loss is dictated by the extent of microwave propagation in a medium. In the following section, the discussions on the amount of microwave propagation with percent tranmittivity have been presented.

For polymeric medium, the % tranmittivity values obtained provides the idea about electromagnetic behavior in the composite material under test, at low microwave electric conductivity. From Fig. 6.9 one can see that ~ 40% transmission loss requires a thickness > 10 mm for virgin PU, whereas, GNCs addition in it improves the magnitude of the transmission loss.

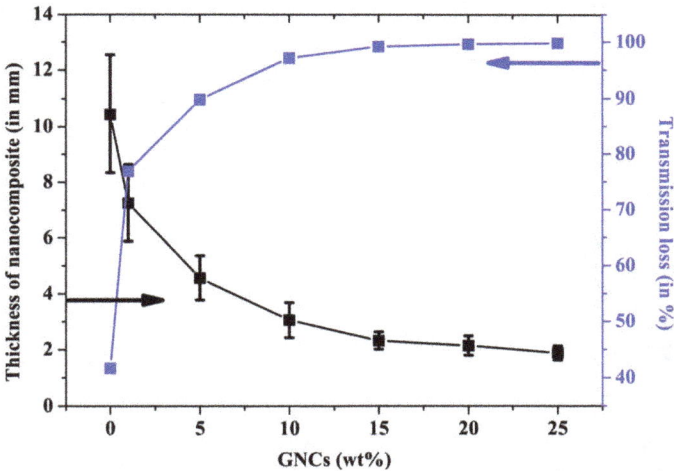

Fig. 6.9 Variation in thickness measured (mm) and transmission loss (%) as a function of weight percent of GNCs in PU. Measurements on base PU are also indicated for comparison.

The thickness of the composite is less than 5 mm, typically, at optimum percolation threshold ~ 5wt % and with addition of GNCs like 25wt %, almost 99% tranmittivity is loss at a lesser thickness value ~ 1–2 mm indicative that addition of GNCs is beneficial because of inherent functionality generating interfacial interactions achieving major loss in tranmittivity @ low thicknesses, relatively.

The embedment of uncorrugated zones in PU with the GNCs flakes is seen including microvoids in Fig. 6.4, in which the amount of filler seems to be insufficient, at this concentration level, to be distributed uniformly. The coverage of GNCs on microvoids gets denser as their weight percent in PU is increased, subsequently.

Interface Polarization Mechanism: Based on the analysis of microwave scattering parameters, the interface polarization mechanism is proposed, showing the amount of dipolar/orientational (OP, radical), atomic (AP), and a number of electron donor (EP) constituents moieties available. They are almost identical in quantity and charge accumulation based polarization remains constant. This leads to a nearly constant real permittivity response for frequency range for any sample. Three shallow relaxation peaks at ~12 (EP), ~10 (AP), and ~9 (OP) GHz are seen for PU in which OP is fading by loading GNCs. This behavior is consistent with vibration spectroscopy discussions presented before.

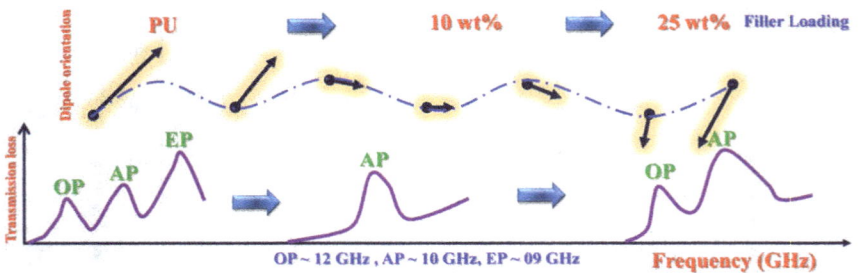

Fig. 6.10 Scheme of interface polarization and variable contribution of polarization modes in PU and GNCs/PU nanocomposites. (OP, radical polarization associated with stereo regular -GNCs-H mode; AP, atomic polarization (urethane amide III δ(CH), at ~1315 cm⁻¹, -N–H- at ~1500 cm⁻¹); EP, electronic polarization indicating doubly bonded oxygen (C=O, -C=N=O-) and nitrogen (-C=N-) sites.

Saturation of radical sites via GNCs hydro-sorption is seen. For EP (12 GHz), at low concentration of GNCs, up to ~ 5wt %, the peak disappeared. This region of loading is, importantly, estimated as the percolation threshold as discussed before in describing σ_{dc}, with efficient dispersion. Following this, the loss peak emergence at high concentration of GNCs is indicative of none equal amount of EP related to doubly bonded oxygen (C=O, -C=N=O-) and nitrogen (-C=N-) sites, whereas, AP (@10 GHz), with PU segmental motion.

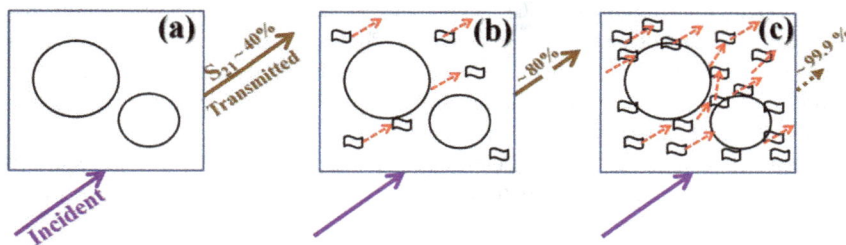

Fig. 6.11 Schematic representation indicating the role of GNCs (hollow rectangles, b, and c) in PU to enhance shielding effectiveness.

Generation of variable strength conducting path (shown by small dashed arrows) possibly couple the incident EM field wiggles to the matrix, arrows at the bottom indicate incident radiation. Whereas, the broken arrows at right (brown) indicate power tranmittivity, S_{21}. The computed transmission loss in percent is represented by a number. Image (a) shows (microvoids) hollow spheres in PU, (b) and (c) low and high wt% GNCs in PU, respectively, showing full GNCs coverage in voids. The conducting path strength is high, leading to dissipation of field wiggles, effectively, at higher GNCs loading urethane amide III δ(CH) (at ~1315 cm^{-1}) and -N−H- (1500 cm^{-1}), acts as a backbone and remains invariant even after GNCs loading. Corresponding interface polarization scheme is shown in Fig 6.11. The associated relaxation time $\tau_{r,OP}$ ~ 60 fs for -GNCs-H stereoregular radical modes. Whereas, host matrix segmental dynamics is having $\tau_{r,AP}$, magnitude, ~ 40-50 fs. The $\tau_{r,OP}$ and $\tau_{r,EP}$ are absent at low weight fraction (up to ~ 5 wt %). The absence is related to aggregates and their interface, having polarization within the periphery of aggregates. Below the percolation threshold, their absence is indicative of homogeneous GNCs dispersion of GNCs in PU having relaxation, occurring via mostly AP which resembles with σ_{dc} data analyzed. The value of σ_{dc} is in the range of 10^{-12}–

10^{-13} S/m, typically, below the percolation threshold, with an estimated activation energy ~ 94 meV of charge carriers pertaining to AP. Beyond the threshold value, dipole and electronic contributions emerge having, respectively, activation energies ~ 93 and 95 meV.

The gradual increase of GNCs in PU provides carrier transport path to incident EM field to get coupled to the host matrix via generating polarization currents as shown in Fig. 6.11_schematically. Image (a) in Fig.6.11_shows PU matrix microvoids with penetration of incident EM field (shown by an arrow at the bottom), having small tranmittivity ~ 40%. Whereas, the addition of GNCs, at low percent, has increased the loss, doubly, by ~ 80%. Hence, it cannot provide an effective conducting path (Fig. 6.11 (b)). At higher weight percent, such as more than 20 wt %, the transmission loss almost reached 99.9% (Figure 6.11 (c)). Due to the dissipation of the wiggling field into the surface, tranmittivity is hindered severely by generating a polarization current via GNCs, consistent with discussion presented earlier. The existence of channels like OP and EP is attributed to polarization related with a homogeneous distribution of GNCs in PU and around the microstructure's interface.

The GNCs prepared by above methodology has been integrated for revealing mechanical and thermal properties in the form of fillers in host epoxy resin matrix. In the following section the utility of GNCs as an effective nanofiller for improving the mechanical and thermal properties of the polymer at low weight fractions have been demonstrated [2].

6.3 Nanocomposite approach for structural engineering

Graphene possesses exceptionally high in-plane elastic modulus (~1 TPa), high strength (~130 GPa) [37] and high specific surface area (>2000 m^2/g) [38], due to its two-dimensional single-atom-thick sheet-like character. It has high thermal conductivity (~5000 W/m-K) [39], thermal stability [40] and excellent electron mobility even at room temperature (~ 10^5 cm^2/Vs) [41]. Such exclusive properties make it an ideal filler material for developing polymer reinforced nano-composites [42,43]. They find multitudes of applications in conducting media [40], active transparent electrodes [44], high strength materials [43,45], electromagnetic interference shielding[46], etc. Further, graphene type nanofillers; such as expanded graphite [47] graphite nanoplatelets [48,49], GO [50,51] and graphene nanoribbons (GNR)[52] have extensively been used as reinforcing element in various polymer matrices with variable weight fractions up to 5 wt%. In such compounded system, the formation of agglomeration as a result of poor dispersion of nanofillers in polymer matrix limits the transfer of its properties to the host medium [53]. Aggregation is ascribed to strong interlayer physical forces like van der Waals interactions between filler sheets and its poor interfacial adhesion with host polymer. The chemical functionalization has been carried out to address these issues [45,54].

Mechanical properties are one of the most deliberated phenomena in epoxy composites due to their widespread applications from aerospace to wind-mill. There are consistent efforts, in this context, to reduce the amount of filler content in epoxy in order to minimize the nanoparticle agglomeration and to achieve superior dispersion and improvements in mechanical properties [43,45,55]. The aggregate performance of nano-composites is a manifestation of the *combo* of various mechanical parameters; like tensile strength-modulus, flexural strength-modulus and fracture toughness, etc. In this, fracture toughness is the most critical parameter, particularly for structural applications. There are a considerable amount of efforts to improve the toughness and to reveal the fracture behavior of epoxy composites [50,53,56]. The advances in mechanical parameters of materials are highly influenced by the fillers physical and chemical properties like geometry, surface area and surface condition i.e. morphology [42,57], surface chemical functionality [58], interface-chemistry [59], and aggregation tendency [53]. In a study, Rafiee et al. [43] compared the mechanical parameters of nano-composites using various fillers like graphene platelets (GPL), single-walled carbon nanotubes (SWCNT) and multi-walled carbon nanotubes (MWCNT) into epoxy. The highest improvement was observed for GPL amongst the other fillers used. Young's modulus, tensile strength, fracture toughness (K_{IC}) and critical strain energy release rate (G_{IC}) of 0.1 wt% GPL composites were found to be increased by ~ 30%, ~40%, ~53% and ~126%, respectively, with respect to pure epoxy. Bortz et al. [50] reported the effect of GO incorporation on properties of epoxy in which the flexural strength and modulus, K_{IC} and G_{IC} showed a monotonic increase with weight fractions between 0.1 and 1 of GO; whereas, the tensile parameters showed improvements, however, inconsistent [50]. At 1 wt% GO loading, the flexural strength and modulus, K_{IC} and G_{IC} were found to improve by 23, 12, 63 and 111%, respectively. In another study, Tang et al. [53] investigated the effect of filler dispersion on properties of graphene/epoxy composites in which poorly dispersed reduced GO composites, showed pedestrian performance compared to highly dispersed composites (~52%) in K_{IC} with 0.2 wt% filler [53].

6.3.1 GNCs as effective nanofiller

In above studies, graphene/its derivatives are found to be superior nanofillers for improving mechanical properties of polymers compared to other allied carbon fillers with filler content typically 0.05 wt% or more [49, 52, 53]. In so far, mechanical properties and fracture mechanisms in epoxy composites at a very low content (< 0.05 wt%) of graphene derivatives have not been explored and not fully understood, even. GNCs, a new graphene derivative, for their nanocomposites have not been reported in the literature. The prepared GNC contained hydroxyl and epoxide polar groups with an exceedingly disordered graphitic backbone having hybrid sp^2 sp^3 phase [26]. It has several advantages over GO; like GNC is synthesized by a simpler route with the less expensive precursor. GNCs have advantages over carbon nanotubes (CNTs); such as (i) they are sheets, offers higher areal content to interact with matrix as compared to rolled CNTs, (ii) they have inherent functionality which reduces the interlayer van der Waals forces between the conjugated sheets and thereby facilitate effective dispersion in polymer by facile sonication.

Current work presents, the effect of GNCs incorporation in epoxy at weight fractions between 0.005 and 2 wt% onto their mechanical properties. The GNCs dispersion is examined by various characterization techniques. The bulk dispersion was explored by the interplay between dispersion-agglomeration phenomena using an optical microscope. The mechanical and thermal properties were studied in detail. Further, the toughening mechanisms and energy absorption through crack propagation were studied using fractography.

GNCs were dispersed into a liquid epoxy using an ethanol as a solvent for nano-composite preparation. Thermo-mechanical properties of epoxy and nano-composites were investigated using dynamic mechanical analyzer (DMA) (PAAR MCR 301 Rheometer) equipped with Rheoplus/32 V3.40 data analyzer software. The measurements were carried out over 25-150°C @ ramp rate of 3 °C/min with 1 Hz frequency. The mechanical tests were performed in a torsion mode on bar specimens with size $50\times12\times4$ mm^3. Tensile, flexural and fracture toughness were measured as per respective ASTM standards using a universal tensile machine (H25KS, Tinius Olsen, USA) equipped with a 5 kN load cell. The tests were done at room temperature and ~ 50% relative humidity. At least six specimens were tested and the mean and standard deviation (SD) are reported.

6.3.2 Dispersibility investigations: homogeneous distribution vs agglomeration and interfacial adhesion of GNCs

Mechanical parameters of nanocomposites are broadly controlled by (a) degree of homogenous dispersion vs. aggregation trend of fillers, (b) stiffness of fillers and (c) matrix/filler interface adhesion [52,52,60]. The GNC/epoxy matrix interfacial bonding plays a pivotal role in tailoring mechanical properties of the prepared nanocomposites.

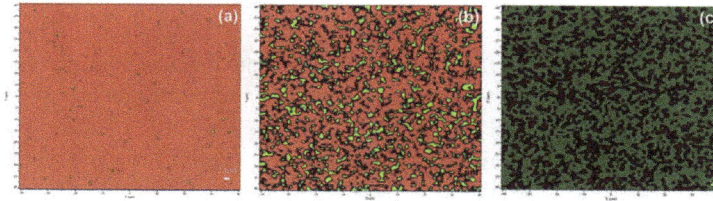

Fig. 6.12 The spectral Raman image of (a) 0.005 wt% and (b) 0.01 wt%, and (c) 0.5 wt%. (scale bar: 2 mm, area 40 40 mm^2). The typical spectral dispersion of GNCs (green) in the epoxy matrix (red) obtained for specimens with GNCs concentration.

6.3.3 Raman mapping of GNCs nanocomposites

Transmission electron microscopy (TEM) and Raman spectral mapping have been used to quantify dispersion of nanofiller. Both GNC and epoxy are carbon-based materials, we found that it was difficult to identify GNCs in epoxy due to poor contrast difference in low wt%. The Raman spectral mapping was successfully used, previously, to evaluate the nanofiller dispersion in the polymer matrix [61]. The same technique has been adopted to assess the GNC dispersion in this present work. The spectral features of Raman spectra collected from each specific spatial coordinates within the selected scan area in the specimen are used as pixels for final imaging.

Thousands of Raman spectra were recorded in order to create an image. For the mapping, nanocomposite samples were cut using a diamond cutter to obtain an ultrafine surface. The Raman spectra were recorded over 1000-1800 cm^{-1}. The spectral variations between 1250-1570 cm^{-1} (range covering D and G features of GNCs) were used to map degree of dispersion of GNCs in nanocomposites. The obtained images consisted of spectral dispersion of epoxy as a background, GNCs and both, as seen in Fig. 6.12. The boundary between GNC and epoxy is indicated by the dark color. For a better understanding of GNC dispersibility into the matrix, spectral images were recorded over large areas ~ 40×40 mm^2. Fig. 6.12 (a) shows such typical image of 0.005 wt% nano-composite. In this image, the epoxy region seems to be dominant with very small islands of GNCs. They are dispersed uniformly in the epoxy matrix. For 0.01 wt% nanocomposites, GNCs (green) are clearly seen to form a *network-centric* structure (Fig. 6.12 (b)), indicating a type of homogeneous dispersion on the surface, due to exfoliation of GNCs into individual layers in the host matrix. At higher GNCs content (0.5 wt%) in the host, GNCs occupied a large area as seen in image (Fig. 6.12 (c)), which may be due to the formation of aggregate (discussed below) owing to the high aspect ratio of GNCs [27].

6.3.4 Optical imaging

Optical micrography was used to obtain more insights into the state of dispersion. Optical micrographs of epoxy and nanocomposites with different concentrations of GNCs are shown in Fig. 6.13. The optical micrograph of pristine epoxy does not show any distinct features (image (a)). The micrographs of the nanocomposites are shown in Fig. 6.13 (b,e,f). In these images several dark island-like structures are clearly visible and attributed to aggregation and the number density of dark islands is increased with a weight fraction of GNCs (Fig. 6.13 b,e,f). Using the micrographs, the following parameters are determined: (a) size of the aggregates and (b) areal density of the aggregates (ADA) (per mm^2). Images were taken at least on three sheet-like specimens (dimension ~ 1 cm^2), to generate aggregate statistics. Nearly, 300-350 images were taken at different sites and the size, as well as the number density of agglomerates, was determined. As observed, the color intensity (black) of aggregates increases with increasing GNCs weight fraction (Fig. 6.13. c-f). With the increase in GNC concentration, the nanoparticle agglomeration increases in the composites. Aggregates with irregular geometrical shapes are well-distributed in the host medium. The average aggregate size is 13 ± 2 mm for 0.01 wt% and 20 ± 4 mm for 0.05 wt% nano-composites. For higher weight fractions the aggregate size decreases slightly with an increase in GNC content up to 2 wt%. The average sizes of GNC aggregates in the composites are given in Table 6.3. The variation in ADA with weight fractions of GNCs is shown in Fig. 6.13. ADA is found to be (3.16 ± 1.09) 10^5 per mm^2, for 0.01 wt%, increased with GNCs weight fraction (Fig. 6.13(g)) and obeyed a power law curve.

Table 6.3: Average size of GNC aggregates with the standard deviation in an epoxy matrix.

Sample	Size of aggregates(in μm)
0.005 wt%	-
0.01 wt%	13 ± 2
0.05 wt%	20 ± 4
0.5 wt%	20 ± 5
1 wt%	09 ± 3
2 wt%	11 ± 4

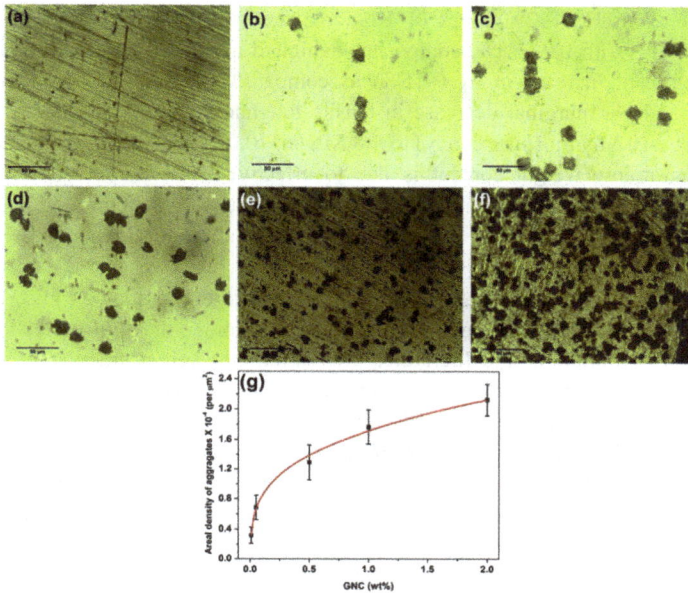

Fig. 6.13 Optical micrographs on the surface of as prepared specimens of (a) neat epoxy and different weight fraction of GNC nanocomposites: (b) 0.01 wt%, (c) 0.05 wt%, (d) 0.5 wt%,(e) 1.0 wt% and (f) 2.0 wt% taken at 20 magnification (scale bar: 50 mm), (g) Areal density of aggregates (measured in number of aggregates/mm²) as a function of GNC wt%. The areal density of aggregates obeyed a power law: ADA α W − W_C^γ, where W ¼ GNCs wt%, W_c critical GNCs wt%, and g is a critical exponent. The curve parameters with the best fit are: W_c ~0.006 and g ¼ 0.3 (correlation factor ¼ 0.025).

From our study, it is concluded that the nano-dispersion of GNCs with the aggregate formation is a challenging task, even at such a low GNC content. One might improve the properties further if the formation of agglomeration can be prevented or minimized.

6.3.5 Mechanical properties of GNCs/nanocomposites

(a) Tensile properties:

The specific tensile modulus (E_S), the ratio between elastic modulus-to-density, and ultimate tensile strength (UTS) of pristine epoxy and nanocomposites are presented in Fig. 6.14 (a). Notably, only with 0.005 wt% of GNCs, nano-composites showed improvements in UTS (~2%) and E_S (~8%) compared to virgin epoxy, though marginal (Fig. 6.14 (a)). With 0.01 wt%, composites exhibited a modest increase in UTS (~4%) and a significant increase in E_S (~15%) as compared to epoxy base (Fig. 6.14 (a)); accompanied by a marginal decrease in tensile toughness (K) (area under stress-strain curve) from ~10 MN/m^2 (pure epoxy) to ~ 8 MN/m^2 (0.01 wt%). The significant increase in modulus at low weight content is due to effective GNC reinforcement owing to chemical interactions between inherent functional groups present in GNCs to that of epoxy. The nano-level dispersion, barring the agglomeration formation, of GNCs in epoxy matrix provides a huge surface area to interact with the matrix. This leads to efficient stress–transfer from matrix to filler. To the best of the author's knowledge, the enhancement in tensile parameters; particularly elastic modulus, with 0.01 wt % GNCs has not been reported in the literature. The minimum weight fraction of nanofiller (graphene) used so far in composite systems is 0.05 wt% as reported by Tang et al. [61]. The advantage of composites containing very low weight fraction of filler, as in the present case, is that the agglomeration of nanofillers in the matrix could be minimized. To further investigate; the reinforcement and agglomeration effect of GNCs in epoxy, tensile properties of the composites with higher weight fractions (up to 2 wt%) are also studied; where the tensile character showed marginal improvement. For example, the composites between 0.05 and 1 wt% showed an increase of 3-10% in E_S as compared to pure epoxy (Fig. 6.14(a)). Whereas, the UTS showed a moderate increase, which is associated with a marginal decrease in K for all the cases. For 2 wt% GNCs, tensile properties do not show any significant improvement; rather strength decreased slightly with comparable modulus values to that of pure epoxy. Maximum increase in tensile properties, in the present case, is exhibited by 0.01 wt% nanocomposites. This is apparently due to uniform dispersion and good chemical interaction of GNC sheets with epoxy. In addition, the GNCs possess large surface area due to their high aspect ratio [58]. Above this concentration, the graphene platelets probably stalk with each other to form aggregates, which resulted in reduced reinforcement effect. In 2 wt% composites, the number of aggregated GNC layers increases to an extent that they form clusters of aggregates (Fig. 6.14). These aggregates seem to affect adversely on the tensile properties, since tensile strength is sensitive to defects produced due to aggregates in nanocomposites [49,53,57].

(b) Flexural properties:

Flexural properties of GNCs/epoxy were investigated by means of three-point bending test. Fig. 6.14 (b) displays flexural strength and modulus of nanocomposites with different GNCs wt. fractions. The load vs. extension curves for pure epoxy and nanocomposites were obtained by flexural measurements.

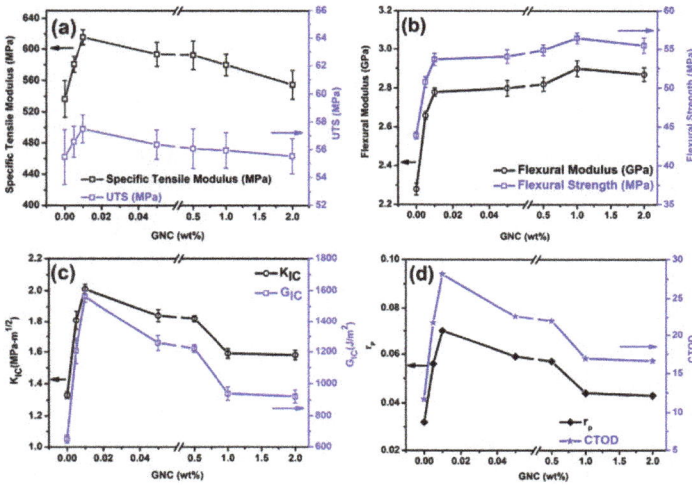

Fig. 6.14 Mechanical properties of epoxy and nanocomposites with various GNCs wt%. (a) specific tensile modulus and UTS vs. GNCs weight fraction, (b) flexural modulus and strength, (c) fracture toughness (K_{IC}) and critical strain energy release rate (G_{IC}) as a function of GNCs weight fraction, (d) Theoretical values of CTOD (d) and plastic zone size (r_p).

Interestingly, significant improvements in flexural strength (~16%) and modulus (~17%) with only 0.005 wt% of GNCs in the matrix (compared to epoxy base) have been observed. Similarly, in case of 0.01 wt% nanocomposites, flexural strength and modulus are increased by ~22% and ~23% respectively. The flexural strength and modulus further increased gradually up to 1 wt% (Fig. 6.14 (b)). The 1 wt% data showed the highest value of flexural strength and modulus, ~29% higher to base epoxy. The results indicated that the optimum flexural properties are obtained with GNCs up to 1 wt%; presumably, due to stiffening of GNCs in epoxy, in spite of forming a large number of aggregates in the matrix. So, unlike tensile properties, flexural or bending properties appeared to be less sensitive to defects in the specimen in terms of GNC aggregates, which is consistent with literature for other nanofillers [42,53,55]. Above a critical weight fraction of filler (in our case ~1 wt%), the flexural properties are also adversely affected; though insignificantly.

(c) Fracture toughness properties:

The K_{IC} and G_{IC} of pure epoxy and nanocomposites are derived using the linear elastic fracture mechanics (LEFM) approach. The K_{IC} and G_{IC} values are plotted with GNC weight fraction in Fig. 6.14(c). The K_{IC} and G_{IC} for epoxy are found to be 1.33 (±0.34) MPa.m$^{1/2}$ and 649 (±25) J/m^2, respectively. The K_{IC} and G_{IC} showed marked improvements, ~36% and ~86% @ 0.005 wt%, and ~51% and ~140% @ 0.01 wt%, compared to well-ordered epoxy. The % increase in K_{IC} and G_{IC} are significantly higher in our case @ 0.01 wt% compared to reported work by Bortz et al. [58] @ 0.1 wt% GO-composites, wherein, the increase is comparable to the reported work by Rafiee et al. [43] @ same wt% platelets based graphene composites. The improvement in notch toughness parameters further provides an indirect clue of *nano-stage* dispersion of GNCs in the host matrix. Such classes of sheets when dispersed uniformly in the matrix reduce the obstacle to the propagation of crack leading to higher energy absorption. The K_{IC} and G_{IC} showed marginal drop-down trends in weight fractions 0.05–2 wt% (Fig. 6.14 (c)); though obtained magnitudes are higher than that of pristine epoxy. The increases in K_{IC} and G_{IC} for obtained weight fractions intervals compared to virgin epoxy are in a range of 29-40% and 40-100%, respectively. The nanocomposites with higher (i.e. 2 wt%) GNCs showed the smallest rise in K_{IC} (~19%) and G_{IC} (~42%) as compared to the virgin host. This is due to the formation of a large number of GNC aggregates discussed previously. These aggregates are acting as defect centres in the polymer matrix. Both, tensile and K_{IC} showed similar trends with respect to filler content (Fig. 6.14 (a) and (c)) (increase in parameters till 0.01 wt%; followed by a decrease). Flexural properties, however, showed an invariable increase with GNC content. It is noteworthy that, in tensile and single-edge-notch-bend (SENB) specimens, fracture occurs through crack initiation and propagation. In flexural test case there is no formation of the crack feature; hence in-complete fracture; although the testing is a destructive technique. The test was terminated at 5% deflection. Therefore, possibly flexural test (bending in absence of pre-crack) is less defect-sensitive. Due to layered GNCs morphology and its ability to improve the mechanical properties remarkably, GNCs could be a promising material for preparation of bio-inspired kind of (seashell [63] as an example) layer-by-layer composites [64,65].

Further, crack tip opening displacement (CTOD) (δ) and plastic zone size (r_p) for the plane strain condition are computed from G_{IC} and K_{IC} strength of epoxy, using equation (3).

$$\delta = \frac{G_{IC}}{\sigma_y} \ and \ r_p = \frac{1}{6\pi} \left(\frac{K_{IC}}{\sigma_y}\right)^2 \qquad (3)$$

The values of CTOD and r_p for epoxy and GNC composites are shown in Fig. 6.14(e). These values increased significantly up to 0.01 wt% compared to host medium. The values show decreasing trends > 0.01 wt%, though the values are higher than that of the virgin values. The GNCs/composites showed larger plastic deformation at the crack tip compared to pure epoxy. The plasticity at the crack tip was further investigated through fractography on SENB specimens. The results are discussed in the subsequent section.

6.3.3 Fracture mechanisms using fractography

The fracture mechanism of the nano-composites can be revealed using fractography analysis. SEM micrographs of the fracture surface of tensile specimens are shown in Fig. 6.15. In 0.01 wt%, the individual sheets are found dispersed with effective uniformity and appear to be coated with epoxy (Fig. 6.15(a)).

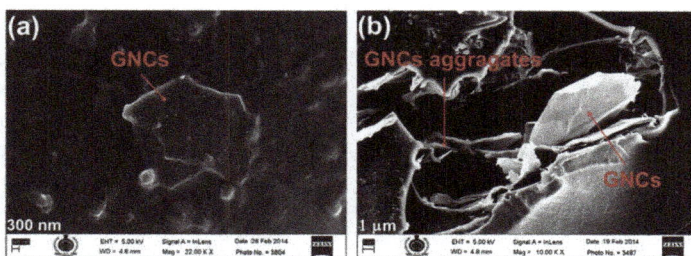

Fig. 6.15 SEM image of tensile fracture surface of nanocomposite (a) 0.01 wt% (scale bar: 300 nm), (b) 2 wt% (scale bar: 1 mm).

This indicates a good interfacial adhesion between epoxy/GNCs interface. At 2 wt% composite, agglomeration of few GNCs layers are distinctly visible in a typical location of the specimen as shown in Fig. 6.15(b). These aggregates are acting as stress concentration zones in composites. Due to the formation of aggregates, the individual conjugated GNC surface may not be in direct contact with epoxy. Hence, effective reinforcement has not been realized, leading to the mechanical properties below expectation, at higher weight fraction nanocomposites as compared to the lower weight fraction counterparts. This fact can be supported by Fig. 6.15(a) which shows a type of *pull-out* feature of a mono-GNC-layer from an aggregate of 2 wt% nanocomposite. The average size of the GNC layer (range 1-10 mm) in nanocomposite is found to be reduced than the pristine GNCs layers (range 10-20 mm)[51]. This is due to the fact that GNCs might undergo breakage of sheets during sonication [45].

Low magnification SEM images are shown in Fig. 6.16 (a) and (b) indicating fracture surface of SENB specimens for epoxy and composites, respectively. The source of crack or stress concentration regime is clearly visible at the middle of the notch in all these micrographs (Fig. 6.16 (a) and (b)). Several tracks of cracks are also visible in the micrographs, propagating in the direction away from the notch front. From the micrographs, it is evident that the fracture occurred mainly due to the action of both normal and shear stresses. As observed in Fig. 6.16 (a) and (b) the appearance of the lines which are not normal to the macroscopic notch, is the indication of shear stresses [66]. At the microscopic level, this is thought to be the mixed-mode (normal and shear) fracture, even though the fracture test was performed using Mode I (tensile stress in the direction orthogonal to the face of the notch). In the notch vicinity or at the crack tip the surface appears to be more ductile compared to the distance away from the notch in all the cases.

The surface appears flat and indicates cleavage-like feature away from the notch, which indicates unstable crack propagation with no gross plastic deformation followed by fracture. Relatively; higher magnification micrographs at the vicinity of the notch are shown in Fig. 6.16 (c) and (d). The *river-bed* marks indicate the formation of micro-cracks and their growth directions in the plane of fracture for epoxy (image (c)) and nanocomposites (image (d)). As evident, the roughness of the nanocomposite surface is much higher than that of neat epoxy which indicates the higher surface area in the former case. Further, the areal density of microcracks in case of nanocomposites is significantly higher than that of neat epoxy.

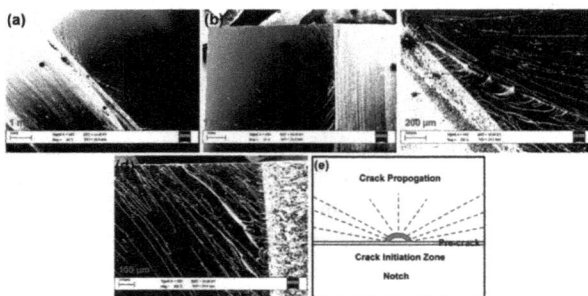

Fig. 6.16 SEM images of the fractured surface of mode 1 fracture toughness specimens for (a) epoxy, (b) 0.01 wt% nanocomposites, at lower magnification (scale bar: 1 mm).

The 3D images of Fig. 6.16 (c) and (d) highlighting the surface roughness obtained by image processing. The microcracks having an undergrowth of parabolic shape bulging out of the crack path indicate the deflection of the crack [49,66]. The twists and tilts in cracks are also distinctly observable in Fig. 6.16(d). These features are more prominent in the fracture surface of nanocomposites with respect to the virgin epoxy. Features clearly suggest that crack deflection is mainly due to obstacles rendered by GNCs in the path of the crack propagation. Higher crack deflection leads to higher energy absorption [50]. So, the nanocomposites are found to absorb more energy to fracture as compared to virgin epoxy. This phenomenon is verified by the higher values of K_{IC} and G_{IC} displayed by the nanocomposites as compared to epoxy (Fig. 6.16 (c)). The possible fracture mechanism in SENB specimen is; schematically, illustrated in Fig. 6.16(e), where the crack initiation zone and its propagation are shown [66].

6.3.4 Thermal and physical properties

Fig. 6.17 (a) presents TGA profiles for GNCs, epoxy, and nano-composites. The weight loss of GNCs between 300 and 850 °C is ~ 50 wt% (inset (ii) of Fig. 6.17(a)), which is related with the removal of oxygen functionalities and thermal degradation of GNCs. In epoxy and nanocomposites, the thermal decomposition processes, mainly, occurred between 376 and 442 °C (Fig. 6.17(a)). Table 6.3 shows, thermal properties achieved via TGA curves. Onset at degradation temperature (T_{onset}) for epoxy is found to be ~ 376 °C. The T_{onset} is increased by ~ 27 °C @ 2 wt % nanocomposites. Similarly, $T_{50\%}$ (temperature @ 50% wt loss) and $T_{d,max}$ (maximum degradation temperature) are increased with increase in GNC wt fraction in the matrix. For 2 wt% GNCs; maximum increase in $T_{50\%}$ and $T_{d,max}$ are found to be ~ 27 °C and ~ 29 °C, respectively (Table 6.3). The residual char yield (CY) @ 600 °C for epoxy, 0.005 and 2 wt% are found to be, respectively, ~6, ~9 and ~15%. This is apparently due to the effective reinforcement of GNCs in an epoxy matrix and their effective bonding with the matrix as discussed earlier.

Tan δ vs. temperature curves for pristine epoxy and GNC composites are presented in Fig. 6.17 (b). The peak of the tan δ curve is ascribed to second order transition or glassy transition temperature (T_g) of the polymer. The pristine epoxy showed a relaxation peak at ~ 85 °C, which corresponds to its T_g. The symmetric, single peak as seen in case of epoxy, interestingly, became asymmetric and broad with the addition of GNCs and resulted into two peaks when curve–fitted with the Gaussian function.

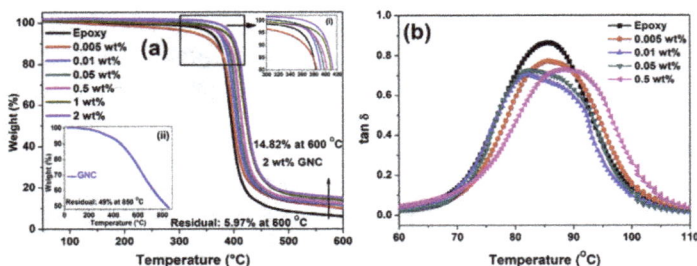

Fig. 6.17 (a) TGA curves of pure epoxy and nanocomposites. The inset (i) shows an expanded area near the T_{onset} and inset (ii) is TGA of GNCs, respectively. (b) tan d as a function of temperature for epoxy and nanocomposites obtained from DMA.

Table 6.4 Thermal parameters of epoxy and nanocomposites obtained from TGA and tan δ curves.

Sample	T_{onset} (°C)	$T_{50\%}$ (°C)	$T_{d,max}$ (°C)	CY (wt%)	T_g	
					T_{g1}	T_{g2}
Pure epoxy	376	396	413	5.97	85.0	
0.005 wt%	378	401	417	9.63	86.4	
0.01 wt%	383	405	418	11.35	81.1	87.9
0.05 wt%	390	408	425	11.55	80.6	90.4
0.5 wt%	395	412	433	12.10	84.0	91.1
1 wt%	400	420	440	13.72	-	-
2 wt%	403	423	442	14.82	-	-

These peaks are assigned to T_{g1} and T_{g2}, which resemble swift chain mobility due to slackly cross-linked domains (at low temperature) and slower chain mobility due to highly cross-linked domains (@ high temperature), respectively [67,68]. The data are presented in Table 6.4. The T_g showed moderate increases in case of nano-composites; particularly the T_{g2}, which increased about 5 °C in case of 0.05 wt%. The increase in T_g with the inclusion of nanofillers is often associated with a hindrance in mobility of polymeric chains [63]. It is reasonable to state that the observed increase in T_g (refer to T_{g2} for composites) is likely due to a good degree of dispersion and robust interfacial adhesion at polymer/GNCs interface, which may cause hindrance in polymer chain mobility. In contrast, the decrease in T_g (refer to T_{g1} for composites) is likely due to disordered polymer networks, which might arise due to filler agglomeration [59]. Molecular dynamics simulation studies have suggested that the surface geometry of the

Thermoset Composite   Materials Research Forum LLC
Materials Research Foundations 38 (2018)   doi: http://dx.doi.org/10.21741/9781945291876

nanofillers influence polymer chain mobility [71]. Therefore, the defected, warped, and the wrinkled surface of nano-carbon may also contribute in decreasing chain mobility (consequently, increase in T_g) through mechanical interlocking with the matrix.

In the similar lines, a composite of MWNT/nylon 6, 6 nanofibers were fabricated and the electrical properties were examined as a function of the filler concentration. The amide functionalized MWNTs were dispersed in formic acid on stabilizing stability for more than 40 h. The MWNTs-suspended in a solution of nylon 6, 6 in formic acid was electrospun to obtain the nanofibers. The I–V studies of the nanofiber found to be sheet improved with increasing filler concentration. [3]

6.4   MWNTs/nylon composite nanofibers by electrospinning

Carbon nanotubes (CNTs) are known for their unique mechanical and electrical properties. Due to which they find applications in electrical devices and materials. Composites of CNTs with a polymer are also used in many devices [72-75]. Properties of CNT composites are dependent on several parameters, such as the filler dispersion, orientation and interfacial bonding etc. [76]. Electrospinning is a unique and efficient tool for producing nanotube–polymer composites and can be used to assemble fibrous polymer sheets with fiber diameters ranging from the nm to μm range. The typical electro spinning set-up consists of a bipolar high voltage source, a syringe injector coupled to a needle (to carry the polymer fluid from the syringe to the spinneret) and a conducting collector to obtain randomly orientated or aligned nanofibers. One can obtain high strength, low weight and low porosity can be achieved along with tailored surface functionality, depending on the polymer type and polymer nanofibers with a wide range of properties, such as electrical conductivity using electrospinning [77-81]. This proves the utility of the polymer nanofibers for a wide variety of applications. Several reports have addressed on conducting polymer composite nanofibers using CNTs as the filler. The degree of dispersion of CNTs in the composite matrix is the key issue. A stable and uniform suspension of nanotubes in the polymer is necessary and for this purpose, several surfactants, such as sodium dodecyl sulfate and triton-X are used [82]. However, this method requires taking out and isolating the surfactants from the solution in order to avoid undesired effects. Nylon shows good chemical resistance and stability. On the other hand, MWNTs have poor degree of dispersion of the in an as-prepared nylon solvent and least studied electrospinning of MWNTs/nylon composites. This section describes fabricated MWNTs/nylon nanofiber composites using an electrospinning technique. Due to the fine dispersion stability of the MWNTs and good solubility of nylon, formic acid was used as the solvent.

6.4.1 Synthesis of composite

Readily available nylon 6, 6 (Sigma Aldrich) was used as the matrix material for the nanofiber and multi-walled nanotubes (MWNTs, Iljin Nanotech Co., Ltd., diameter ~15–25 nm) were used. Initially, the raw MWNTs were functionalized [84] with sulfuric and nitric acid solution (95% H_2SO_4: 65% HNO_3=3:1) by sonicating it for ~4–7 h. Followed to this MWNTs were collected on a PTFE (polytetrafluoroethylene) membrane (pore size ~1 μm) and were neutralized. The functionalization of amine group was achieved through a route chemical method using thionyl chloride ($SOCl_2$) and ethylenediamine. In a typical synthesis, 0.5 g of the functionalized MWNTs was added in 200 ml of $SOCl_2$ solution and sonicated for 5 min. The mixture was refluxed for 24 h at 50–70 °C, filtered and washed with anhydrous tetrahydrofuran (THF) to remove any unreacted $SOCl_2$. After this residue was stirred in 200 ml of ethylenediamine for 3 days at 50–70 °C and filtered through a 1 μm membrane. On filtration, the residue was washed with anhydrous ethyl alcohol to remove the residual ethylenediamine [84]. The functionalized MWNTs were then dried at ~70 °C under atmospheric conditions. This is followed by characterization such as Fourier transform infrared (FTIR) spectroscopy.

Subsequently, functionalized MWNTs were dispersed in formic acid then nylon was added and heated to allow for dissolution until the viscosity of the solution reached its optimum level. The nylon concentration is varied 10 to 15 wt.% and dispersed MWNTs/nylon solution with a nanotube concentration ranging from 1 to 20 wt.% was obtained using this process. The MWNTs/nylon solution was spun into fibers using an electrospinning system (NanoNC Co, Korea). Fig. 6.18(a) shows a schematic of electrospinning process.

Initially, a syringe was filled with an MWNTs/nylon solution and assigned to flow rate of 0.1–0.5 ml/h. The MWNTs/nylon nanofibers were sprayed on the ITO coated glass for a 3 min period. The thickness of the nanofiber sheet was found to be ~1 μm. The metal coated glass electrode with an area of ~0.25 cm^2 was placed on the MWNTs/nylon fiber sheet sprayed on the ITO glass. Fig. 6.18(b) shows a schematic diagram of the analysis. Following this their textural morphology was studied by environment scanning electron microscopy (ESEM) and transmission electron microscopy (TEM). MWNTs/nylon nano fibers were studied for electrical properties of the as a function of the filler content.

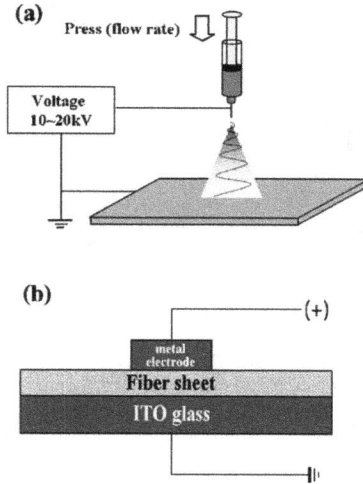

Fig. 6.18 (a) Schematic diagram of the electrospinning process, and (b) schematic diagram of the electrical measurement on the MWNTs/nylon non-woven mat.

6.4.2 Characterizations

(a) Dispersion of MTCNTs:

The dispersion of the CNTs in the polymer composites is the crucial issue. Different approaches are used for dispersion of CNTs in polymer solvents. Among various method an acid treatment is used to disperse the CNTs in solvents such as DMF [86], NMP, and water. The good dispersion of the acid-treated MWNTs in DMF attributed to the strong coulombic attraction interaction between the positively charged amide groups ($-NR2$) and the negatively charged MWNTs. [85,86]. For acid-treated MWNTs are well dispersed in water since it attached with functional groups such as carbonyl, hydroxyl, and carboxyl, which contain oxygen. Due to the hydrophilic interaction between the acid-treated MWNTs and water molecules is increased and an electrostatic repulsive force is induced between the MWNTs. As a result, the inter-tube dispersion is very efficient. A dispersion of MWNTs in an aqueous solution can be improved using these two interactions. Such forces are at pH of 7 [87]. Dispersed MWNTs can be precipitated around pH 7. In former case formic acid since the nylon is soluble in formic acid.

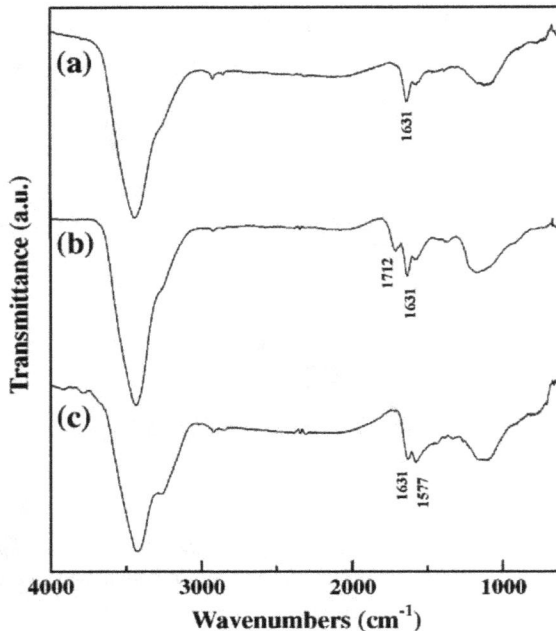

Fig. 6.19 FT-IR spectra for (a) pristine MWNTs, (b) carboxylic functionalized MWNTs, *and (c) amide functionalized MWNTs.*

(b) FTIR of functionalized MWNTs:

Fig 6.19 shows FTIR spectra of (a) pristine MWNTs, (b) carboxylic functionalized MWNTs and (c) amide functionalized MWNTs. In case of MWNTs, the peak at ~1631 cm^{-1} was assigned to the C-C stretching vibration mode associated with sidewalls while for acid treatment of the MWNTs, an additional peak was observed at ~1712 cm^{-1} due to a C-O stretch of the carboxyl group. With amine treatment of the carboxylated MWNTs, the emergence of a new peak at ~1577 cm^{-1} was observed. It is assigned to the in-plane N–H molecular vibrations of the amine group [88]. It is believed that a substitution reaction occurs and a –NH group replaces the –OH group of the carboxylated MWNTs after amide functionalization to form the –CO–NH functional group so peak at ~1712 cm^{-1} is disappeared on amine treatment.

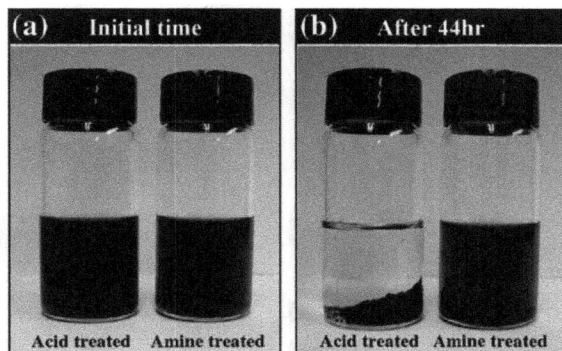

Fig. 6.20 Images of the MWNTs dispersed in a formic acid solution (sonication ~30 min). The bottle on the left side contained carboxylic functionalized MWNTs and the bottle on the right side contained amide functionalized MWNTs. (a) Initial status and (b) after 44 h.

Fig. 6.20 shows images of the dispersion of carboxylated MWNTs and the amide functionalized MWNTs in the formic acid solution in (a) initial stage and (b) after 44 h. The figure shows that the amide functionalized MWNTs were more stable in formic acid than the carboxylated MWNTs. At initial stage, both had good dispersion properties but the amide functionalized MWNTs dispersion shows more stability after more than 40 hrs. In contrast to carboxylated MWNTs agglomerate, precipitated and settle to the bottom of the solution for the same period. For the acid-treated MWNTs, it appears that the molecular interactions between the acid molecules and carboxyl group (on the surface of the MWNTs) are weaker than the coulombic attraction between the positively charged amide groups ($-NR_2$) and the negatively charged MWNTs in DMF suspension. Furthermore, the electrostatic repulsive force between the acid-treated MWNTs in formic acid was also weak because of low pH. However, for the amide functionalized MWNTs, the amine-terminated MWNTs have a stronger interaction with the formic acid molecules than the carboxylated MWNTs. Hence, the amide functionalized MWNTs showed better stability in formic acid than carboxylated MWNTs.

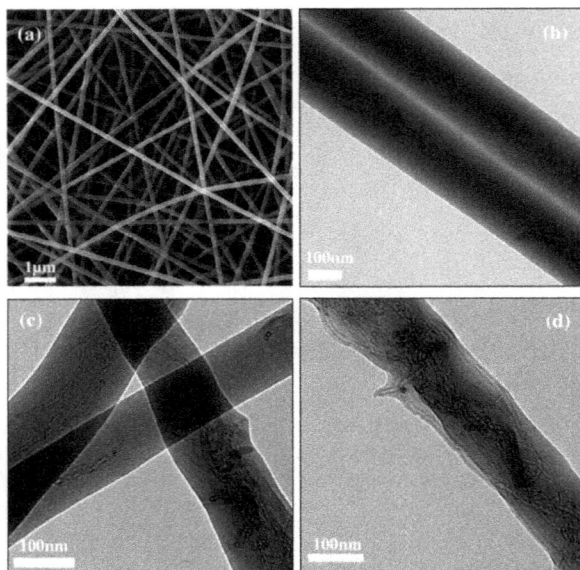

Fig. 6.21 SEM images of the (a) only nylon fibers with a 15 wt.% nylon solution. (b), (c) and (d) are TEM images of the MWNTs/nylon composite nanofibers with 2 wt. %, 10 wt.%, and 20 wt.% of MWNTs in the nylon fibers respectively. All the samples were electrospun at 18 kV with a needle and a collector distance of 10 cm. The solution flow rate was 0.4 ml/h.

(c) Morphological analysis:

Fig. 6.21(a) shows the SEM images it is observed that nylon nanofibers are randomly oriented in the non-woven mat and have a diameter of ~ 150–200 nm. Fig. 6.21(b), (c), and (d) shows the TEM images of the MWNTs/nylon spun nanofibers containing 2, 10 and 20 wt.% MWNTs, respectively. The TEM images show that as increase in wt.% of the MWNTs in the nylon 6, 6 fibers, it shows increases in density. In another experiment, a fluid blocked the outlet spinneret as a result of the high concentration of MWNTs in the fluid.

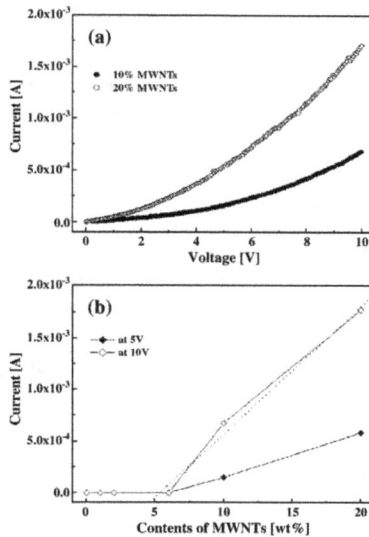

Fig. 6.22 (a) I–V characteristics for the nylon nanofibers loaded with (i) 10 wt.% and (ii) 20 wt.% of MWNTs. (b) The plot of the current as a function of the filler wt.% at (i) 10 V and (ii) 5 V.

6.4.3 I–V characteristic of the nanofiber composite

Fig. 6.22(a) shows a typical I–V characteristic of the nanofiber composite mat containing 10 wt.% and 20 wt.% MWNTs. The contact resistance was ignored. The I-V characteristics of the nanofiber mat showed a nonohmic behavior. Fig. 6.22(b) gives a plot of the MWNT loading in the nylon 6, 6 fibers as a function of current (A). No substantial variation in the current up to ~6 wt.% loading in the fiber is observed. But, at ~10 wt.% MWNTs, the current was ~0.15 and ~0.68 mA at an applied voltage of ~5 and 10V, respectively. The current increased from 0.59mA to 1.77mA with increasing MWNT loading in the fiber from 10 wt.% to 20 wt.%. The observed trend in an increase of current proposes that the CNT loading boosts the electron conduction path. Moreover, there is the possibility of "bridging" between the nylon phases with increasing carbon nanotube loading. As a result of the effective concentration of charge carrier increases which, in effect, reduce hopping distance of conduction electron and favor percolation phenomenon. The MWNTs incorporated in the nylon/MWNTs composite nanofibers with different concentrations were confirmed by TEM (Fig. 6.4 (b–d)). The nylon/MWNTs composite nanofibers containing 20 wt.% MWNTs contained more closely packed MWNTs. The close packing of MWNTs with covalent interlocking with the matrix allows electron charge transfer from the matrix to the loaded MWNTs through chemical bonding between the filler and the matrix. For MWNTs in nylon 6, 6 nanofibers, charge can be transferred from the nylon matrix to the MWNTs through the chemical bonding between the nanotubes and matrix. As a result, the MWNTs can facilitate electron transport and increase the electrical conductivity of the nanofibers [89].

A correlation has been established between the dispersibility of the thin multiwall carbon nanotubes (t–MWCNTs) in the composite, achieved via the chemical and mechanical dispersion routes, to their field emission parameters. The pristine t–MWCNTs (0.1 wt %) have been soaked in terpineol solution ($C_{10}H_{18}O$, (R)–2–(4–Methyl–3–cyclohexenyl)–2–propanol)) followed by the sonication and admixing with the ethyl cellulose polymer to obtain the raw–composite. In the chemical route, the t–MWCNTs have been –COOH functionalized to obtain better dispersed and stabilized suspension in the terpineol. The presence of –COOH affects the hydrogen bonding capacity of the t–MWCNTs which in turn stimulates the Lewis basicity of the medium and assist dispersion process. In the mechanical route, the dispersibility of the t–MWCNTs in the ethyl cellulose has been enhanced, via untanglement from the shear force, using a three-roll milling technique. Moreover, the superior dispersibility of the t–MWCNTs has been observed for the composite synthesized by the combination of the chemical and mechanical dispersion routes. The field emission parameters have been influenced by the degree of dispersion of the t–MWCNTs in the composite. Details of the synthesis routes and field emission analysis are presented [4].

6.5 Carbon nanotube composite: Dispersion routes and field emission parameters

Carbon nanotubes (CNTs) considered as a promising field electron emission source for the field emission displays (FEDs) and vacuum electronic devices [90-92]. In the field emission quantum mechanical tunnelling process which takes place under an applied electric field the CNTs have simple and economical cathode designing. It has various advantages such as low turn-on–field, high and uniform density of emission sites and low–cost production technology in contrast to the Spindt type micro–tips [93,94]. The fabrication of CNT–field emitters with greater field emission parameters are reported in many reports [95]. Further, the composite based techniques for high-performance CNT–emitters, proved to be advantages as they can be screen–printed on a large area, have good adhesion properties with cost-effective production [96]. In general, synthesis of CNT–nanocomposites did by a) surface modification (functionalization) of the CNTs [97], b) homogeneous dispersion and c) integration to desired epoxy/polymer matrix. On the other hand, the dispersion of the CNTs in the nanocomposites is a major issue to be resolved for nanotube–reinforced polymers. Due to their large surface area (> 1000 m^2–g^{-1}), which is several orders of magnitude larger than the surface of conventional fillers, CNTs exhibit a strong tendency to form agglomerates. The functionalization enhances dispersibility and interfacial bonding of the nanotubes in the hybrid material system [98]. In order to have tuned properties of CNT-nanocomposite, the homogeneous dispersion of CNTs in the polymer matrix or an untanglement of the agglomerate is very important. A considerable amount of work related to the functionalization of CNTs and synthesis of CNT–nanocomposites has been carried out [98-102]. Different reports related to functionalization are available e.g. graphitic carbon nanofibers functionalized by reactive linker molecules derived from diamines and triamines has been reported [100]. Secondly, a strong acid surface modification of the nanotubes has been suggested in which the CNTs were treated with triethylenetetramine (TETA) solution in UV/O_3 environment [100]. Further plasma surface modification and ultrasonication time effect on the mechanical properties of multiwall carbon nanofibers–polycarbonate has been investigated [101]. MWCNTs etched with ethanol, methanol and water vapors have been studied [103]. It can be concluded that in most cases mechanical properties of the CNT–nanocomposites were explored and in general, not much attention has been paid to studies of the field emission parameters of the CNT–nanocomposites synthesized by dispersing the CNTs via various dispersion routes.

In section, a relationship has been established between the dispersibility of the t–MWCNTs in the composite towards their field emission parameters. For dispersion studies, chemical and mechanical dispersion routes are discussed. For the chemical dispersion method, the t–MWCNTs has been –COOH functionalized by the H_2O_2 treatment, whereas, for the mechanical dispersion route the shear–mixing of the t–MWCNTs have been carried out to fabricate the CNT–composite cathode layers. Study of field emission analysis revealed that the combination of the chemical and mechanical dispersion method shows excellent field emission parameters.

6.5.1 Synthesis of thin multiwall carbon nanotube composite

Readily available thin multiwall carbon nanotubes (t–MWCNTs (Iljin-Nanotech, Korea), purity 95 %, diameter: 5–7 nm, length ~ 10 μm) used as the starting material. For functionalization, initially, pristine t–MWCNTs were soaked in H_2O_2 solution followed by the sonication of the solution for about 12 h. Further, this t–MWCNTs have been filtered and characterterized using the Fourier transform infrared (FTIR) technique to confirm the functionalization. Fig. 6.23 is the synthesis scheme for the t–MWCNT–composites.

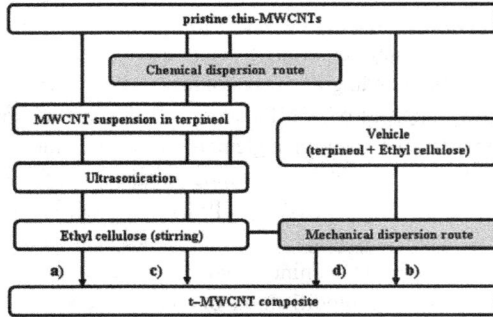

Fig. 6.23 Synthesis scheme for the thin multiwall carbon nanotube composite.

From above scheme, we can see the four different methods to synthesize the t–MWCNT–composite. In route (a) pristine t–MWCNT and α- terpineol (($C_{10}H_{18}O$, (R)–2–(4–Methyl–3–cyclohexenyl)–2–propanol)) solution was admixed and sonicated for ~12 h and then stirred with ethyl cellulose for about 1 h. These samples were labeled as the *raw–composites*. In method (b) use of the mechanical dispersion route(MDR) was used. In this, pristine t–MWCNTs along with the ethyl cellulose and terpineol was premixed for a period of 10–15 min. Followed to this, the composite was subjected to the three roll milling (calendar) process (Kyong Yong Machinery Co.Ltd. Korea) for a short dwell (residence) time ~ 15 min to obtain fine dispersion of the pristine t–MWCNTs in the polymer composite under the enormous shear forces. The viscosity of the composite has been monitored constantly. In the method (c), the chemical dispersion route (CDR) has been adopted, in which the pristine t–MWCNTs have been functionalized by refluxing them in a hydrogen peroxide (H_2O_2) solution for 12 h and filtered. The filtered nanotubes followed the same process sequence as described in method (a). These samples were designated as the *CDR–composites*. In method (d), initially, the chemical functionalization of the nanotubes has been carried out followed by a three roll milling of the composite at the end of the process sequence. Since, method (d) was the combination of method (b) and (c), the samples synthesized by this method were designated as the *CMDR–composites*. Thus, the composite was synthesized using the pristine, functionalized, shear mixed and the combination of functionalized/shear mixed t–MWCNTs. In all the cases the weight content of the t–MWCNTs was kept constant i.e. 0.1 wt%. The obtained composite was screen printed (printing area: 1×1 cm^2) on a

Indium Tin Oxide (ITO) coated (thickness 200 nm) glass substrate and kept in the oven for ~ 30 min for drying. The volatile organic components were removed by drying them at ~ 450°C for 10 min. The field emission measurements were carried out in the diode configuration with a fixed inter-electrode distance ~200 μm. The applied voltage sweep of 50–700 V has been employed and all measurements have been carried out at a vacuum level of ~ 10–7 Torr.

6.5.2 Characterization

(a) Dispersibility test for the pristine and H_2O_2 functionalized t–MWCNTs:

Chemical dispersion route and its FTIR:

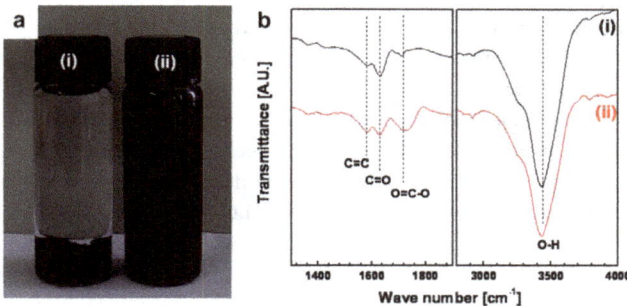

Fig. 6.24 (a) Recorded digi-cam photograph for (i) pristine t-MWCNTs and (ii) H_2O_2-treated t-MWCNTs dispersed in phenol (t-MWCNTs, 0.1 wt%). The photograph was recorded after 20 days. (b) Recorded Fourier transform infrared (FTIR) spectra for (i) pristine t-MWCNTs and (ii) H2O2-treated t-MWCNTs.

Fig. 6.24 (a) is the recorded digital photograph for (i) pristine t–MWCNTs and (ii) H_2O_2–treated–t–MWCNTs suspended in a terpineol solution. The photograph was snapped after about twenty days. From Fig. 6.24(a), the huge difference was observed for the suspension of the pristine and H_2O_2–treated– t–MWCNTs in the solution. It is observed that after dispersing the raw t-MWCNTs and H_2O_2–treated–t–MWCNTs in the terpineol solution. The nanotubes have observed to be gradually sedimentation from the upper part of the solution within 30 min. However, the suspension of H_2O_2–treated–t–MWCNTs showed better stability up to twenty days. It was studied with the FTIR technique. Fig. 6.24(b) shows the recorded FTIR spectra for the (i) pristine t–MWCNTs and (ii) H_2O_2–

treated–t–MWCNTs. The FTIR spectrum in (i) shows characteristics bands at ~ 1580, 1630, 1725 and 3433 cm^{-1}, which corresponds to the –C=C, –C=O, –O=C–O and –OH vibrations, respectively [104]. In spectrum (ii) the bands at ~ 1580 and 1725 cm^{-1} has been modified due to the H_2O_2 treatment. The observed modifications, in the spectrum, indicate the molecular vibrations associated with the –COOH functional group attached to the side walls of the nanotubes. The functionalization of nanotubes could positively contribute to the dispersion of the t–MWCNTs. From Fig. 6.24(a) one can see that the colloidal property of –COOH functionalized t–MWCNT suspension in the terpineol solution improved significantly as compared to the pristine t–MWCNT suspension in terpineol. The improved dispersibility and stability of the t–MWCNTs in the terpineol may be due to high viscosity of the terpineol and is associated with interaction between branched –OH group of terpineol and the polar –COOH functional of t–MWCNTs via π-π* stacking mechanism. The π-π* interaction between –COOH–terpineol and presence the bulky –OH group in both structures may hinder the hydrogen bonding tendency of the t–MWCNTs with the medium. As a result, the Lewis basicity (i.e. availability of a free electron pair) of terpineol medium is stimulated towards the –COOH functionalized t–MWCNTs [105]. However, the absence of such mechanism for the pristine t–MWCNTs. Thus, surface–functionalized tubes showed a good degree of dispersion and stabilization in the terpineol medium. Moreover, it is also crucial to functionalize t–MWCNTs without significantly affecting their properties. In the present study, the functionalization of t–MWCNTs was carried out using the H_2O_2 reagent which is 'mild' as compared with the HNO_3/H_2SO_4. The nanotube etching (shortening) rate for the H_2O_2 functionalization process is ~ 130 nm h^{-1}, whereas, for the HNO_3/H_2SO_4 acid treatment the nanotube etching rate is ~ 200 nm h^{-1} [106]. Retaining a high aspect ratio of the t–MWCNTs is of special interest towards an improvement of the field emission performance of a CNT. In general, the mean field enhancement factor,γ_m and turn–on–field depend on the aspect ratio of the CNTs, for grater aspect ratio CNTs one could achieve the higher magnitude of mean field enhancement factor, γ_m and lower turn-on–field [106]. Thus, the precaution was taken not to cut short t–MWCNTs by the functionalization process.

(b) Mechanical dispersion method: Three roll milling (Calendaring) process

This method has significant processing challenges. Firstly a high amount of nano-scale dispersion is required including high aspect ratios compounded with property enhancements. In case of t–MWCNTs, sufficient amount of shear stresses is required to be applied so entanglement and dispersion in the ethyl cellulose matrix occur uniformly without damaging the nanotubes via substantially reducing their length. The three roll milling technique involved pure shear forces subjected to the composite as compared

Thermoset Composite
Materials Research Foundations **38** (2018)

Materials Research Forum LLC
doi: http://dx.doi.org/10.21741/9781945291876

with the other types of milling process which relies on compressive impact as well as shear stress [108]. Fig. 6.25 shows the general configuration scheme of a three-roll mill.

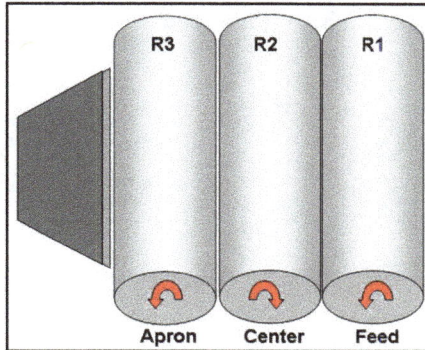

Fig. 6.25 Schematic diagram showing the general configuration of a three-roll mill.

Between the feed and the center rolls, the intense shear mixing occurs. The gap between the adjacent rolls and mismatch in the angular velocity between the rolls results in high shear rates. The composite material flows under and over the adjacent rollers due to surface tension. The material is collected by placing a collector blade in contact with the apron roll. The three roll mill yield in locally very high shear forces within a short residence time ~ 15 min. The calendaring process, in the present work, has been carried out by utilizing a commercially available laboratory scale three-roll mill, for the dispersion of t–MWCNTs. They consist of three stainless steel rolls that are ~ 80 mm in diameter. The angular velocity mismatch between the adjacent rolls has been fixed with a condition ω_3 (feed) = $3\omega_2$ (center) = $9\omega_1$ (apron). The setting of the mill was electronically controlled and the process enables the precise monitoring. The effects of roller gap setting in the calendaring process have been comprehensively reported in light of the structure and multi-functional properties of CNT nanocomposites [109]. Using a pressure gauge control system, the gap setting is adjusted, accurately, to maintain a specified value at ~ 20 μm.

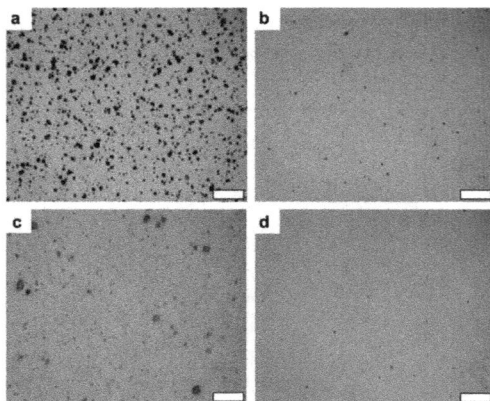

Fig. 6.26 Recorded optical micrographs for (a) raw-composite, (b) MDR-composite, (c) CDR-composite and (d) CMDR-composite (scale bar: 20 µm).

Morphological study of composites:

The optical microscopy images are shown in Fig. 6.26(a)-(d) for the composite samples prepared by the mentioned methods (a)-(d). In Fig. 6.26(a), dark spots can be seen for the raw– composites where the size of the spot and areal density is varied. This implies that t–MWCNT have aggregated to a large extent in the raw–composite. In part (b) for the MDR–composite, excluding a small amount of t–MWCNT aggregation spots, uniform distribution of the spot size as well as the areal density of the spots. This is indicative of three roll milling method has effectively dispersed and untangled the nanotubes in the ethyl cellulose matrix. The part (c) optical image CDR–composite, shows the very good aggregation of nanotubes in the composite as well which is a contrast to uniform dispersibility of the t–MWCNTs part (d) is seen for the CMDR–composite. The reduction in agglomeration of nanotubes is indicative of well-conserving ability of dispersion even reinforced them in the ethyl cellulose polymer. The presence of O– and EtO– in the backbone of ethyl cellulose offers stearic hindrance to the bulky –COOH functionalities present onto the surface of the t–MWCNTs. During the mechanical dispersion process, in addition to primary chemical dispersion, further agglomeration of the nanotubes might have hindered. The knead–vortex has been experienced by the composite material when it passes through the rolls. The vortex formation enables the final entanglement and dispersion of the nanotubes that occurs in the area between the rolls. Due to untanglement from the shear force, the shear mixing could provide a good amount of dispersion of the nanotubes in the composites. The amount of disentanglement depends on the type of polymer composite and the surface modifications of the nanotube fillers [108]. The – COOH surface modified CNTs positively participates in the mechanical dispersion process. As compared with the composites synthesized by other routes, the composite obtained by the method (d) has a high dispersibility of the t–MWCNTs. The field emission parameters have been measured for the synthesized composites.

6.3.3 Field emission parameters for the t–MWCNT–composite

Fig. 6.27 is the plot of variation in the measured current density (mAcm^{-2}) as a function of applied electric field (V µm^{-1}) for the composites prepared by the methods (a)-(d). Each plot is the average of five recorded profiles. The inset shows the corresponding Fowler–Nordheim (F-N) plot for log (I–V^{-2}) as a function of V^{-1}.

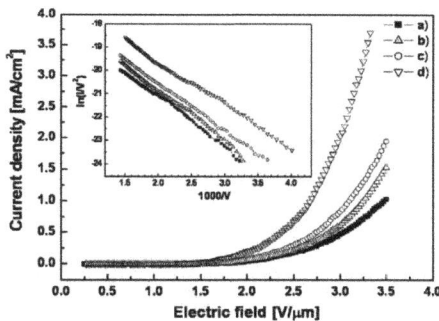

Fig. 6.27 Variation in field emission current density (measured in mA cm^{-2}) as a function of applied electric field (V µm⁻1) for (a) raw-, (b) MDR-, (c)CDR- and (d) CMDR-composite. Inset is the corresponding Fowler–Nordheim plot.

From Fig. 6.27 one can see that the magnitude of the current density monotonically increases with increase in the applied electric field, however, no major variations in the current density was observed up to ~ 2 Vµm^{-1}, for the plots (a)-(d). Moreover, the J–E behavior typically follows the F–N tunneling mechanism at the low electric field, from 2–2.5 Vµm^{-1}, and current densities. With the subsequent increase in the field, from 2.5–3 Vµm^{-1}, and increased current density (> 0.5 mA–cm^{-2}), the plot (d) exhibit a marked current density variation that sharply deviates from the plots (a)-(c). With further increase in the applied electric field, from 3–3.5 $V\mu m^{-1}$, and current density, the J–E characteristics yield to the electron tunneling behavior governed by the F–N theory. For each plot, the magnitude of the turn-on–electric field (measured in $V\mu m^{-1}$) was estimated at a current density ~10 μAcm^{-2}, where as, the magnitude of measured current density, J, is coated at 3 V μm^{-1}. Table 6.5 shows the field emission parameters enlisted for the composite synthesized by various methods. One can see that, for the raw–composite, the turn-on– the field is ~ 1.698 $V\mu m^{-1}$ and decreased up to ~ 1.649 $V\mu m^{-1}$ for the MDR–composite.

Table 6.5 Field emission parameters for the carbon nanotube composite synthesized by various methods.

	Turn-on-field $(V\ \mu m^{-1})$ $(@10\ \mu A\ cm^{-2})$	Current density $(\mu A\ cm^{-2})$ $(@\ 3\ V\ \mu m^{-1})$	Mean field enhancement factor	Area (m^2)
(a)	1.968	467.8	6.881.7	5.9×10^{-18}
(b)	1.969	613.7	6760.3	8.7×10^{-18}
(c)	1.574	824.4	7123.9	8.5×10^{-18}
(d)	1.374	2030.2	7851.4	1.2×10^{-17}

The magnitude of the turn-on–field lowered down further, ~ 1.574 $V\mu m^{-1}$, for the CDR–composite. And the turn-on–field has achieved the lowest value ~ 1.374 $V\mu m^{-1}$, for the composite synthesized by the combination of the chemical and mechanical dispersion routes i.e. the CMDR–composite. Thus, significant improvement in the turn-on–field parameter has been achieved for the CMDR–composite. Furthermore, the magnitude of current density, J, (measured at 3 $V\mu m^{-1}$) is ~ 467.8 μAcm^{-2} for the raw–composite and gradually observed to be increased from 613.7 $\mu A\ cm^{-2}$ to 824.4 μAcm^{-2} to finally 2030.2 μAcm^{-2} for the MDR–, CDR–, and CMDR–composites, respectively. As compared with the current density, J, of the raw-composite, the magnitude of current density is almost twice and ~ 4.5 times for the CDR– and CMDR–composite, respectively. In general, it is difficult to extract physically meaningful parameters from a fit of the F–N equation to the I–V plots [109]. This is especially true for I–V plots obtained from randomly distributed multiple CNT emitting tips such as those on the t–MWCNTs–composite cathode layers. However, a qualitative estimation can be made by computing the mean field enhancement factor, γ_m, for the CNT field emitters, using the generally accepted F–N theory [110] in terms of the voltage–based notion:

$$J_{loc} = \frac{\alpha \beta^2 V^2}{(t_f)^2 \phi} \exp\left[-\frac{V_{fb\phi^{3/2}}}{\beta V}\right] \dots \tag{4}$$

where V is the applied voltage, β is the conversion factor for the voltage-to-barrier-field (the field at the CNT tip that determines whether FE occurs), J_{loc} is the *local* current density (not the *macroscopic* current density, as used in this paper), ϕ is the local work function, a and b are universal constants (with a=1.541433 × 10^{-6} A eV V^{-2}, b=6.830890 × 10^9 eV$^{3/2}$Vm^{-1}), and ϑ_F and t_F are mathematical functions known as the Nordheim functions, evaluated at the Fermi level. Tabulations of the special field emission elliptic functions ϑ_F and t_F are given in the literature [111]. The value of ϑ_F is typically ~ 0.95 for an array of MWCNT composite cathode layers [111]. For a parallel plate arrangement with the plate separated by a gap of thickness Δ, we have $\gamma = \beta\Delta$. The field enhancement factor, γ is the ratio of the *barrier–field* to the *turn–on–field*.

From Fig. 6.27, an F–N plot has been determined (shown as inset) and its slope was measured in order to estimate the mean field enhancement factor, γ_m, for the raw– as well as the other composites. The detailed process is described elsewhere [111]. The slope correction factor, s, has been taken as ~ 1 and the local work-function, ϕ, of the nanotubes was assumed to be a constant ~ 5 eV. It can be seen that, the equation (4) itself is degenerate, because it describes two mathematical parameters (offset a and slope b) and three canonical physical quantities such as, curvature of the emitter tip, d, conversion factor, β, and work function, ϕ. Thus, only a few comments could be made on the details of the emission physics from the plot of variations in current density with the applied electric field, since the electronic properties of the tips cannot be easily separated from the geometric field–enhancement factor. Nevertheless, one could extract an average of the distribution of the field emission parameters. The mean field enhancement factor, γ_m, computed for the raw–composite is 6881.7 and decreased down to 6760.3 for the MDR– composite and thereafter, a gradual increase was observed from 7123.9 to 7851.4 for the CDR– and CMDR–composite, respectively. Thus, the computed mean–field– enhancement factor, γ_m, increases by ~ 14 % for the CMDR–composite as compared to the raw cathode layers. For the raw composite, the computed geometric area participating in the electron field emission process is 5.9× 10^{-18} m^2 and increases gradually from 8.5 × 10^{-18} to 8.7 × 10^{-18} and up to ~ 1.2 × 10^{-17} m^2 for the MDR–, CDR–, and CMDR– composites, respectively. The active area of the field emission is increased approximately two times for the CMDR–composites as compared with the raw–composite.

In general, field emission parameters, measured for an array of emitters, depending on the position of the individual emitter on the cathode surface. One can find that superior field emission properties have been obtained for the CNT–composite synthesized by the combination of the chemical– and mechanical dispersion routes. This indicates that CNTs that were dispersed well in the nanocomposite when printed onto the cathode layers have an optimum position to give favorable field emission properties. It is interesting to note that, the weight content of the t–MWCNTs in the composite was 0.1wt% which is the order of magnitude smaller as compared with the other reports [112] presented before [113].

Summary

In summary of EMI analysis of microwave data provided a basic insight into the behavior of EM field in heterogeneous ponderable media. The C–H, N–H, and C-O nonplanar bonds in PU, decisively, played a pivotal role in modifying transport parameters thereby getting hydrosorpted with sp^3 C–H sites of GNCs. The percolation threshold (@ ~ 5 wt %) is in good agreement with literature report [173]. It suggests superior dispersibility of GNCs in PU. In the host, the active modes of polarization are modified by GNCs, especially, at the skin void. Loaded GNCs provide transport path to EM wiggles by producing mobile charge carriers along amide III polymer segments near microvoids. The coupling occurs via various modes of polarization resulting in dramatic enhancement in tranmittivity loss. The loss improves, for PU, from ~ 40 to 99.9% with a reduction in coating thickness from a centimeter scale down to the millimeter, @ 25 wt % GNCs. The GNCs/PU nanocomposites are promising for building shielding patterns, especially, at short range projectile tracking.

To summarize for GNC/epoxy composites, we have demonstrated mitigation in mechanical as well as thermal properties at low weight fractions of GNCs in host epoxy. They are exfoliated into individual sheets in epoxy at lower weight fractions (<0.05 wt%) levels. GNCs constituted micron-sized aggregates in the epoxy matrix at wt. fractions > 0.01 wt%. The areal aggregate density increases with GNC wt. fraction, which adversely affected the mechanical properties. The mechanical parameters; such as, for tensile, flexural and fracture toughness showed significant enhancements due to GNC incorporation irrespective of its content in epoxy. Optimum reinforcement is obtained @ 0.01 wt% nanocomposites; presumably, due to the cooperative effects of large surface area, uniform dispersion and good interfacial adhesion. The thermal and physical properties are enhanced considerably with GNC. Mechanisms for fracture energy absorption in nanocomposites are mainly governed by the formation of more number of microcracks and their deflection as compared to neat epoxy. On account of these properties, GNC could be promising nanofiller for structural applications.

In carbon nanotube studies, Non-woven mats of nylon nanofibers loaded with MWNTs were fabricated using an electrospinning process. Formic acid was used as the solvent. No surfactants were used to help functionalize the MWNTs. Moreover, the –NH$_2$ terminated carboxylated MWNTs dispersion was stable for more than 40 h compared with the carboxylated nanotubes. The observed improvement in the –NH$_2$ functionalized MWNTs was attributed to the strong intra-molecular forces that overcome the inter-tube coulombic attraction. The filler concentration was varied from 0 to 20 wt.% and the I–V characteristics were examined. The I–V characteristics were found to be non-ohmic and improved with increasing filler concentration in the nylon nanofiber. This increase was attributed to the enhancement of the electron conduction process by MWNT loading.

In field emission studies, the parameters of the composite, synthesized via various dispersion techniques, have been influenced by the dispersibility of the t–MWCNTs in the composite. The typical turn-on–field, (measured at $10\mu A$ cm^{-2}) for the raw–composite, decreases from its virgin value of 1.698 to 1.649 Vμm^{-1}, whereas, the magnitude of current density (at 3 V μm^{-1}) increases form 467.8 to 613.7 μA cm^{-2} for the composite synthesized by the mechanical dispersion route. The magnitude of turn-on–field continues to decrease and attains a local minimum at 1.374 Vμm^{-1} for the composite synthesized by a combination of the chemical and mechanical dispersion routes. Whereas, the magnitude of current density continues to increase. With respect to the raw-composite, approximately, two and four-time increase in the current density has been observed i.e. 824.4 and 2030.2 μA cm^{-2} for the composite synthesized by the chemical and combination of both the routes. The mean-field–enhancement factor computed for the composites increases from its initial values of 6881.7 up to 7851.4 for the composite synthesized by the combination of the methods. The geometric field emission area computed for the raw–composite is 5.9×10^{-18} m^2 and gradually increases up to ~ 1.2×10^{-17} m^2 for the composite synthesized via a combination of chemical- and mechanical dispersion technique. Thus superior field emission parameters have been obtained for the composite synthesized by the combination of the chemical and mechanical dispersion routes. The presence of –COOH on the t–MWCNTs hinders the hydrogen donation and improves the dispersibility of the nanotubes through the π–π* stacking interactions with the terpineol medium. Furthermore, the steric hindrance offered by the legends O- and EtO- to the –COOH maintains the dispersibility of t–MWCNTs in the ethyl cellulose. Thus the functionalization positively contributes to the dispersion of the t–MWCNTs. Whereas, the mechanical dispersion of the t–MWCNTs in the composite, using calendaring technique, offers good untanglement from the knead–vortex shear force. Both dispersion techniques are scalable and can easily be integrated into synthesis process of the composite.

References

[1]   A. Kumar, P. S. Alegaonkar, Impressive transmission mode electromagnetic interference shielding parameters of graphene-like nanocarbon/polyurethane nanocomposites for short range tracking countermeasures, ACS Appl. Mater. Interfaces, 7 (2015) 14833–42. https://doi.org/10.1021/acsami.5b03122

[2]   D. K. Chouhan, S. K. Rath, A. Kumar, P. Alegaonkar, S. Kumar, G. Harikrishnan, T. U. Patro, Structure-reinforcement correlation and chain dynamics in graphene oxide and laponite-filled epoxy nanocomposites, J. Mater. Sci. 50 (2015) 7458-72. https://doi.org/10.1007/s10853-015-9305-5

[3]   J. S. Jeong, S. Y. Jeon, T. Y. Lee, J. H. Park, J. H. Shin, P. S. Alegaonkar, A.S. Berdinsky, and J. B. Yoo, Fabrication of MWNTs/nylon conductive composite nanofibers by electrospinning, Diamond and related materials, 15 (2006) 1839-1843. https://doi.org/10.1016/j.diamond.2006.08.026

[4]   J. H. Park, P. S. Alegaonkar, S. Y. Jeon, and J. B. Yoo, Carbon nanotube composite: Dispersion routes and field emission parameters. Composites Science and Technology, 68 (2008), 753-759. https://doi.org/10.1016/j.compscitech.2007.08.030

[5]   P. P. Kuzhir, A. G. Paddubskaya, M. V. Shuba, S. A. Maksimenko, S. Bellucci, Electromagnetic shielding efficiency in Ka-band: carbon foam versus epoxy/carbon nanotube composites. J. Nanophoton 6(2012) 061715-20. https://doi.org/10.1117/1.JNP.6.061715

[6]   R. Rohini, S. Bose, Electromagnetic interference shielding materials derived from gelation of multiwall carbon nanotubes in polystyrene/poly(methyl methacrylate) blends, ACS Appl. Mater. Interfaces 6 (2014)11302–10. https://doi.org/10.1021/am502641h

[7]   S. Maiti, N. K. Shrivastava, S. Suin, B. Khatua, Polystyrene/ mwcnt/graphite nanoplate nanocomposites: efficient electro-magnetic interference shielding material through graphite nano-plate–MWCTs–graphite nanoplate networking, ACS Appl. Mater. Inter, 5(2013)4712–24. https://doi.org/10.1021/am400658h

[8]   S. K. Rath, S. Dubey, G. S. Kumar, S. Kumar, A. Patra, J. Bahadur et al., Multi-walled cnt-induced phase behaviour of poly (vinylidene fluoride) and its electro-mechanical properties, J. Mater. Sci. 49 (2014)103–113. https://doi.org/10.1007/s10853-013-7681-2

[9]   A. Singh A. Carbon nanotubes based nanocomposite for electromagnetic wave
       absorption and dynamic structural strain sensing. Ind. J. Pure Appl. Phys.
       51(2013)439–43.

[10]   S. H. Park, P. T. Theilmann, P. M. Asbeck, P. R. Bandaru, Enhanced
       electromagnetic interference shielding through the use of functionalized carbon-
       nanotube-reactive polymer composites, IEEE Trans. Nanotechnol.
       9(2010)464–69. https://doi.org/10.1109/TNANO.2009.2032656

[11]   M. H. Al-Saleh, W. H. Saadeh, U. Sundararaj, EMI shielding effectiveness of
       carbon based nanostructured polymeric materials: a comparative study, Carbon
       60(2013)146–56. https://doi.org/10.1016/j.carbon.2013.04.008

[12]   M. Arjmand, T. Apperley, M. Okoniewski, U. Sundararaj, Comparative study of
       electromagnetic interference shielding proper-ties of injection molded versus
       compression molded multi-walled carbon nanotube/polystyrene composites,
       Carbon 50(2012)5126–34. https://doi.org/10.1016/j.carbon.2012.06.053

[13]   Y. Yang, M. C. Gupta, K. L. Dudley, R. W. Lawrence, Novel carbon nanotube-
       polystyrene foam composites for electromagnetic interference shielding, Nano
       Lett., 5(2005)2131–34. https://doi.org/10.1021/nl051375r

[14]   R. Kumar, S. R. Dhakate, T. Gupta, P. Saini, B. P. Singh, R. B. Mathur, Effective
       improvement of the properties of light weight carbon foam by decoration with
       multi-wall carbon nanotubes, J. Mater. Chem. A 1(2013) 5727–35.
       https://doi.org/10.1039/c3ta10604g

[15]   B. Yuan, L. Yu L, L. Sheng, K. An, X. Zhao, Comparison of electromagnetic
       interference shielding properties between single- wall carbon nanotube and
       graphene sheet/polyaniline composites, J.Phys. D: Appl. Phys, 45 (2012) 235108-
       14. https://doi.org/10.1088/0022-3727/45/23/235108

[16]   Z. Liu, G. Bai, Y. Huang, Y. Ma, F. Du, F. Li, et al. Reflection and absorption
       contributions to the electromagnetic interference shielding of single-walled carbon
       nanotube/polyur-ethane composites, Carbon 45 (2007)821–27.
       https://doi.org/10.1016/j.carbon.2006.11.020

[17]   Y. Huang, N. Li, Y. Ma, F. Du, F. Li, X. He et al. The influence of single-walled
       carbon nanotube structure on the electromagnetic interference shielding efficiency
       of its epoxy composites, Carbon 45(2007)1614–21.
       https://doi.org/10.1016/j.carbon.2007.04.016

[18]   T. Kuilla, S. Bhadra, D. Yao, N. H. Kim, S. Bose, J. H. Lee, Recent Advances in graphene based polymer composites, Prog. Polym. Sci. 35(2010)1350−75. https://doi.org/10.1016/j.progpolymsci.2010.07.005

[19]   V. K. Singh, A. Shukla, M. K. Patra, L. Saini, R. K. Jani, S. R. Vadera, Microwave absorbing properties of a thermally reduced graphene oxide/nitrile butadiene rubber composite, Carbon 50(2012)2202−08. https://doi.org/10.1016/j.carbon.2012.01.033

[20]   J. Liang, Y. Wang, Y. Huang, Y. Ma, Z. Liu, J. Cai et al. Electromagnetic interference shielding of graphene/epoxy composites, Carbon 47(2009)922−25. https://doi.org/10.1016/j.carbon.2008.12.038

[21]   V. Eswaraiah, V. Sankaranarayanan, S. Ramaprabhu, Functionalized graphene−pvdf foam composites for EMI shielding, Macromol. Mater. Eng. 296(2011) 894−98. https://doi.org/10.1002/mame.201100035

[22]   Z. Chen, C. Xu, C. Ma, W. Ren, H. M. Cheng, Lightweight and flexible graphene foam composites for high-performance electromagnetic interference shielding. Adv. Mater. 25(2013)1296− 1300. https://doi.org/10.1002/adma.201204196

[23]   H. B. Zhang, Q. Yan, W. G. Zheng, Z. He, Z. Z. Yu, Tough graphene−polymer microcellular foams for electromagnetic inter-ference shielding. ACS Appl. Mater. Inter. 3(2011) 918−24. https://doi.org/10.1021/am200021v

[24]   C. Basavaraja, W. J. Kim, D. Y. Kim, S. H. Do, Synthesis of polyaniline-gold/graphene oxide composite and microwave absorption characteristics of the composite films. Mater. Lett. 65(2011)3120−23. https://doi.org/10.1016/j.matlet.2011.06.110

[25]   M. H. AlSaleh, U. Sundararaj, X-Band EMI shielding mechanisms and shielding effectiveness of high structure carbon black/polypropylene composites, J. Phys. D: Appl. Phys. 46(2013)035304-8. https://doi.org/10.1088/0022-3727/46/3/035304

[26]   A. Kumar, S. Patil, A. Joshi, V. Bhoraskar, S. Datar, P. Alegaonkar, Mixed phase, sp2−sp3 bonded, and disordered few layer graphene-like nanocarbon: Synthesis and characterizations. Appl. Surf. Sci. 271(2013)86-92. https://doi.org/10.1016/j.apsusc.2013.01.097

[27]   P. Larkin, Infrared and Raman spectroscopy: principles and spectral interpretation; elsevier: Amsterdam, 2011.

[28]   B. Stuart, Infrared Spectroscopy; Wiley: Hoboken, NJ, USA, 2005. https://doi.org/10.1002/0471238961.0914061810151405.a01.pub2

[29]   D. Stauffer, A. Aharony, Introduction to Percolation Theory; CRC Press: Boca Raton, FL, USA, 1994.

[30]   S. Obukhov, First order rigidity transition in random rod networks, Phys. Rev. Lett. 74(1995) 4472-75. https://doi.org/10.1103/PhysRevLett.74.4472

[31]   O. Regev, P. N. ElKati, J. Loos, C. E. Koning, Preparation of conductive nanotube–polymer composites using latex technology. Adv. Mater. 16(2004)248–51. https://doi.org/10.1002/adma.200305728

[32]   Z. Ounaies, C. Park C, K. Wise, E. Siochi, J. Harrison, Electrical properties of single wall carbon nanotube reinforced polyimide composites. Compos. Sci. Technol. 63(2003)1637–1646. https://doi.org/10.1016/S0266-3538(03)00067-8

[33]   J. Sandler, J. Kirk, I. Kinloch, M. Shaffer, A. Windle, Ultra-low electrical percolation threshold in carbon-nanotube-epoxy composites. Polymer 44(2003)5893–99. https://doi.org/10.1016/S0032-3861(03)00539-1

[34]   Agilent, PAN Microwave Network Analyzer, Catalogue and Product Note E8364B, 2009

[35]   A. Jonscher, Dielectric relaxation in solids, Ed. 1st, Chelsea Dielectrics Press: London, 1983.

[36]   N. Li, Y. B. Huang, F. Du, X. He, X. Lin, H. Gao, Y. Ma et al. Electromagnetic interference (EMI) shielding of single-walled carbon nanotube epoxy composites. Nano Lett. 6(2006)1141–45. https://doi.org/10.1021/nl0602589

[37]   C. Lee, X. Wei, J. W. Kysar, J. Hone, Measurement of the elastic properties and intrinsic strength of monolayer graphene, Science 321(2008)385-388. https://doi.org/10.1126/science.1157996

[38]   M. D. Stoller, S. Park, Y. Zhu, J. An, R. S. Ruoff, Graphene-based ultracapacitors. Nano Lett. 8(2008)3498-502. https://doi.org/10.1021/nl802558y

[39]   A. A. Balandin, S. Ghosh, W. Bao, I. Calizo, D. Teweldebrhan, F. Miao F, C. N. Lau, Superior thermal conductivity of single-layer graphene, Nano Lett. 8(2008)902-907. https://doi.org/10.1021/nl0731872

[40]   Y. H. Kahng, S. Lee, W. Park, G. Jo, M. Choe, J. H. Lee, H. Yu, T. Lee, K. Lee, Thermal stability of multilayer graphene films synthesized by chemical vapor deposition and stained by metallic impurities, Nanotechnol 23(2012) 075702-08. https://doi.org/10.1088/0957-4484/23/7/075702

[41]   X. Du, I. Skachko, A. Barker, Andrei EY. Approaching ballistic transport in suspended graphene, Nat. Nanotechnol 3(2008)491-495. https://doi.org/10.1038/nnano.2008.199

[42]   S. Stankovich, D. A. Dikin, G. A. Dommett, K. M. Kohlhaas, E. J. Zimney, E. A. Stach, et al Graphene-based composite materials, Nature 442(2006)282-286. https://doi.org/10.1038/nature04969

[43]   M. A. Rafiee, J. Rafiee, Z. Wang, H. Song, Z. Z. Yu, N. Koratkar, Enhanced mechanical properties of nanocomposites at low graphene content, ACS Nano3(2009)3884-90. https://doi.org/10.1021/nn9010472

[44]   S. Watcharotone, D. A. Dikin, S. Stankovich, R. Piner, I. Jung, G. H. Dommett, et al. Graphene-silica composite thin films as transparent conductors, Nano Lett 7(2007)1888-92. https://doi.org/10.1021/nl070477+

[45]   Y. Yang, W. Rigdon, X. Huang, X. Li, Enhancing graphene reinforcing potential in composites by hydrogen passivation induced dispersion, Sci. Rep. 3(2013)2086-2090. https://doi.org/10.1038/srep02086

[46]   A. G. D'Aloia, F. Marra, A. Tamburrano, G. De Bellis, M. S. Sarto, Electromagnetic absorbing properties of graphene–polymer composite shields, Carbon 73(2014)175-84. https://doi.org/10.1016/j.carbon.2014.02.053

[47]   A. Yasmin, J. J. Luo, I. M. Daniel, Processing of expanded graphite reinforced polymer nanocomposites, Composites Science and Technology66(2006)1182-1189. https://doi.org/10.1016/j.compscitech.2005.10.014

[48]   A. Yasmin, I. M. Daniel, Mechanical and thermal properties of graphite platelet/epoxy composites, Polymer 45(2004)8211-9. https://doi.org/10.1016/j.polymer.2004.09.054

[49]   I. Zaman, T. T. Phan, H. C. Kuan, Q. Meng, L. T. La, L. Luong, O. Youssf, J. Ma, Epoxy/graphene platelets nanocomposites with two levels of interface strength, Polymer 52(2011)1603-11. https://doi.org/10.1016/j.polymer.2011.02.003

[50]   D. R. Bortz, E. G. Heras, I. Martin-Gullon, Impressive fatigue life and fracture toughness improvements in graphene oxide/epoxy composites, Macromolecules, 45(2011)238-45. https://doi.org/10.1021/ma201563k

[51]   C. Bao, Y. Guo, L. Song, Y. Kan, X. Qian, Y. Hu, In situ preparation of functionalized graphene oxide/epoxy nanocomposites with effective reinforcements, J. Mater.Chem. 21(2011)13290-8. https://doi.org/10.1039/c1jm11434d

[52] M. A. Rafiee, W. Lu, A. V. Thomas, A. Zandiatashbar, J. Rafiee, J. M. Tour, et al. Graphene nanoribbon composites. ACS Nano. 4(2010)7415-20. https://doi.org/10.1021/nn102529n

[53] L. C. Tang, Y. J. Wan, D. Yan, Y. B. Pei, L. Zhao, Y. B, Li, et al. The effect of graphene dispersion on the mechanical properties of graphene/epoxy composites, Carbon 60(2013)16-27. https://doi.org/10.1016/j.carbon.2013.03.050

[54] Y. J. Wan, L. C. Tang, D. Yan, L. Zhao, Y. B. Li, L. B. Wu, J. X, Jiang, G. Q. Lai, Improved dispersion and interface in the graphene/epoxy composites via a facile surfactant-assisted process, Compos. Sci.Technol. 82(2013)60-68. https://doi.org/10.1016/j.compscitech.2013.04.009

[55] F. Yavari, M. A. Rafiee, J. Rafiee, Z. Z. Yu, N. Koratkar, Dramatic increase in fatigue life in hierarchical graphene composites, ACS applied materials & interfaces, 2(2010)2738-2743. https://doi.org/10.1021/am100728r

[56] M. Naebe, J. Wang, A. Amini, H. Khayyam, N. Hameed, L. H. Li, et al. Mechanical property and structure of covalent functionalised graphene/epoxy nanocomposites, Sci. Rep., 4(2014)4375-4379. https://doi.org/10.1038/srep04375

[57] A. J. Crosby, J. Y. Lee, Polymer nanocomposites: the "nano" effect on mechanical properties, Polym. Rev. 47(2007)217-229. https://doi.org/10.1080/15583720701271278

[58] T. Ramanathan, A. A. Abdala, S. Stankovich, D. A. Dikin, M. Herrera-Alonso, R. D. Piner et al. Functionalized graphene sheets for polymer nanocomposites. Nat. Nanotechnol, 3(2008)327-335. https://doi.org/10.1038/nnano.2008.96

[59] F. W. Starr, T. B. Schrøder, S. C. Glotzer, Molecular dynamics simulation of a polymer melt with a nanoscopic particle, Macromolecules 35 (2002)4481-4492. https://doi.org/10.1021/ma010626p

[60] J. Liang, Y. Huang, L. Zhang, Y. Wang, Y. Ma, T. Guo, Y. Chen, Molecular-level dispersion of graphene into poly (vinyl alcohol) and effective reinforcement of their nanocomposites, Adv. Funct. Mater. 19(2009)2297-2302. https://doi.org/10.1002/adfm.200801776

[61] S. A. Shojaee, A. Zandiatashbar, N. Koratkar, D. A. Lucca, Raman spectroscopic imaging of graphene dispersion in polymer composites, Carbon 62(2013)510-513. https://doi.org/10.1016/j.carbon.2013.05.068

[62] J. Zhang, B. Zhang, Q. Xue, Z. Wang, Ultra-elastic recovery and low friction of amorphous carbon films produced by a dispersion of multilayer graphene,

Diamond Relat. Mater. 23(2012)5-9.
https://doi.org/10.1016/j.diamond.2011.12.011

[63]   X. Li, P. Nardi, Micro/nanomechanical characterization of a natural
       nanocomposite material—the shell of Pectinidae, Nanotechnology 15(2003)211-
       219. https://doi.org/10.1088/0957-4484/15/1/038

[64]   T.U. Patro, H. D. Wagner, Layer-by-layer assembled pva/laponite multilayer free-
       standing films and their mechanical and thermal properties, Nanotechnology
       22(2011)455706-12. https://doi.org/10.1088/0957-4484/22/45/455706

[65]   P. Podsiadlo, A. K. Kaushik, E. M. Arruda, A. M. Waas, B. S. Shim, J. Xu, et al.
       Ultrastrong and stiff layered polymer nanocomposites, Science318(2007)80-83.

[66]   E. S. Greenhalgh, Failure Analysis and Fractography of Polymer Composites, Ed.
       first, Woodhead Publishing, New Delhi, 2009.

[67]   S. K. Rath, V. K. Aswal, C. Sharma, K. Joshi, M. Patri, G. Harikrishnan, et al.
       Mechanistic origins of multi-scale reinforcements in segmented polyurethane-clay
       nanocomposites, Polymer, 55(2014)5198-5210.
       https://doi.org/10.1016/j.polymer.2014.08.035

[68]   D. K. Chouhan, S. K. Rath, A. Kumar, P. Alegaonkar, S. Kumar, G. Harikrishnan,
       T. U. Patro, Structure-reinforcement correlation and chain dynamics in graphene
       oxide and laponite-filled epoxy nanocomposites, J. Mater. Sci. 50(2015)7458-72.
       https://doi.org/10.1007/s10853-015-9305-5

[69]   M. El Achaby, Y. Essamlali, N. El Miri, A. Snik, K. Abdelouahdi, A. Fihri, M.
       Zahouily, A. Solhy, Graphene oxide reinforced chitosan/polyvinylpyrrolidone
       polymer bio-nanocomposites, J. Appl. Polym Sc., 22 (2014) 131-136.
       https://doi.org/10.1002/app.41042

[70]   S. Ganguli, A. K. Roy, D. P. Anderson, Improved thermal conductivity for
       chemically functionalized exfoliated graphite/epoxy composites, Carbon
       46(2008)806-817. https://doi.org/10.1016/j.carbon.2008.02.008

[71]   G. D. Smith, D. Bedrov, L. Li, O. A. Byutner, A molecular dynamics simulation
       study of the viscoelastic properties of polymer nanocomposites, J. Chem. Phys.
       117(2002) 9478-89. https://doi.org/10.1063/1.1516589

[72]   Jae-Hong Park, Gil-Hwan Son, Jin-San Moon, Jae-Hee Han, Alexander S.
       Berdinsky, D.G. Kuvshinov, Ji-Beom Yoo, Chong-Yun Park, Screen printed
       carbon nanotube field emitter array for lighting source application J. Vac. Sci.
       Technol., B 23 (2005) 749-753. https://doi.org/10.1116/1.1851535

[73] Y.J. Jung, G.H. Son, J.H. Park, Y.W. Kim, Alexander S. Berdinsky, J.B.Yoo, C.Y. Park, Fabrication and properties of under-gated triode with CNT emitter for flat lamp. Diamond Relat. Mater. 14 (2005) 2109-12. https://doi.org/10.1016/j.diamond.2005.07.029

[74] Shuying Yang, Karen Lozano, Azalia Lomeli, Heinrich D. Foltz, Robert Jones, Electromagnetic interference shielding effectiveness of carbon nanofiber/LCP composites. Compos., A 36 (2005) 691-97.

[75] J.S. Moon, J.H. Park, T.Y. Lee, Y.W. Kim, J.B. Yoo, C.Y. Park, J.M. Kim, K.W. Jin, Transparent conductive film based on carbon nanotubes and PEDOT composites. Diamond Relat. Mater. 14 (2005) 1882-87. https://doi.org/10.1016/j.diamond.2005.07.015

[76] C. Park, Z. Ounaied, K.A.Watson, R.E. Crooks, J. Smith Jr., S.E. Lowther, J.W. Connell, E.J. Siochi, J.S. Harrison, T.L. St. Clair, Dispersion of single wall carbon nanotubes by in situ polymerization under sonication. Chem. Phys. Lett. 364 (2002) 303-08. https://doi.org/10.1016/S0009-2614(02)01326-X

[77] E.J. Ra, K.H. An, K.K. Kim, S.Y. Jeong, Y.H. Lee, Anisotropic electrical conductivity of MWCNT/PAN nanofiber paper. Chem. Phys. Lett. 413 (2005) 188-93. https://doi.org/10.1016/j.cplett.2005.07.061

[78] K.Frank, Y. Gogotsi, A. Ali, N. Naguib, H. Ye, G. Yang, C. Li, P. Willis, Electrospinning of continuous carbon nanotube-filled nanofiber yarns. Adv. Mater. 15 (2003) 1161-65. https://doi.org/10.1002/adma.200304955

[79] D.I. Cha, H.Y. Kim, K.H. Lee, Y.C. Jung, J.W. Cho, B.C. Chun, Electrospun nonwovens of shape-memory polyurethane block copolymers. J. Appl. Polym. Sci. 96 (2005) 460-65. https://doi.org/10.1002/app.21467

[80] M.M. Bergshoef, G.J. Vancso, Transparent nanocomposites with ultrathin, electrospun nylon-4, 6 fiber reinforcement. Adv. Mater. 11 (1999) 1362-65. https://doi.org/10.1002/(SICI)1521-4095(199911)11:16<1362::AID-ADMA1362>3.0.CO;2-X

[81] Y. Wang, Y. Xia, Dynamic tensile properties of E-glass, Kevlar49 and polyvinyl alcohol fiber bundles. , J. Mater. Sci. Lett .19 (2002) 583-86. https://doi.org/10.1023/A:1006730312279

[82] Valerie C. Moore, Michael S. Strano, Erik H. Haroz, Robert H. Hauge,Richard E. Smalley, Individually suspended single-walled carbon nanotubes in various surfactants. Nano Lett. 3 (2003) 1379-82. https://doi.org/10.1021/nl034524j

[83] J. Liu, A.G. Rinzler, H. Dai, J.H. Hafner, R.E. Smalley, Fullerene pipes. Science 280 (1998)1253-56. https://doi.org/10.1126/science.280.5367.1253

[84] Tae Young Lee, Ji-Beom Yoo, Lee TY, Yoo JB. Adsorption characteristics of Ru (II) dye on carbon nanotubes for organic solar cell. Diamond Relat. Mater. 14 (2005) 1888-90. https://doi.org/10.1016/j.diamond.2005.08.055

[85] H. Hou, J.J. Ge, J. Zwng, Q. Li, D.H. Reneker, A. Greiner, S.Z.D. Cheng, Electrospun polyacrylonitrile nanofibers containing a high concentration of well-aligned multiwall carbon nanotubes. Chem. Mater. 17 (2005) 967-73. https://doi.org/10.1021/cm0484955

[86] Jie Liu, Michael J. Casavant, Michael Cox, D.A.Walters, P. Boul,Wei Lu,A.J. Rimberg, K.A. Smith, Daniel T. Colbert, Richard E. Smalley, Controlled deposition of individual single-walled carbon nanotubes on chemically functionalized templates. Chem.Phys. Lett. 303 (1999) 125-29. https://doi.org/10.1016/S0009-2614(99)00209-2

[87] K. Esumi, M. Ishigami, A. Nakajima, K. Sawada, H. Honda, Chemical treatment of carbon nanotubes. Carbon 34 (1996) 279-81. https://doi.org/10.1016/0008-6223(96)83349-5

[88] T. Ramanathan, F.T. Fisher, R.S. Rouff, L.C. Brinson, Amino-functionalized carbon nanotubes for binding to polymers and biological systems. Chem. Mater. 17 (2005) 1290-95. https://doi.org/10.1021/cm048357f

[89] R. Haggenmueller, F. Du, J.E. Fischer, K.I. Winey, Interfacial in situ polymerization of single wall carbon nanotube/nylon 6, 6 nanocomposites. Polymer 47 (2006)2381-88 https://doi.org/10.1016/j.polymer.2006.01.087

[90] R.H Baughman, A.A. Zakhidov, W.A.de Heer. Carbon nanotubes–the route toward applications. Science 297(2002) 787-92. https://doi.org/10.1126/science.1060928

[91] W.B Choi, D.S.Chung, J.H Kang, H.Y.Kim, Y.W.Jin, I.T Han,Y.H Lee,J.E. Jung, N.S.Lee, G.S.Park, J.M.Kim. Fully sealed, high–brightness carbon–nanotube field–emission display. Appl Phys Lett 75(1999) 3129-31. https://doi.org/10.1063/1.125253

[92] H.J Kim, J.J.Choi, J.H.Han, J.H.Park,J.B Yoo. Design and field Emission test of carbon nanotube pasted cathodes for traveling–wave tube applications. IEEE Trans Electron Devices 53(2006) 2674-80. https://doi.org/10.1109/TED.2006.884076

[93] C.A.Spindt, I.Brodie, L. Humphrey, E. Westerberg . Physical properties of thin-film field emission cathodes with molybdenum cones. J Appl Phys 47(1976) 5248-63. https://doi.org/10.1063/1.322600

[94] J.M.Bonard, H.Kind, T.Stockli. Field emission form carbon nanotubes: the first fiveyears. Solid State Electronics 45(2001) 893-14. https://doi.org/10.1016/S0038-1101(00)00213-6

[95] H.Cui,O. Zhou, B.R.Stoner. Deposition of aligned bamboo–like carbon nanotubes viamicrowave plasma enhanced chemical vapor deposition. J Appl Phys 88 (2000) 6072-74. https://doi.org/10.1063/1.1320024

[96]N.S Lee,D.S. Chung, I.T Han, J.H.Kang, Y.S.Choi, H.Y Kim, S.H.Park,Y.W. Jin, W.K.Yi, M.J.Yun, J.E.Jung, C.J. Lee, J.H. Yoo, S.H.Jo, C.G.Lee, J.M Kim. Application of carbonnanotubes to field emission displays. Diamond and Relat Maters 10(2001) 265-70. https://doi.org/10.1016/S0925-9635(00)00478-7

[97] A.Hirsch, Functionalization of single–walled carbon nanotubes. Angew Chem Int Ed 41(2002) 1843-59. https://doi.org/10.1002/1521-3773(20020603)41:11<1853::AID-ANIE1853>3.0.CO;2-N

[98] J.Zhu, H. Peng, F.Rodriguez-Macias, J.L.Margrave, V.N. Khubashesku, A.M. Imam, K.Lozano, E.V Barrera. Reinforcing epoxy polymer composites through covalent integration of functionalized nanotubes. Adv Funct Mater 14(2004) 643-48. https://doi.org/10.1002/adfm.200305162

[99] J.L.Matthew, J. Vergne, E.D. Mowles,W.H. Zhong,D.M. Hercules,C.M. Lukehart CM. Surface functionalization and characterization of graphitic carbon nanofibers (GCNFs). Carbon 43(2005)2883–93. https://doi.org/10.1016/j.carbon.2005.06.003

[100] M.L.Sham, J.K. Kim. Surface functionalities of multi–wall carbon nanotubes after UV/Ozone and TEAT treatments. Carbon 44(2006) 768-77. https://doi.org/10.1016/j.carbon.2005.09.013

[101] P.He , Y. Gao , J. Lian , L. Wang, D. Qian , J. Zhao , W. Wang ,M.J. Schulz,X.P. Zhou, D.Shi. Surface modification and ultrasonication effect on the mechanical properties of carbon nanofiber/polycarbonate composites. Composites: Part A 37(2006) 1270-75. https://doi.org/10.1016/j.compositesa.2005.08.008

[102] G.Yu, J. Gong, S.Wang, D. Zhu, S.He, Z. Zhu. Etching effects of ethanol on multiwalled carbon nanotubes. Carbon 44(2006) 1218-24. https://doi.org/10.1016/j.carbon.2005.10.050

[103] J.S.Moon, P.S. Alegaonkar, J.H. Han, T.Y Lee, J.B Yoo, J.M. Kim. Enhanced field emission properties of thin-multiwalled carbon nanotubes: role of SiOx coating. J Appl Phys 100(2006) 1043031-37. https://doi.org/10.1063/1.2384795

[104] B.Kim, Y.H.Lee, J.H Ryu, K.D.Suh. Enhanced colloidal properties of single-wall carbon nanotubes in _-terpineol and Texanol. Colloids and Surfaces A 273(2006)161–64. https://doi.org/10.1016/j.colsurfa.2005.08.024

[105] J.Liu, A.G. Rinzler, H. Dai, J.H.Hafner, R.K. Bradley, P.J. Boul, A. Lu, T. Iverson, K. Shelimov, C.B. Huffman, F. Rodriguez-Macias, Y.S. Shon, T.R. Lee, D.T. Colbert, R.E.Smalley. Fullerene pipes. Science 280(1998) 1253-56. https://doi.org/10.1126/science.280.5367.1253

[106] Y.Cheng, O. Zhou. Electron field emission from carbon nanotubes. Comptes Rendus Physique 4(2003) 1021-33. https://doi.org/10.1016/S1631-0705(03)00103-8

[107] F.H.Gojnu, M.H.G.Wichmann, U.Kopke, B.Fiedler, K. Schulte . Carbon nanotubereinforced epoxy-composites; enhanced stiffness and fracture toughness at low nanotube content. Comp Sci and Technol.64(2004) 2363-71. https://doi.org/10.1016/j.compscitech.2004.04.002

[108] E.T.Thostenson, T.W.Chou. Processing–structure–multi–functional property relation ship in carbon nanotube/epoxy composites. Carbon 44(2006) 3022-29. https://doi.org/10.1016/j.carbon.2006.05.014

[109] A.Yasmin, J.LAbot, I.M Daniel. Processing of clay/epoxy nanocomposites by shear Mixing. Scr Mater 49(2003) 81-86. https://doi.org/10.1016/S1359-6462(03)00173-8

[110] R.H. Nordheim, L.W.Fowler Electron emission in intense electric field. Proc Roy Soc London A 119(1928) 173-81. https://doi.org/10.1098/rspa.1928.0091

[111] D. Temple . Recent progress in field emitter array development for high performance applications. Mater Sci Eng R 24(1999) 185-239. https://doi.org/10.1016/S0927-796X(98)00014-X

[112] A.S.Berdinsky, A.V.Shaporin,J.B. Yoo, J.H.Park, P.S.Alegaonkar, J.H. Han, G.H.Son . Field enhancement factor for an array of MWNTs in CNT paste. Appl Phys A 83(2006) 377-83. https://doi.org/10.1007/s00339-006-3482-7

[113] P.C.P.Watts,S.M. Lyth, E.Mendosz, S.R.P.Silva. Polymer supported carbon nanotube arrays for field emission and sensor devices. Appl Phys Lett 89(2006) 1031131-33. https://doi.org/10.1063/1.2345615

Thermoset Composite   Materials Research Forum LLC
Materials Research Foundations 38 (2018) doi: http://dx.doi.org/10.21741/9781945291876

Chapter 7

Conductive Thermoset Composites

Halima Khatoon, Sajid Iqbal, Sharif Ahmad[*]

Jamia Millia Islamia, New Delhi, India

hkn.nasir02@gmail.com, saj143frnd@gmail.com, sharifahmad_jmi@yahoo.co.in[*]

Abstract

Conductive thermoset composites (CTC) have gained much attention among scientists and technologists due to their excellent electrical, mechanical, and thermal properties. These properties of CTC make them a superior candidate in various applications like EMI shielding, anti-corrosive coatings, electronic packaging, LED's, etc. In view of this, the present chapter highlights the significance of CTC along with the historical background of some important thermoset polymers, different types of CTC and their methods of synthesis. In addition to this, the applications of CTC are described. Moreover, the problems associated with CTC and their solutions are discussed in detail.

Keywords

Thermoset Polymers, Conductive Fillers, Composites, Properties, Applications

Contents

7.1 Introduction

Polymers are a versatile class of materials, playing a significant role in everyday life. Their importance can be visualized in terms of safety, comfort, economy and reduction in noise pollution within the society [1]. They have widely replaced the use of metals, alloys and other materials. So far no material has been found that can substitute polymers. They have wide applications ranging from daily needs in engineering, biomedical, defence, space, etc. Polymers, based on structure and properties are mainly classified as (i) Thermosets and (ii) Thermoplastics.

Thermosets are the prepolymer resins that are present in liquid (soft or viscous) form before their curing and once they get cured, their structure cannot be reshaped by heating or melting or in other words one can say that they retain their strength and shape even after heating [2]. This property makes them advantageous over other form of plastic i.e. thermoplastics, which soften on heating [3]. Because of this, thermoset polymers are widely used for the production of permanent structure in solid shapes. Commercially, wide range of thermosetting polymers are available in the form of polyurethanes, epoxy, vulcanized rubber, phenolic, polyesters, polyamides, etc. [4]. These polymers are widely being used in different industrial applications. However, some time they lack the required thermo-physical, mechanical and electrical properties. Thus, to overcome these drawbacks thermoset polymers are being modified in the form of their blends, organic-inorganic hybrid and composites [5].

Conductive thermoset composites (CTC) are processed by dispersing the conductive particles such as gold, silver, palladium, nickel, copper, graphite, carbon fibre, conducting polymers, etc., within the insulating thermoset polymer matrix [6–8]. The presence of these conductive particles enhances the physico-mechanical, thermal, electrical and other properties. The dispersion of the conductive particles in thermoset composites not only improve the value-aided properties but also reduces the cost of production. In CTC, the thermoset polymer act as a matrix while the conductive fillers act as a reinforcing material that governs the load bearing and other physico-mechanical properties. It is noted that approximately 75% of composites are based on thermoset polymer matrices. The importance of composite materials was realized since the age of early civilization. The people of that era were used to make buildings by using mud and straw. In the past few years, especially the conducting polymers (CP) based thermoset polymer composites are in high demand, because of their high conductivity, good environmental stability, and low-cost manufacturing processes of CP [9]. It combines the processibility of thermoset matrix and conductivity of the conducting polymers. Thus, the CTC attains the best mechanical, thermal and electrical properties.

In view of this, the present chapter highlights the significance of CTC along with the historical backgrounds of some important thermoset polymers, different types of CTC and their methods of synthesis. In addition to this, the properties and applications of CTC have been described. Moreover, the problems associated with CTC and their solutions are also discussed in detail.

7.2   Historical background of thermoset polymers

The first thermoset polymer "vulcanised rubber" was discovered in 1839 by Goodyear's. After that, Baekeland reported phenolic resin in 1909. He also introduced the moulding process and use for production of these polymers, for which he got a patent. Some important thermoset polymers and their year of discovery are listed below [10–12]:

- *Natural Rubber-Goodyear's-1839*

- *Phenolic Resin-Baekeland-1909*

- *Unsaturated Polyester-Carleton ellis-1930*

- *Epoxy-Dr. Pierre Castan-1936*

- *Polyurethane-Otto Bayer-1938*

- *Polybenzoxazine-Higginbottom-1980s*

Epoxy: Epoxy resin is one of the most important commercial thermoset which finds various applications in the field of paints, adhesives, coatings, electronics, etc. [13]. It contains one or more than one α-epoxy groups at terminal, cyclic, or internal in their

Materials Research Forum LLC
doi: http://dx.doi.org/10.21741/9781945291876

backbone. After the thermosetting reaction, it is converted to a tough and solid material. The α-epoxy (1, 2-epoxy) is the most common and simplest type of epoxy while the diglycidyl ether of bisphenol A (DGEBA) is considered as the most widely used commercial epoxy resin [14]. It (DGEBA) can be easily synthesized according to the following reaction scheme 7.1.

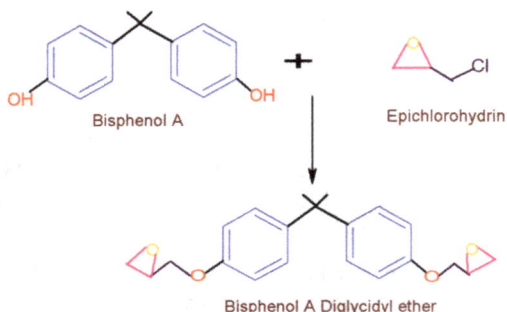

Scheme 7.1 Synthesis of Bisphenol A Diglycidyl ether.

Polyesters: Polyesters can be classified in two forms i.e. (i) thermoplastics and (ii) thermosets. These polymers contain ester functional groups in the main chain. Unsaturated polyesters (UP) are the thermoset polymers, generally prepared by the condensation reaction of polyols and polycarboxylic acids (scheme 7.2) [15]. They have wide applications in naval construction, offshore applications, waterlines, building construction, paint and coating industries due to their excellent rigidity, high strength, excellent dielectric properties and good chemical resistance [16].

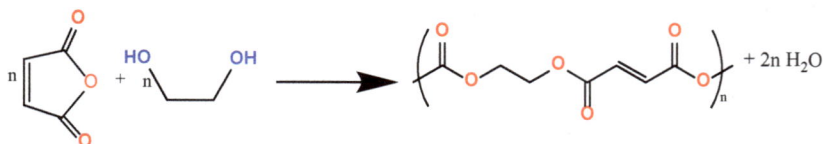

Scheme 7.2 General synthesis of unsaturated polyester resin.

Materials Research Forum LLC

doi: http://dx.doi.org/10.21741/9781945291876

Polyurethane: Polyurethane is the most versatile existing polymer, comprises of hard (NCO) and soft (OH) segments with the polar urethane group in the main backbone of polymer chain. It was first discovered by Otto Bayer in 1938. After this, a continuous interest grew. The most commonly used commercial polyurethane fibre called as Lycra was first developed by DuPont [18]. They are found in many forms like films, plastics, foams, elastomers, coatings, etc. They exhibit excellent chemical resistant, tear resistance, physico mechanical, and good processability [17]. They can be synthesised by step growth polymerization between the diol and di-isocyanates. A general reaction for the synthesis of PU is shown in scheme 7.3.

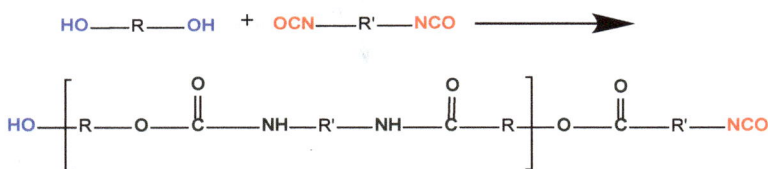

Scheme 7.3 General reaction for the synthesis of polyurethanes.

Polybenzoxazine- A new thermoset polymer: Polybenzoxazine has emerged as a novel type of thermoset polymer as an alternative for epoxies, phenolics, polyimides, etc. They have amine and hydroxyl moiety in their polymeric chain. They can be synthesised by the ring opening polymerization of benzoxazine monomer. The benzoxazine monomer, obtained by the reaction of bisphenol A, amine and formaldehyde, can polymerizes to form polybenzoxazines as shown in scheme 7.4. Polybenzoxazines provide excellent characteristics such as high heat resistance, flame retardance, good stability, low water absorption, good electrical property and high flexibility [19]. These newly developed resins possess unique features, namely (i) no change in volume upon curing, (ii) low water absorption, (iii) high T_g value, (iv) high char yield, (v) no strong acid catalysts required for curing, and (vi) no by-products formation.

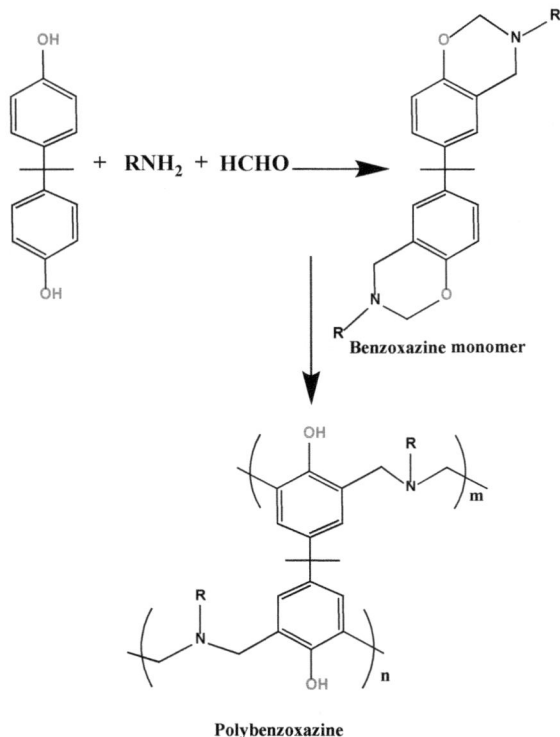

Scheme 7.4. Synthesis of benzoxazine monomer and polybenzoxazine therefrom.

7.3   Method of Composite processing

Thermoset composites can be processed mainly by two broad methods (i) wet forming method and (ii) hot melt prepreg method.

The wet forming method is used for the processing of the thermoset polymer resins in fluid state and their curing is performed by external heat and pressure. The Melt compounding method, Hand layup, Bag molding, Filament winding and Pultrusion are some examples of wet forming methods [20]. Out of these the hand layup is the most common and cost effective method. In this process, the reinforcing agents are placed by hand in a mold and then the thermoset resin is applied through a roller or a brush. Fig. 7.1 represents a typical model for the hand layup method. This method can be used for small to large-scale production required for boats, storage tanks, showers and tubs. Qin et al. [21] have synthesized graphene nanoplatelete coated carbon fibers/epoxy (GnP-CF/epoxy) composite by prepreg and hand layup methods. For the fabrication of composite, the prepregs were first warmed-up to room temperature. Unidirectional composites were obtained by laying up twelve layers by hand. The prepreg stacks were sandwiched between two Teflon-lined steel plates. The thickness of the final composites was controlled by a spacer made of high temperature tape. This assembly was inserted in a nylon vacuum bag, in which primary vacuum was created, and placed into the autoclave for curing. The thermal program was run initially at 2 h at 75 °C and followed by next 2 h at 125 °C. The pressure during cure was maintained at 0.59 MPa to form CF/epoxy composites.

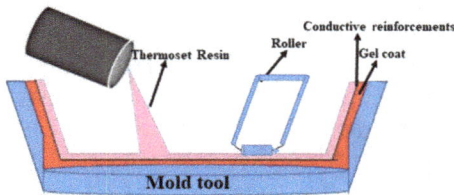

Fig. 7.1 Hand layup model for CTC formation.

Bag molding is another wet forming method widely used for composite formulation. In this process, the composite materials are subjected to vacuum in order to remove the air bubbles from the laminate. Post this stage; the material may be subjected to atmospheric pressure while it undergoes curing process in an oven. Fig 7.2 Shows the process model for bag molding synthesis.

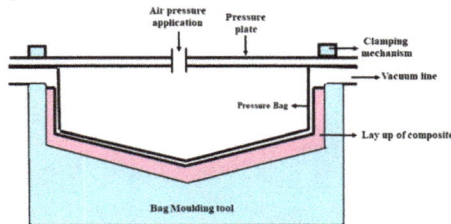

Fig. 7.2 Bag molding model for CTC formation.

Hot melt prepreg method is another useful method for the processing of CTC. It is a two-stage process in which, the prepreg is formed by the impregnation of conductive fillers into the thermoset resins followed by the curing of prepreg composites at high temperature and pressure. Toshio et al. [22] have used a hot-melt prepreg method for the synthesis of aligned CNT/epoxy composites. For this, first they have prepared the CNT/epoxy prepreg and then convert it into composites.

Apart from these, sonication, stirring or milling are some of the other methods to produce conductive thermoset composites. In these methods, a conductive filler and the insulating thermoset polymers are first mixed separately with a suitable solvent, then their solutions were stirred or sonicated and then mixed together followed by solvent evaporation to form composites. However, these methods are difficult to scale up for commercial uses because of the high viscosity of the composites, which allows only the low loading of conductive fillers. Therefore, there is an urgent need to develop a new method which can be used for the production of CTC. The *in situ* method can be considered as a new and advance method that may be used to solve these problems, as it is a single stage process, which requires less or no solvent. In recent years, CTC are being synthesised by *in situ* methods. In this method, generally the thermoset monomers are polymerized in the presence of conductive fillers or the conductive fillers are modified/functionalised (e.g. GO, rGO, CNT, Ag, Au, silica etc.) polymerised (conducting polymers) in presence of thermoset polymers. Fig. 7.3 represents a general representation for the *in situ* method for the processing of CTC. In the literature, reported by She et al. this method was used to synthesise reduced graphene oxide/polystyrene (rGO/PS) composites [23]. They have used a two steps *in situ* process, in the first step GO/PS composite was synthesized and in the next step rGO/PS composite was obtained by the reduction of GO to rGO using hydrazine hydrated followed by thermal reduction in a box resistance furnace at 200 °C for 12 h.

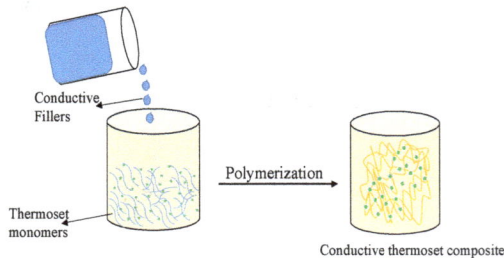

Fig. 7.3 In situ polymerization method for CTC formation.

7.4   Different types of CTC

A number of conductive thermoset composites have been reported by many researchers. The conductive nanofillers like conducting polymers, graphene, CNT, MWCNTs, clay, gold, silver, etc. are being used to prepare CTC by many researchers. However, the present chapter highlights the use of various types of conductive particles and their conductive composite in different thermoset (epoxy, polyurethane, polyester and polybenzoxazines) matrices.

An exhaustive literature survey on different CTCs are discussed below, discussing various aspects of these materials.

7.4.1   Epoxy Based CTC

For the past several decades, epoxies are being used in various plastics industries. They have a wide range of applications in the field of adhesives, paints and coatings. The literature revealed that more than 90% of epoxy thermoset polymer resins are mainly obtained from bisphenol A (BPA) [24]. However, there is a growing demand to develop such epoxies which have value aided properties and applications [25]. Recently, epoxy based conductive composites are considered as an engineered technique used to improve the overall properties. In a work reported by Bao et al. the surface modification of GO to functionalised graphene oxide (FGO) was carried out by using hexachlorocyclotriphosphazene (HCTP) and glycidol [26]. Later the FGO/epoxy nanocomposite was developed through *in situ* polymerization. The composite showed good mechanical and electrical conductivity than that of plain epoxy. An increase of 38% in hardness and 113% in storage modulus was noted. The electrical conductivity was found to be of 6 orders higher than that of the plain. A thermoset epoxy clay nanocomposite was synthesized by Costas et al. [27]. To improve the mechanical and thermal properties of the composite, they modified the clay with diprotonated forms of polyoxypropylene diamines of type α, ω-[NH$_3$CHCH$_3$CH$_2$(OCH$_2$CHCH$_3$)xNH$_3$]$^{2+}$ with x"2.6, 5.6, and 33.1. It was evaluated that the diamine intercalated clay epoxy nanocomposite reduced the plasticising effect, which improve the mechanical and thermal properties and also reduces the cost and time for the synthesis of nanocomposite.

7.4.2   Polyurethane based CTC

Polyurethanes, due to their versatile properties, have attracted a tremendous interest in research and development. Lyudmyla et al. [28] have synthesized conductive MWCNT filled polyurethane thermoset composite. They have investigated the chemical interaction (wander Vaal or covalent) of MWCNT with PU, their glass transition dynamics and laser interferometric creep rate spectroscopy. The composite shows 2-3 fold increase in modulus and tensile strength due to the covalent interfacial interaction between MWCNT and PU. Mcclory et al. [29] have synthesized conductive thermoplastic nanocomposites of MWCNT (1wt %) loaded PU via an addition polymerization reaction. They noticed that only 0.1 wt.% addition of MWCNT in PU exhibited a remarkable enhancement in stiffness, strength and toughness. The tensile strength, % elongation and increase in young's modulus was found to be 397, 302 and 561% respectively than that of plain PU. Berns et al. [30] prepared a new conductive composite based on thermoset PU and conductive ionic liquid (dialkylimidazolium salts) and studied the mechanical and conduction properties. They have also established a correlation between the specific conductivity and mechanical properties. Jaaoh et al. [31] synthesized a conductive composite of PANI/PU and studied their electrostrictive behavior and vibration energy harvesting performance. A detailed review on conducting polymer based polyurethane conductive composite was recently published by Ahmad et al. [32].

7.4.3 Polyester based CTC

Apart from epoxy, phenolic resin and polyurethane, unsaturated polyester (UPE) is another very important thermoset polymer matrix that have been used for the preparation of conductive thermoset composites [15]. In a recent work reported by Jin and co-workers, a heat conductive material of steel/PE sandwich composite was designed for the application in the field of electronics, electrical and aerospace [33]. The prepared composite exhibit high thermal conductivity, chemical stability, excellent corrosion resistance and electrical insulation behaviour. Further, the alignment of the steel wire in the composite was also studied on the thermal transmission behavior. It was observed that the longitudinal alignment of steel led to increase in the thermal transmission speedily with the increased loading of steel wire. However, it increases gently when steel wire was placed in transverse direction. A reduced graphene oxide based PE (rGO/PE) conductive nanocomposite was synthesized by Bora et al. [34]. The thermal, mechanical, electrical and antibacterial properties were evaluated and an increase of 123% in tensile strength and 87% in Young's modulus at 3 wt. % loading of the rGO was reported. Electrical conductivity of 3.7×10^{-4} S/cm was observed for 3% rGO/PE composite.

7.4.4 Polybenzoxanines based CTC

Polybenzoxanies possess excellent thermoset properties due to which it acts as a potential alternative for epoxy bismaleimides, esters and polyamides [35]. However, it has some drawbacks too, like it cures at relatively high temperature (190 °C) and brittle in nature [35]. Thus, to overcome this and to improve other property there is an urgent need to modify them in the form of blends and composite.

In recent years polybenzoxazine based conductive composites have gained much attention due to versatile properties like excellent thermal stability, high yield, high glass transition, water resistant, high versatility of molecular design and no shrinkage after curing [36]. A boron nitride filled polybenzoxazine composite with very high thermal conductivity of 32.5 W/mK was synthesized by Ishida et al. [37]. The excellent conductivity was attributed to the formation of conductive networks with low thermal resistance along the conductive paths.

A benzoxazine functionalised MWCNT (MWCNT-FBz) was successfully synthesized by Wang et al. through Diels-alder reaction and later the composite of MWCNT-PBz was also fabricated [38]. The prepared composite showed good electrical conductivity of 7×10^{-5} S/cm which is due to the compatibility between benzoxazine-based resins and the benzoxazine-functionalized MWCNTs.

7.5 Properties of CTC

As compared to virgin conducting nanoparticles and insulating thermoset polymer matrix, the conductive thermoset composites are exhibiting more stable and superior physico-mechanical, thermal and electrical properties. The properties of some CTC are discussed below.

7.5.1 Thermal properties

Conductive thermoset composites show a unique thermal property called positive temperature coefficient (PTC) in which they exhibit a sharp insulator to conductor transition [39]. CTC show a large PTC due to the thermal expansion of thermoset polymers and conductive pathway of conducting fillers. The principle of PTC is generally based on three factors (i) percolation threshold, (ii) conduction mechanism and (iii) the thermo-mechanical properties of the composite. The CTC shows the increase in volume and decrease in conductive phase with the increasing temperatures and hence the conductance of composite decreases with increase in temperature [40].

7.5.2 Mechanical properties

Generally the mechanical properties of the CTC depends on three factors (i) strength and modulus of the conductive filler, (ii) strength and chemical stability of the insulating thermoset polymer resins and (iii) types of interaction between the filler and resin. In a literature reported by Joseph et al., superior mechanical properties of sisal fibre reinforced several thermoset polymers matrix, such as epoxy, polyester and phenol-formaldehyde, composites have been investigated [41]. It was further found that among these conductive thermoset composites, phenolic type polymer matrices show better mechanical properties, in terms of tensile and flexural strength, than epoxy and polyester resins which can be attributed to the high interfacial bonding in phenolic composites. Minoo et al. have improved the mechanical properties of the epoxy by dispersing 0.1wt% of graphine oxide (GO) [42]. The flexural strength, storage modulus and glass transition temperature of composite was studied and found to be higher than that of plain epoxy, which can be due to the better dispersion and strong bonding between the GO and epoxy.

7.5.3 Electrical properties

J.F Feller et al. have compared the electrical properties of short carbon fibers (SCF) reinforced polyester and polyepoxy conductive polymer composites (CPC) [43]. The SCF with polyepoxy exhibit better property than that of polyester based composite. Both the CPC show a PTC effect between 90-160°C. In another report, Tomohiro et al. have discussed the electrical properties of PANI based conductive thermoset composite, in which PANI was used as a conductive filler, DBSA and PTSA as a dopant and DVB as a crosslinking agent [44]. The resulting composite shows an excellent electrical property with the conductivity of 1 S/cm.
Properties of some of the conductive thermoset composites are tabulated in table 7.1.

7.6   Applications of conductive thermoset composites

Nowadays, new conductive thermoset materials are in high demand in developing industries due to their fast cooling ability. Among other conductive thermoset materials, polymer composites have following merits:

(a) Light weight.

(b) Ease of synthesis.

(c) Good electrical insulation and electrical conduction.

(d) Corrosion resistant.

Some of the important applications of conductive thermoset polymer composite in various industrial areas is illustrated in Fig. 7.4, and are discussed in the following sections:

Table 7.1 Properties of conductive thermoset polymer.

S. No.	System	Matrix	Filler	Thermal Properties	Conductivity $(S\ cm^{-1})$	Ref.
1.	Epoxy/PTh	Epoxy	PTh	First degradation occur at 384 °C.	0.87	[6]
2.	Epoxy/CF	Epoxy	CF	-	25	[44]
3.	Epoxy/PEDOT	Epoxy	PEDOT	First degradation occur at 323 °C.	0.7	[45]
4.	Polyurethane/ Graphite	Polyurethane	Graphite	-	0.1	[46]
5.	Graphene/CNT/ polyurethane	polyurethane	GCNT	-	0.01	[47]
6.	Epoxy/CNT	Epoxy	CNT	-	1	[48]
7.	Phenol formaldehyde/rGO	Phenol formaldehyde	rGO	First degradation occur at 212 °C.	0.03	[49]
8.	Epoxy/CNT	Epoxy	CNT	First degradation occur at 383 °C.	1.5×10^{-5}	[50]
9.	Phenol formaldehyde/GO	Phenol formaldehyde	GO	First degradation occur at 372 °C.	7	[51]
10.	phenolic foam (PF)/MWCNTs	phenolic foam	MWCNT	First degradation occur at 299 °C.	1.3×10^{-2}	[52]
11.	Polyurethane/ Carbon Fibre (CF)	Polyurethane	CF	-------	10^{-5}	[53]
12.	Epoxy/Graphene Nanoplatelets (GNPs)	Epoxy	GNPs	First degradation occur at 384 °C.	10^{-3}	[54]

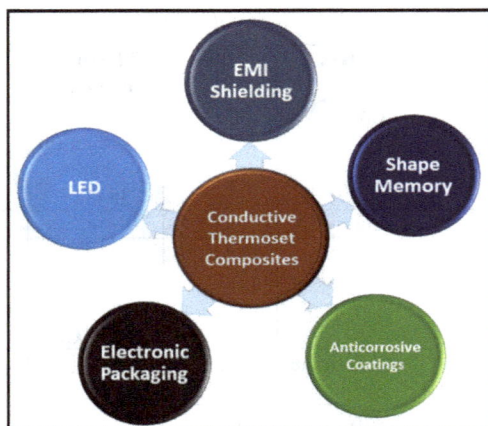

Fig. 7.4 Applications of conductive thermoset composites.

7.6.1 Electromagnetic interference (EMI) shielding

Electromagnetic interference (EMI) shielding is an undesired electromagnetic (EM) waves caused by the excessive use of electronic devices such as mobile phones, television, computers, ACs, etc. and other industrial equipment's. This is not only a big problem for the electronic industries but also affects the human health very badly causing cancer, birth defects, memory loss, etc. [55]. Therefore, it must be overcome by developing better shielding materials at low cost. For this purpose, Aal et al. [56] have fabricated carbon black reinforced epoxy composites (CB/EP) and tested its EMI shielding properties. The author found that the composite showed superior EMI shielding properties. Moreover, the synthesized composite also showed excellent antistatic properties. Liu et al. [57] prepared single-walled carbon nanotube (SWCNT)/polyurethane (PU) composite. The authors found that the synthesized composite showed tremendous EMI shielding properties and are light in weight. Huang et al. [58] fabricated the SWCNT reinforced epoxy composite and investigated its EMI shielding properties. It was observed that with the addition of 15 wt.% SWCNT the said composite showed excellent absorption properties (20-30 dB) in the X-band (8-12.4 GHz) region. Liang et al. [59] have synthesized graphene/epoxy composites for EMI shielding. The composite showed very low threshold value of 0.52 vol.%. The synthesized composite showed outstanding shielding effectiveness of -21 dB in the X-band region. Hoang et al. [60] prepared multiwalled carbon nanotubes dispersed polyurethane (MWCNT/PU) composite and investigated its EMI shielding ability. The author found that the composite with 22 wt.% of MWCNT showed excellent EMI shielding properties with an average shielding effectiveness of 20 dB. The superior shielding effectiveness was attributed due to the high conductivity of the composite (10 S m^{-1}). Later on, Gupta et al. [61] have synthesized multiwalled carbon nanotubes (MWCNTs) based polyurethane (PU) composites by solvent casting technique. The synthesized composite showed excellent EMI shielding properties with maximum absorption of -41.6 dB. The enhanced shielding properties was attributed due to the excellent conductivity (7.9 Scm^{-1}) of the composite than that of pristine PU (10^{-14} Scm^{-1}). Verma et al. [62] have synthesized graphene dispersed commercial polyurethane (PUG) composite and studied its EMI shielding properties. The authors found that the synthesized composite showed outstanding EMI shielding properties of 21 dB in the X-band region with a sample thickness of 3 mm. The superior properties of the composites may be attributed to the high electrical conductivity and increased electrical polarization.

7.6.2 Anti-corrosive coatings

Corrosion is an undesirable phenomenon, which causes great loss to the economy and society. This cannot be completely eradicated but can be overcome by developing some good materials such as corrosion inhibitors, paints and coatings. For this instance, Ahmad et al. [63] have formulated PANI/alkyd composite for high performance anticorrosive coatings. The uniform dispersion of PANI provides higher strength to the films that result in better corrosion protection. Mostafaei et al. [64] have reported the synthesis of Epoxy/polyaniline(PANI)-ZnO composite coatings and investigated its corrosion resistance properties. The corrosion resistance properties of the composites have been tested in 3.5 wt.% NaCl solution and found to be excellent corrosion protective coatings. The composite coating was found to be uniform, crack free, and compact. The PANI/ZnO acted as a physical barrier, which restricted the penetration of water molecules and other corrosive ions leading to the formation of superior corrosion protective coatings. Zhu et al. [65] have fabricated carbonyl iron particles (CIP)/silica/PU composite and studied its thermal, mechanical and anticorrosive properties. The CIP/silica/PU composite showed excellent corrosion resistance properties than that of CIP/PU composite. The excellent corrosion resistance performance was due to the superior barrier effect of silica shell. Wei et al. [66] have synthesized CNT reinforced PU composites and explored its anticorrosive performance in 3.0 wt.% NaCl solution. The author found that the corrosion protection efficiency of the synthesized coating was found to be very high (97.7%). This was attributed to the homogeneous dispersion of CNT in PU matrix, which provides an excellent barrier effect and improves the corrosion protective performance. Zhang et al. [67] have prepared polyvinylpyrrolidone/reduced graphene oxide (PVP-rGO) composites and investigated its corrosion protective performance. The composite with 0.7 wt.% rGO showed excellent corrosion protective performance (0.3 mm/year) than that of pristine epoxy resin (1.3 mm/year). Ahmad et al. [68] have reported the synthesis of polythiophene dispersed epoxy (PTh/EP) composites and studied its anticorrosive properties. It was observed that the synthesized composite coating showed superior corrosion protective performance (E_{corr}/I_{corr}: -0.30301 V/3.1587x10^{-9} Acm^{-2}) than that of pristine epoxy-PA coating (-0.45505 V/9.3862 x 10^{-7} Acm^{-2}).

7.6.3 Shape memory application

Shape memory polymers (SMPs) have attracted wide attention due their high processability, superior recoverability. Moreover, they are light in weight and are of low cost. These properties of SMPs lead to their use in intelligent medical devices, smart textiles, sensors and actuators [69]. For this persistence, Sahoo et al. [70] have synthesized polypyrrole-reinforced polyurethane composite by chemical oxidative polymerization and explored its shape memory behaviour. It was observed that the composite showed outstanding shape recovery of 80-90% at applied voltage of 40 V. Oh et al. [71] have fabricated graphene dispersed polyurethane composite for shape memory application. The author found that the composite showed exceptional shape memory behaviour with slight modification of MeOH. The excellent shape memory properties of the synthesized composites was due to the formation of covalent bonds between graphene and polyurethane. Raja et al. [72] have prepared composites of polyurethane, polylactide and modified carbon nanotubes (PU/PLA/modified CNT) and investigated its thermal, mechanical and electroactive shape memory properties. The said composite showed superior recoverability of 95% in 5 sec under an applied voltage of 40V, which was very much higher than those of pristine PU/PLA and modified CNT (15 sec and 40 sec). The superior shape memory properties was attributed due to the excellent thermal and electrical conductivities. Lu et al. [73] have synthesized polymer composite using epoxy, CNT and boron nitride (BN) and studied its shape memory performance. The authors found that with the addition of equivalent amount (4 wt.%) of CNT and boron nitride showed remarkable shape recovery within 60 sec. This was maybe due to the synergistic effect of CNT and BN.

7.6.4 Other applications

Lu et al. [74] fabricated polyurethane/alumina (PU/Al$_2$O$_3$) and polyurethane/carbon fibre (PU/CF) composites. The author found that the PU/CF composite showed superior conductivity than that of PU/Al$_2$O$_3$ (10 times) and pristine PU (50 times). The author also observed that the PU/Al$_2$O$_3$ composite absorbed more moisture (water molecules) than that of PU/CF composite; hence, an increase in dielectric loss was observed. This proved that the PU/CF composite displayed great potential as electronic packaging material. Lu et al. [75] have synthesized polyaniline(PANI)/epoxy composites by *in-situ* polymerization and studied its different properties. The synthesized composite showed excellent dielectric constant (3000) and low dielectric loss of 0.5 at 10 kHz. The superior dielectric properties of the composites was attributed due to the homogeneous dispersion of PANI into the epoxy polymeric matrix and found to be a potential candidate for capacitor applications. Du et al. [76] have reported the synthesis of epoxy/graphite composite for proton exchange membrane (PEM) fuel cells. The composite showed high conductivity (200-500 S cm^{-1}), flexural modulus (2×10^4 MPa) and strength (72 MPa) with the addition of 50 wt.% carbon content. These properties of the composite led to their application in bipolar plate for fuel cells. Pramanik et al. [77] fabricated hyperbranched poly(ester amide)/polyaniline (HBPEA/PANI) composite and studied its antistatic properties. The synthesized composite showed high tensile strength (7.2-12.25 MPa), impact resistance (>100 cm) and scratch hardness (8.5-10 kg). Further, the addition 10 wt.% of PANI displayed low sheet resistance (7.37×10^5 Ω/sq) than that of pristine polymer (2.94×10^7 Ω/sq) and found to suitable for antistatic applications. The decrease in sheet resistance may be attributed to the formation of conductive network between PANI and HBPEA. Cho et al. [78] have synthesized polyamide (PA)/reduced graphene (rGO) composite through melt blending method. The titanate-coupling agent (TCA) was used to modify rGO to enhance the chemical compatibility. The TCA/rGO/PA composite displayed superior conductivity (53%) than that of PA/rGO and found to be promising for LED heat sink due to its excellent thermal dissipation.

7.7 Problems and solution associated with CTC

As the thermoset and their composites are highly cross-linked they are difficult to recycle [79]. The major problem associated with the CTC is their permanent and intractable three-dimensional structure which hinders its recycling or reprocessing by usual methods [80]. Previously the CTC waste was generally disposed in landfill, which made a huge pressure on the environment [81]. Thus, to reduce the pressure on environment, many countries especially European country have made strict rules and regulations on the disposal of CTC waste to make the environment greener and cleaner. Hence, it is necessary to introduce some efficient recycling routes which must be already considered at the beginning of the design stage to recycle CTC with high value.

There are several other methods to recycle CTC:

- The methods of assembling should be as easy as the dismantling at the end of their life.

- Standardize the polymers used

- Choosing materials which can be recycle easily

- Incompatible polymers should be avoided

Literature reveals that there are four main methods for recycling the CTCs that have been used since the early 90s of 20th century. These are (i) mechanical recycling, (ii) pyrolysis process, (iii) fluidized bed process, and (iv) chemical recycling [82]. In a mechanical recycling process, the CTC can be recycled mainly for fillers and fibres of wide ranged lengths, however, the obtained fibres are not long and of high value. Kouparistsas et al. have used this process to grind composites to recycle them [83]. Pyrolysis is another most common thermal process in which the composites are recycles over high temperature by which some liquid products, some useful solid residues, long fibres with high modulus and a mixture of gas releases [84]. Another thermal process is the fluidized bed process that can produce clean fibres and fillers with good energy recovery. A successful recycling of glass fibre composites was carried out by Pickering et al. using fluidized-bed process at a fluidizing velocity of 1.3 m/s and a bed temperature of 450°C [85]. Chemical recycling is not as effective as mechanical and thermal processes because the solution or solvent used in this process are not feasible to all composites. Thus to develop an effective method for chemical recycling of thermoset CF/Epoxy composites Li et al. in a mixed solution of H_2O_2 and DMF [86]. They have used a two-steps chemical process (shown in Fig. 7.5,) to obtain the CF and epoxy. For this, first they treated the composite in acetic acid at 120°C for 30 min, which helped in the expansion of layered surface and to accelerate the degradation process. After that in the next step, a solution mixture of H_2O_2 and DMF was used to further oxidising it. After these chemical modifications, finally, the composites were degraded and the clean CF was obtained and the rest degraded by-products were obtained as a deep yellow liquid.

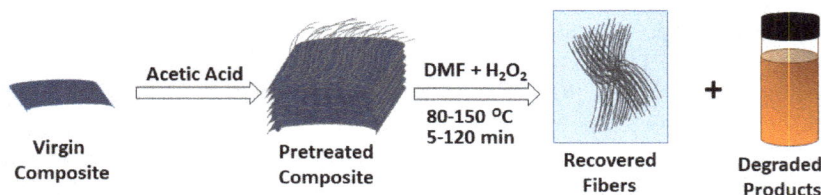

Fig. 7.5 Chemical recycling of CF/Epoxy thermoset composite (Adapted from ref no 86).

Further, they have also demonstrated the mechanism of decomposition of CF/EP composites in DMF and H_2O_2. As shown in Fig. 7.6 the decomposition process is mainly divided into five steps: initiation, diffusion, oxidative decomposition, dissolution and separation. In the initiation process, the H_2O_2 produces active free radicals that further diffuses to the surface of the CF/EP composites. These free radicals are then transfer and oxidative decomposition takes place. Finally, the decomposed products were dissolved in DMF, which will be easy to dissolve and remove. Thus, this process helps in the decomposition of 90 % epoxy resin and clean CF have been obtained.

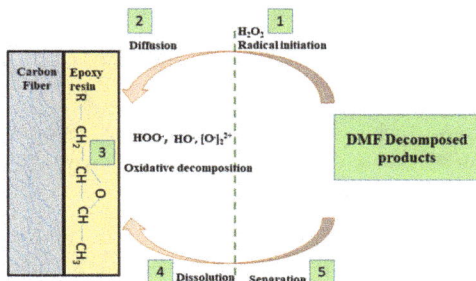

Fig. 7.6 Decomposition mechanism of recycling of CF/Epoxy composite (Adapted from ref no 86).

Some other methods like reversible cross-linking with covalent chemistry or supramolecular interactions have also been practiced to alter the thermoset into thermoplastics.

Conclusion

The present book chapter highlight the significance of conductive thermoset composite (CTC) along with their historical backgrounds of some important thermoset polymer matrix, types of CTC and methods of synthesis of CTC. Moreover, the problems associated with CTC and their solutions have been discussed in detail. In addition to this, the applications of CTC have also been discussed.

Acknowledgment

One of the author, Halima Khatoon, is highly thankful to UGC-MANF for providing financial assistance.

Thermoset Composite   Materials Research Forum LLC
Materials Research Foundations **38** (2018)   doi: http://dx.doi.org/10.21741/9781945291876

References

[1]   Hassan Namazi, Polymers in our daily life, Bioimpact. 7 (2017) 73–74.

[2]   R.J.J.W. Jean-Pierre Pascault, Henry Sautereau, Jacques Verdu, 2002
        Thermosetting polymers, New york, CRC.

[3]   Thermoset Composites - An Introduction, (2001) 1–3.
        https://www.azom.com/article.aspx?ArticleID=301

[4]   A. Dotan, Biobased Thermosets, (2013), Handbook of thermoset plastics, Elsevier
        Inc.,800.

[5]   V.K. Thakur, M.K. Thakur, Processing and characterization of natural cellulose
        fibers/thermoset polymer composites, Carbohydr. Polym. 109 (2014) 102–117.
        https://doi.org/10.1016/j.carbpol.2014.03.039.

[6]   O. Zabihi, A. Khodabandeh, S.M. Mostafavi, Preparation, optimization and
        thermal characterization of a novel conductive thermoset nanocomposite
        containing polythiophene nanoparticles using dynamic thermal analysis, Polym.
        Degrad. Stab. 97 (2012) 3–13.
        https://doi.org/10.1016/j.polymdegradstab.2011.10.022.

[7]   R. Rohini, P. Katti, S. Bose, Tailoring the interface in graphene/thermoset polymer
        composites: A critical review, Polym. 70 (2015) A17–A34.
        https://doi.org/10.1016/j.polymer.2015.06.016.

[8]   O. Shepelev, S. Kenig, H. Dodiuk, Nanotechnology Based Thermosets, Third Edit,
        Elsevier Inc., 2013. https://doi.org/10.1016/B978-1-4557-3107-7.00016-6.

[9]   C. Li, H. Bai, G. Shi, Conducting polymer nanomaterials: electrosynthesis and
        applications., Chem. Soc. Rev. 38 (2009) 2397–2409.
        https://doi.org/10.1039/b816681c.

[10]   H. Dodiuk, S.H. Goodman, Introduction, Handb. Thermoset Plast. (2013) 1–12.
        https://doi.org/10.1016/B978-1-4557-3107-7.00001-4.

[11]   H. Ishida, overview and historical background of polybenzoxazine research, in:
        Handb. Benzoxazine Resins, 2011: pp. 3–69.

[12]   I.H. Updegraff, Unsaturated Polyester Resins, in: Handb. Compos., Springer,
        Boston, MA, 1982, 17-37. https://doi.org/10.1007/978-1-4615-7139-1_2

[13]   Sharif Ahmad, S.M. Ashraf, E. Sharmin, F. Zafar, Studies on ambient cured
        polyurethane modified epoxy coatings synthesized from a sustainable resource,
        Prog. Cryst. Growth Charact. Mater. 45 (2002) 83–88.

[14] W. Brostow, S.H. Goodman, J. Wahrmund, Epoxies, Handbook of Thermoset Plastics 2014. https://doi.org/10.1016/B978-1-4557-3107-7.00008-7.

[15] J.M. Sadler, F.R. Toulan, G.R. Palmese, J.J. La Scala, Unsaturated polyester resins for thermoset applications using renewable isosorbide as a component for property improvement, J. Appl. Polym. Sci. 132 (2015) 1–11. https://doi.org/10.1002/app.42315.

[16] A. Kandelbauer, G. Tondi, O.C. Zaske, S.H. Goodman, Unsaturated Polyesters and Vinyl Esters, Third Edit, Elsevier Inc., 2013, 111-172. https://doi.org/10.1016/B978-1-4557-3107-7.00006-3.

[17] C. Zhang, Y. Li, R. Chen, M.R. Kessler, Polyurethanes from Solvent-Free Vegetable Oil-Based Polyols, ACS Sus. Chem. Engg. 10 (2014) 2465-2476. https://doi.org/10.1021/sc500509h.

[18] H. Janik, M. Sienkiewicz, J. Kucinska-Lipka, Polyurethanes, Handbook of thermoset plastics, Elsevier Inc., Waltham MA, 2014.

[19] I.H. Hathaikarn Manuspiya, Polybenzoxaines-based composites for increased dielectric constant, in: Handb. Benzoxazine Resins, 2011: pp. 621–637.

[20] M. Biron, Thermoset and composite, Elsevier Ltd., Brooklyn NY USA 2003, 560.

[21] W. Qin, F. Vautard, L.T. Drzal, J. Yu, Mechanical and electrical properties of carbon fiber composites with incorporation of graphene nanoplatelets at the fiber-matrix interphase, Compos. Part B Eng. 69 (2015) 335–341. https://doi.org/10.1016/j.compositesb.2014.10.014.

[22] T. Ogasawara, S.Y. Moon, Y. Inoue, Y. Shimamura, Mechanical properties of aligned multi-walled carbon nanotube/epoxy composites processed using a hot-melt prepreg method, Compos. Sci. Technol. 71 (2011) 1826–1833. https://doi.org/10.1016/j.compscitech.2011.08.009.

[23] N. Wu, X. She, D. Yang, X. Wu, F. Su, Y. Chen, Synthesis of network reduced graphene oxide in polystyrene matrix by a two-step reduction method for superior conductivity of the composite, J. Mater. Chem. 22 (2012) 17254. https://doi.org/10.1039/c2jm33114d.

[24] R. Auvergne, S. Caillol, G. David, B. Boutevin, J.P. Pascault, Biobased thermosetting epoxy: Present and future, Chem. Rev. 114 (2014) 1082–1115. https://doi.org/10.1021/cr3001274.

[25] S. Zhao, M.M. Abu-Omar, Synthesis of Renewable Thermoset Polymers through Successive Lignin Modification Using Lignin-Derived Phenols, ACS Sustain.

Chem. Eng. 5 (2017) 5059–5066.
https://doi.org/10.1021/acssuschemeng.7b00440.

[26]   X.Q. and Y.H. Chenlu Bao, Yuqiang Guo, Lei Song, Yongchun Kan, In situ preparation of functionalized graphene oxide/epoxy nanocomposites with effective reinforcements, J. Mater. Chem. 21 (2011) 13290–13298.

[27]   C.S. Triantafillidis, P.C. LeBaron, T.J. Pinnavaia, Thermoset epoxy-clay nanocomposites: The dual role of α,ω-diamines as clay surface modifiers and polymer curing agents, J. Solid State Chem. 167 (2002) 354–362.
https://doi.org/10.1006/jssc.2001.9541.

[28]   L. V. Karabanova, R.L.D. Whitby, V.A. Bershtein, A. V. Korobeinyk, P.N. Yakushev, O.M. Bondaruk, A.W. Lloyd, S. V. Mikhalovsky, The role of interfacial chemistry and interactions in the dynamics of thermosetting polyurethane-multiwalled carbon nanotube composites at low filler contents, Colloid Polym. Sci. 291 (2013) 573–583. https://doi.org/10.1007/s00396-012-2745-4.

[29]   C. McClory, 1, T. McNally, 1, G.P. Brennan, 2, J. Erskine, Thermosetting Polyurethane Multiwalled Carbon Nanotube Composites, J. Appl. Polym. Sci. 105 (2007) 1003–1011.

[30]   B. Berns, H. Deligöz, B. Tieke, F. Kremer, Conductive composites of polyurethane resins and ionic liquids, Macromol. Mater. Eng. 293 (2008) 409–418. https://doi.org/10.1002/mame.200700405.

[31]   D. Jaaoh, C. Putson, N. Muensit, Enhanced strain response and energy harvesting capabilities of electrostrictive polyurethane composites filled with conducting polyaniline, Compos. Sci. Technol. 122 (2016) 97–103.
https://doi.org/http://dx.doi.org/10.1016/j.compscitech.2015.11.020.

[32]   H. Khatoon, S. Ahmad, A Review on Conducting Polymer Reinforced Polyurethane Composites, J. Ind. Eng. Chem. 53 (2017) 1-22.
https://doi.org/10.1016/j.jiec.2017.03.036.

[33]   F.Q. Yang, Y.F. Jin, X. Qian, Simulation of Thermal Conductivity of Sandwich Composite, Appl. Mech. Mater. 278-280 (2013) 523–526.
https://doi.org/10.4028/www.scientific.net/AMM.278-280.523.

[34]   C. Bora, P. Bharali, S. Baglari, S.K. Dolui, B.K. Konwar, Strong and conductive reduced graphene oxide/polyester resin composite films with improved mechanical

strength, thermal stability and its antibacterial activity, Compos. Sci. Technol. 87 (2013) 1–7. https://doi.org/10.1016/j.compscitech.2013.07.025.

[35] K.S. Santhosh Kumar, C.P. Reghunadhan Nair, Polybenzoxazine-new generation phenolics, Handb. Thermoset Plast. (2013) 45–73. https://doi.org/10.1016/B978-1-4557-3107-7.00003-8.

[36] S. Matsumura, A.R. Hlil, C. Lepiller, J. Gaudet, D. Guay, Z. Shi, S. Holdcroft, A.S. Hay, Ionomers for proton exchange membrane fuel cells with sulfonic acid groups on the end-groups: Novel branched poly(ether-ketone)s, Am. Chem. Soc. Polym. Prepr. Div. Polym. Chem. 49 (2008) 511–512. https://doi.org/10.1002/pola.

[37] H. Ishida, S. Rimdusit, Very high thermal conductivity obtained by boron nitride-filled polybenzoxazine, Thermochim. Acta. 320 (1998) 177–186. https://doi.org/10.1016/S0040-6031(98)00463-8.

[38] Y.H. Wang, C.M. Chang, Y.L. Liu, Benzoxazine-functionalized multi-walled carbon nanotubes for preparation of electrically-conductive polybenzoxazines, Polymer (Guildf). 53 (2012) 106–112. https://doi.org/10.1016/j.polymer.2011.11.040.

[39] X. S. Yi, G. Wu, Y. Pan, Properties and applications of filled conductive polymer composites, Polym. Int. 44 (1997) 117–124. https://doi.org/10.1002/(SICI)1097-0126(199710)44:2<117::AID-PI811>3.0.CO;2-L.

[40] G. Boiteux, J. Fournier, D. Issotier, G. Seytre, G. Marichy, Conductive thermoset composites: PTC effect, Synth. Met. 102 (1999) 1234–1235. https://doi.org/10.1016/S0379-6779(98)01432-5.

[41] K. Joseph, S. Varghese, G. Kalaprasad, S. Thomas, L. Prasannakumari, P. Koshy, C. Pavithran, Influence of interfacial adhesion on the mechanical properties and fracture behaviour of short sisal fibre reinforced polymer composites, Eur. Polym. J. 32 (1996) 1243–1250. https://doi.org/10.1016/S0014-3057(96)00051-1.

[42] M. Naebe, J. Wang, A. Amini, H. Khayyam, N. Hameed, L.H. Li, Y. Chen, B. Fox, Mechanical Property and Structure of Covalent Functionalised Graphene/Epoxy Nanocomposites, Sci. Rep. 4 (2014) 1–7. https://doi.org/10.1038/srep04375.

[43] J.F. Feller, I. Linossier, Y. Grohens, Conductive polymer composites: Comparative study of poly(ester)-short carbon fibres and poly(epoxy)-short carbon

fibres mechanical and electrical properties, Mater. Lett. 57 (2002) 64–71.
https://doi.org/10.1016/S0167-577X(02)00700-0.

[44] T. Yokozeki, T. Goto, T. Takahashi, D. Qian, S. Itou, Y. Hirano, Y. Ishida, M. Ishibashi, T. Ogasawara, Development and characterization of CFRP using a polyaniline-based conductive thermoset matrix, Compos. Sci. Technol. 117 (2015) 277–281. https://doi.org/10.1016/j.compscitech.2015.06.016.

[45] O. Zabihi, Preparation and characterization of toughened composites of epoxy/poly(3,4-ethylenedioxythiophene) nanotube: Thermal, mechanical and electrical properties, Compos. Part B Eng. 45 (2013) 1480–1485. https://doi.org/10.1016/j.compositesb.2012.09.029.

[46] B. Redondo-Foj, P. Ortiz-Serna, M. Carsí, M.J. Sanchis, M. Culebras, C.M. Gómez, A. Cantarero, Electrical conductivity properties of expanded graphite-polycarbonatediol polyurethane composites, Polym. Int. 64 (2015) 284–292. https://doi.org/10.1002/pi.4788.

[47] M. Verma, S.S. Chauhan, S.K. Dhawan, V. Choudhary, Graphene nanoplatelets/carbon nanotubes/polyurethane composites as efficient shield against electromagnetic polluting radiations, Compos. Part B Eng. 120 (2017) 118–127. https://doi.org/10.1016/j.compositesb.2017.03.068.

[48] N. Yamamoto, R. Guzman de Villoria, B.L. Wardle, Electrical and thermal property enhancement of fiber-reinforced polymer laminate composites through controlled implementation of multi-walled carbon nanotubes, Compos. Sci. Technol. 72 (2012) 2009–2015. https://doi.org/10.1016/j.compscitech.2012.09.006.

[49] X. Zhao, Y. Li, J. Wang, Z. Ouyang, J. Li, G. Wei, Z. Su, Interactive oxidation-reduction reaction for the in situ synthesis of graphene-phenol formaldehyde composites with enhanced properties, ACS Appl. Mater. Interfaces. 6 (2014) 4254–4263. https://doi.org/10.1021/am405983a.

[50] E. Ivanov, R. Kotsilkova, E. Krusteva, E. Logakis, A. Kyritsis, P. Pissis, C. Silvestre, D. Duraccio, M. Pezzuto, Effects of processing conditions on rheological, thermal, and electrical properties of multiwall carbon nanotube/epoxy resin composites, J. Polym. Sci. Part B Polym. Phys. 49 (2011) 431–442. https://doi.org/10.1002/polb.22199.

[51] F.Y. Yuan, H. Bin Zhang, X. Li, H.L. Ma, X.Z. Li, Z.Z. Yu, In situ chemical reduction and functionalization of graphene oxide for electrically conductive

phenol formaldehyde composites, Carbon N. Y. 68 (2014) 653–661. https://doi.org/10.1016/j.carbon.2013.11.046.

[52] Q. Li, L. Chen, X. Li, J. Zhang, X. Zhang, K. Zheng, F. Fang, H. Zhou, X. Tian, Effect of multi-walled carbon nanotubes on mechanical, thermal and electrical properties of phenolic foam via in-situ polymerization, Compos. Part A Appl. Sci. Manuf. 82 (2016) 214–225. https://doi.org/10.1016/j.compositesa.2015.11.014.

[53] I.S. Gunes, G.A. Jimenez, S.C. Jana, Carbonaceous fillers for shape memory actuation of polyurethane composites by resistive heating, Carbon N. Y. 47 (2009) 981–997. https://doi.org/10.1016/j.carbon.2008.11.053.

[54] N.J. Yi Wang, Jinhong Yu, Wen Dai, Yingze Song, Dong Wang, Liming Zeng, Enhanced Thermal and Electrical Properties of Epoxy Composites Reinforced With Graphene Nanoplatelets, Polym. Compos. 36 (2015) 556–565.

[55] S. Iqbal, S. Ahmad, Recent development in hybrid conducting polymers: Synthesis, applications and future prospects, J. Ind. Eng. Chem. 60 (2017) 53–84. https://doi.org/10.1016/j.jiec.2017.09.038.

[56] N.M. Barkoula, B. Alcock, N.O. Cabrera, T. Peijs, Fatigue properties of highly oriented polypropylene tapes and all-polypropylene composites, Polym. Polym. Compos. 16 (2008) 101–113. https://doi.org/10.1002/pc.

[57] Z. Liu, G. Bai, Y. Huang, Y. Ma, F. Du, F. Li, T. Guo, Y. Chen, Reflection and absorption contributions to the electromagnetic interference shielding of single-walled carbon nanotube/polyurethane composites, Carbon N. Y. 45 (2007) 821–827. https://doi.org/10.1016/j.carbon.2006.11.020.

[58] Y. Huang, N. Li, Y. Ma, F. Du, F. Li, X. He, X. Lin, H. Gao, Y. Chen, The influence of single-walled carbon nanotube structure on the electromagnetic interference shielding efficiency of its epoxy composites, Carbon N. Y. 45 (2007) 1614–1621. https://doi.org/10.1016/j.carbon.2007.04.016.

[59] J. Liang, Y. Wang, Y. Huang, Y. Ma, Z. Liu, J. Cai, C. Zhang, H. Gao, Y. Chen, Electromagnetic interference shielding of graphene/epoxy composites, Carbon N. Y. 47 (2009) 922–925. https://doi.org/10.1016/j.carbon.2008.12.038.

[60] A.S. Hoang, Electrical conductivity and electromagnetic interference shielding characteristics of multiwalled carbon nanotube filled polyurethane composite films, Adv. Nat. Sci. Nanosci. Nanotechnol. 2 (2011). https://doi.org/10.1088/2043-6262/2/2/025007.

[61]   T.K. Gupta, B.P. Singh, S. Teotia, V. Katyal, S.R. Dhakate, R.B. Mathur, Designing of multiwalled carbon nanotubes reinforced polyurethane composites as electromagnetic interference shielding materials, J. Polym. Res. 20 (2013) 32–35. https://doi.org/10.1007/s10965-013-0169-6.

[62]   M. Verma, P. Verma, S.K. Dhawan, V. Choudhary, Tailored graphene based polyurethane composites for efficient electrostatic dissipation and electromagnetic interference shielding applications, RSC Adv. 5 (2015) 97349–97358. https://doi.org/10.1039/C5RA17276D.

[63]   J. Alam, U. Riaz, S. Ahmad, High performance corrosion resistant polyaniline/alkyd ecofriendly coatings, Curr. Appl. Phys. 9 (2009) 80–86. https://doi.org/10.1016/j.cap.2007.11.015.

[64]   A. Mostafaei, F. Nasirpouri, Epoxy/polyaniline–ZnO nanorods hybrid nanocomposite coatings: Synthesis, characterization and corrosion protection performance of conducting paints, Prog. Org. Coatings. 77 (2014) 146–159. https://doi.org/10.1016/j.porgcoat.2013.08.015.

[65]   J. Zhu, S. Wei, I.Y. Lee, S. Park, J. Willis, N. Haldolaarachchige, D.P. Young, Z. Luo, Z. Guo, Silica stabilized iron particles toward anti-corrosion magnetic polyurethane nanocomposites, RSC Adv. 2 (2012) 1136–1143. https://doi.org/10.1039/C1RA00758K.

[66]   H. Wei, D. Ding, S. Wei, Z. Guo, Anticorrosive conductive polyurethane multiwalled carbon nanotube nanocomposites, J. Mater. Chem. A. 1 (2013) 10805. https://doi.org/10.1039/c3ta11966a.

[67]   Z. Zhang, W. Zhang, D. Li, Y. Sun, Z. Wang, C. Hou, L. Chen, Y. Cao, Y. Liu, Mechanical and anticorrosive properties of graphene/epoxy resin composites coating prepared by in-situ method, Int. J. Mol. Sci. 16 (2015) 2239–2251. https://doi.org/10.3390/ijms16012239.

[68]   A.G. and S.A. Neha Kanwar Rawat, Influence of microwave irradiation on various properties of nanopolythiophene and their anticorrosive nanocomposite coatings, RSC Adv. (2014) 1–12. https://doi.org/10.1039/c2ra21135a.

[69]   Q. Meng, J. Hu, A review of shape memory polymer composites and blends, Compos. Part A Appl. Sci. Manuf. 40 (2009) 1661–1672. https://doi.org/10.1016/j.compositesa.2009.08.011.

[70]   J.W.C. Nanda Gopal Sahoo, Yong Chae Jung, Nam Seo Goo, Conducting Shape Memory Polyurethane-Polypyrrole Composites for an Electroactive Actuator,

Macromol. Mater. Eng. (2005) 1049–1055.
https://doi.org/10.1002/mame.200500211.

[71] S.M. Oh, K.M. Oh, T.D. Dao, H. Il Lee, H.M. Jeong, B.K. Kim, The modification of graphene with alcohols and its use in shape memory polyurethane composites, Polym. Int. 62 (2013) 54–63. https://doi.org/10.1002/pi.4366.

[72] M. Raja, S.H. Ryu, A.M. Shanmugharaj, Thermal , mechanical and electroactive shape memory properties of polyurethane (PU)/ poly (lactic acid) (PLA)/ CNT nanocomposites, Eur. Polym. J. 49 (2013) 3492–3500.
https://doi.org/10.1016/j.eurpolymj.2013.08.009.

[73] H. Lu, Y. Yao, W. Min, J. Leng, D. Hui, Composites : Part B Significantly improving infrared light-induced shape recovery behavior of shape memory polymeric nanocomposite via a synergistic effect of carbon nanotube and boron nitride, Compos. PART B. 62 (2014) 256–261.
https://doi.org/10.1016/j.compositesb.2014.03.007.

[74] X. Lu, G. Xu, Thermally Conductive Polymer Composites for Electronic Packaging, J. Appl. Polym. Sci. 65 (1997) 2733–2738.
https://doi.org/10.1002/(SICI)1097-4628(19970926)65:13%3C2733::AID-APP15%3E3.0.CO;2-Y.

[75] J. Lu, K.S. Moon, B.K. Kim, C.P. Wong, High dielectric constant polyaniline/epoxy composites via in situ polymerization for embedded capacitor applications, Polymer (Guildf). 48 (2007) 1510–1516.
https://doi.org/10.1016/j.polymer.2007.01.057.

[76] L. Du, S.C. Jana, Highly conductive epoxy/graphite composites for bipolar plates in proton exchange membrane fuel cells, J. Power Sources. 172 (2007) 734–741.
https://doi.org/10.1016/j.jpowsour.2007.05.088.

[77] S. Pramanik, J. Hazarika, A. Kumar, N. Karak, Castor Oil Based Hyperbranched Poly (ester amide)/ Polyaniline Nano fi ber Nanocomposites as Antistatic Materials, Ind. Eng. Chem. Res. 52 (2013) 5700–5707.

[78] E.C. Cho, J.H. Huang, C.P. Li, C.W. Chang-Jian, K.C. Lee, Y.S. Hsiao, J.H. Huang, Graphene-based thermoplastic composites and their application for LED thermal management, Carbon N. Y. 102 (2016) 66–73.
https://doi.org/10.1016/j.carbon.2016.01.097.

[79] Y. Zhang, A.A. Broekhuis, F. Picchioni, Thermally self-healing polymeric materials: The next step to recycling thermoset polymers?, Macromolecules. 42 (2009) 1906–1912. https://doi.org/10.1021/ma8027672.

[80] M.M. Obadia, B.P. Mudraboyina, A. Serghei, D. Montarnal, E. Drockenmuller, Reprocessing and Recycling of Highly Cross-Linked Ion-Conducting Networks through Transalkylation Exchanges of C-N Bonds, J. Am. Chem. Soc. 137 (2015) 6078–6083. https://doi.org/10.1021/jacs.5b02653.

[81] J. Palmer, O.R. Ghita, L. Savage, K.E. Evans, Successful closed-loop recycling of thermoset composites, Compos. Part A Appl. Sci. Manuf. 40 (2009) 490–498. https://doi.org/10.1016/j.compositesa.2009.02.002.

[82] S.J. Pickering, Recycling Technologies For Thermoset Composite Materials, Adv. Polym. Compos. Struct. Appl. Constr. ACIC 2004. 37 (2004) 392–399. https://doi.org/10.1016/B978-1-85573-736-5.50044-3.

[83] C.E. Kouparitsas, C.N. Kartalis, P.C. Varelidis, C.J. Tsenoglou, C.D. Papaspyrides, Recycling of the fibrous fraction of reinforced thermoset composites, Polym. Compos. 23 (2002) 682–689. https://doi.org/10.1002/pc.10468.

[84] A. Torres, I. De Marco, B.M. Caballero, M.F. Laresgoiti, J.A. Legarreta, M.A. Cabrero, A. González, M.J. Chomón, K. Gondra, Recycling by pyrolysis of thermoset composites: characteristics of the liquid and gaseous fuels obtained, Fuel. 79 (2000) 897–902. https://doi.org/10.1016/S0016-2361(99)00220-3.

[85] S.J. Pickering, R.M. Kelly, J.R. Kennerley, C.D. Rudd, N.J. Fenwick, A fluidised-bed process for the recovery of glass fibres from scrap thermoset composites, Compos. Sci. Technol. 60 (2000) 509–523. https://doi.org/10.1016/S0266-3538(99)00154-2.

[86] P. Xu, J. Li, J. Ding, Chemical recycling of carbon fibre/epoxy composites in a mixed solution of peroxide hydrogen and N,N-dimethylformamide, Compos. Sci. Technol. 82 (2013) 54–59. https://doi.org/10.1016/j.compscitech.2013.04.002.

Chapter 8

Waterborne Thermosetting Polyurethane Composites

Sashivinay Kumar Gaddam [a], Pothu Ramyakrishna [b], Aditya Saran [c], Rajender Boddula [d,*]

[a] Polymers & Functional Materials Division, Indian Institute of Chemical Technology, Uppal Road, Tarnaka, Hyderabad-500007, Telangana state, India.

[b] College of Chemistry and Chemical Engineering, Hunan University, Changsha 410082, P.R. China

[c] Department of Microbiology, Marwadi University, Rajkot, Gujarat-360003.

[d] CAS Key Laboratory of Nanosystem and Hierarchical Fabrication, National Center for Nanoscience and Technology, Beijing 100190, PR China

research.raaj@gmail.com

Abstract

This chapter provides a background of waterborne thermosetting polyurethane composite synthesis, physico-chemical properties and their applications as coatings, adhesives and printing inks. The reinforcement of waterborne polyurethane dispersion (PUD) matrix with different inorganic nanofillers develops cross-linking networks and leads to their high modulus, strength, durability, and resistance towards weather and chemical attacks. Thereby, a brief survey on different functionalization methods of nanofillers for the development of advanced waterborne PUD thermosetting composites with specific properties, including shape memory, fire retardancy, corrosion resistance and antimicrobial activity, is given.

Keywords

Waterborne Polyurethanes, Thermoset, Composites, Metal/Metaloxide Nanoparticles, Carbohydrates, Clays, Carbon Nanomaterials

Contents

8.1 Introduction

Based on their thermal behavior, polymers are mainly classified as thermoplastics and thermosets. Thermoplastics are linear polymers and upon heating beyond their melting point or glass transition temperature (T_g), polymer chains are free to move and the polymer can flow. On the other hand, thermosets are crosslinked polymers and upon heating, they retain their polymer network structure and thus polymer cannot flow and cannot be reshaped after polymerization. Therefore, a thermoset can be defined as polymer that forms a three-dimensional networked structure through covalent chemical bonds that penetrate the whole polymer backbone [1]. Thermosetting polymers play an important role in many industrial applications due to their high strength, durability, thermal and chemical resistances as provided by their high crosslinked network structure [2–4]. Typical examples of thermosetting polymers are polyurethanes (PU), polyesters, epoxies, formaldehydes and vinyl esters [5]. There is a large demand for tailoring the properties of thermosetting polymers in order to improve their high-performance industrial applications. Therefore, a variety of fillers are often introduced into the polymer matrix to develop thermosetting composite materials. These fillers are often nano-particles (nano clay, metal oxides, carbon materials and cellulose nanocrystals) and strongly influence the properties of composites at very low volume fractions.

Since its inception, polyurethane (PU) has found extensive applications as coatings, adhesives, sealants, inks, paints, and foams because of its excellent abrasion resistance, chemical resistance, high toughness and low film forming temperatures [6, 7]. Generally, polyurethanes represent an important class of thermoplastics and thermosets usually contain a significant number of urethane linkages and are prepared via polyaddition reaction between the petroleum-based polyol and diisocyanates [8, 9]. Thermosetting PUs are widely used in metallurgy, automobiles, mill run, shoes, and printing industries [10]. In order to meet the rising demands of their applications, PU thermoset composites are prepared by filling of PU matrix with inorganic functional fillers. Recently, growing threats from environmental pollution, depletion of the world crude oil feedstocks and gradual increment in crude oil prices have forced the PU industry to develop sustainable materials [11, 12]. In sustainable thermosetting materials, PUs are developed using bio-based polyols those derived from renewable resources. However, recent advances in PU manufacturing suggest that introduction of waterborne polyurethane dispersions (PUDs) is more advantageous than their solvent-borne counter parts in terms of nontoxicity, nonflammability, and low or no content of volatile organic compounds (VOC) such as solvents [13, 14].

Waterborne PUDs have many applications as coatings, adhesives, primers, pint additives, deformers, associate thickeners, pigment paints, bio-elastomers, etc. [15, 16]. However, most of the waterborne PUDs are developed from linear polyols and are considered as thermoplastic polymers. These thermoplastic PUDs have poor thermo-mechanical properties, water, and solvent resistance as compared to cross-linked solvent borne PUs [17]. Hence, there are many methods developed for the purpose of improving the properties of PUDs and most of these methods involved hybridizing the polymer with inorganic fillers such as inorganic oxides (silica, titania, ZnO, etc), metal nanoparticles (Ag and Au), carbon materials (graphene, GO, CNTs), clay and carbohydrate materials (cellulose, starch). By all these methods, cross-linked networks should be developed in PU matrix and thus obtained PUDs considered as waterborne thermosetting PU composites.

8.2 PUD thermosetting composites

8.2.1 Inorganic oxide based PUD thermosetting composites

8.2.1.1 Silica-based PUD thermosetting composites

Silica nanoparticles can be introduced into the polymer systems by blending, *in-situ* polymerization or by a sol-gel process [18]. The sol-gel process is an extensively used method, involves the polycondensation of silanol groups of silica (nano & fumed) or aminosilylated chain extenders, which develops the siloxane cross-linking networks in the polymer matrix [19, 20]. The inclusion of silica nanoparticles into the PUD matrix enhances the performance of polymer for hydrophobic surface coatings with high flexibility, improved thermo-mechanical, optical and electrical properties. Furthermore, PU-silica hybrid materials also exhibit improved water, chemical and corrosion resistance and antifouling performance with low film formation time [21, 22].

Jeon et al. developed waterborne polyurethane-silica hybrids by a sol-gel process [23]. The PU prepolymer was prepared using PTMG, IPDI, and DMPA and capped at one end with APTES. After neutralization, NCO terminated PU ionomer was mixed with nanosilica (Aerosil 200) in different ratios (2, 5 and 10 wt%). The NCO terminated PU ionomer was dispersed in water and further chain extended with DETA. The sol-gel process took place in dispersion step. Ethoxy groups of APTES undergo hydrolysis and gives silanol groups which can subsequently undergo polycondensation with silanol groups on the nanosilica to develop a three-dimensional siloxane cross-linking networks in PU matrix. It was found that, with an increase in nanosilica content, thermo-mechanical properties and resistance to water were enhanced without impairing the film transparency.

Marine biofouling is a serious issue to all immersed substrates in sea water such as ship hulls, pipe lines, offshore platforms, and bridges. Rahman et al. developed a waterborne polyurethane-silane protective coating for marine fouling by a sol-gel process [24]. The prepolymer was developed from PTMG, H_{12}MDI, and DMPA and after neutralization; prepolymer was capped at one end with TMSiP-EDA, which can act as both chain extender and crosslinker. The NCO terminated PU ionomer was dispersed in water and further chain extended with EDA. The methoxy groups of TMSiP-EDA were hydrolyzed and condensed to develop siloxane cross-linking networks in PU matrix. The thermal stability, tensile strength and Young's modulus of PUD-silane coatings were enhanced with an increase in TMSiP-EDA content from 1.75 to 5.17 mole% and they exhibited superior anti marine fouling properties than conventional PUD coatings.

It is a challenge to achieve satisfactory adhesion of a printing ink on the surface of low-energy plastic films used for food packaging [25]. The introduction of hydrophobic monomers into polymer chains enhances the adhesion to the low-energy substrate surface [26]. The non-polar silane chain extenders exhibit surface activity and migrate to the surface of PUDs, offer low surface energy with high hydrophobicity [27, 28], and hence improve the wettability and adhesion of PUDs on the substrates. Lei et al. developed waterborne PUDs for aqueous printing ink applications, using a diamine silane chain extender (AEAPTMS) in different ratios (1, 2, 3 & 4 wt%) [29]. The increase of

AEAPTMS content enhances the hydrophobicity, tensile strength, glass transition temperature (T_g), water resistance and thermal stability of PUD films. The PUD showed good adhesion on OPP film (Fig. 8.1) with 2% AEAPTMS loading and it was also observed that having two amine groups AEAPTMS could improve the properties of PUDs greatly than APTES which had single amine group.

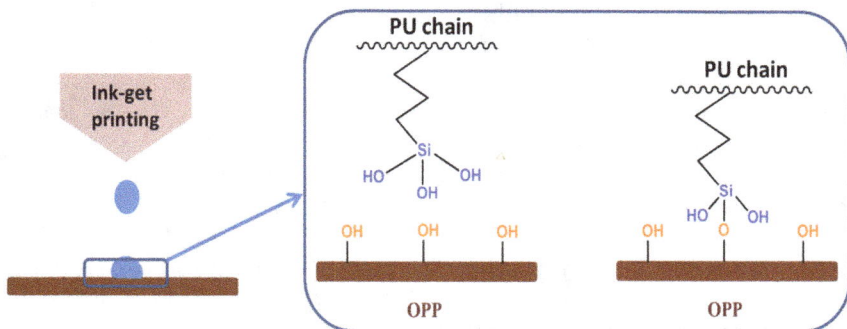

Figure 8.1 Schematic representation of adhesion of PUDs on OPP films.

Apart from the incorporation of silica nanoparticles, some of the researchers attempted to introduce more cross-linking networks using different synthetic approaches to develop PUD-silane thermosets. Dendrimers and hyperbranched polymers have attracted considerable attention due to their remarkable benefits of low solution viscosity, excellent miscibility and having a large number of functional end groups [30-32]. Introduction of these dendrimers or hyperbranched polymers into PUD-silane nanocomposites enhance the miscibility of nanosilica, which can fill in the cavities of polymer and this can effectively prevent the aggregation of nanosilica [33]. Jena et al. developed waterborne polyurethane-silica hybrid coatings through a sol-gel process [34]. A hyperbranched polyester polyol was prepared by single-step melt condensation reaction from DMPA and glycerol and was used to synthesize waterborne PUD with H_{12}MDI and DMPA. APTES used as silane chain extender and TEOS solution was used as silica sol, added to the PUD in different concentrations (2, 5 and 7 wt%). The increased siloxane network density enhanced the glass transition temperature (T_g), thermal stability, hydrophobicity, water resistance and abrasion resistance of PUD films.

Recently, "Click" chemistry becomes a powerful tool in PU synthesis because of its advantages such as high selectivity, quantitative yield, short reaction time, mild reaction conditions and absence of by-products [35, 36]. Click chemistry can combine the particular groups or compounds together by triazole rings. Sun et al. introduced Click chemistry mechanism in PUD-silane composite synthesis [37]. The functionalized nano silica with silylated azide coupling agent was added to the alkyne-functionalized waterborne PUD in different concentrations (1, 3, 5 and 10 wt%) in the presence of Cu(I) catalyst. The triazole rings were developed by the reaction of azide and alkyne groups and the enhanced thermal stability, hardness, hydrophobicity and weather resistance were reported as increased in silica content.

More recently, "Thiol-ene" coupling (TEC) chemistry has emerged as a green methodology in plant oil based PU synthesis with targeted properties [38]. The double bonds of plant oils can be converted into thiol functionality through TEC with functional thiols [39, 40]. In their study, Fu et al. fabricated bio-based waterborne PU-silica composites through a sol-gel process [41]. This group developed castor oil based thiol monomers including carboxylated castor oil (MACO) and silylated castor oil (MSCO) through TEC chemistry. By varying the MSCO content, different PUD-silane composites were developed with different silica content (1.7, 2, 2.3, 2.5 and 2.7 wt%) using MACO as an internal emulsifier in place of DMPA. With increased silica content, the thermal stability and surface properties of PUD films were enhanced with decrease in their transparency.

UV-initiated photopolymerization has also attracted much attention in PUD synthesis to develop the cross-linked networks. UV-curable chemistry offers excellent characteristic properties including independent reaction temperature, controllable irradiation wavelength, irradiation time, and the intensity of UV-light [42]. Moreover, this route is eco-friendly and energy efficient which meets with the requirement of "green" chemistry. Zhang et al. developed UV-curable waterborne PU-silica composites using vinyl functionalized TEOS sol [43]. The functionalization of TEOS sol was achieved by γ-Methacryloxypropyltrimethoxysilane (KH-570). Different quantities of vinyl functionalized silica (TEOS sol; 5, 10, 15, 17.5 and 20 wt%) were mixed with HEMA end capped PUD, and the final composite films were obtained by UV curing of latex films using a photoinitiator (Irgacure 2959). The incorporation of functionalized silica improved the hardness, thermal stability and water resistance (Fig. 8.2) of composite films without disturbing their transparency.

Shape memory polymers respond to temperature have traditionally been used as temperature sensors, actuators and have also been used in medical applications including sutures, catheters and stents [44-46]. Lee et al. studied the effect of weight ratios of vinyl modified nano silica (0.5, 1, 2 & 5 wt%) on the shape memory properties of UV-curable waterborne polyurethane-silica nanocomposites [47]. The vinyl terminated PU prepolymer was reacted with nano silica modified with allyl isocyanate and dispersion films were cured under UV light in the presence of a photoinitiator. The incorporated vinyl modified nano silica particles significantly enhanced the thermo-mechanical properties and 99% shape recovery and hysteresis to cyclic loadings were obtained with 1% loading of the modified silica.

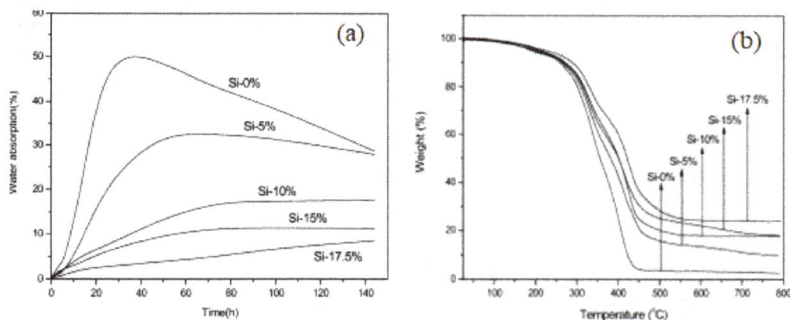

Figure 8.2 Improved water resistance (a) and thermal stability (b) of PUD films with different silica contents.

Polyimide is an important class of heterocyclic polymers with excellent heat resistance and superior mechanical and durable properties [48, 49]. So, the thermal properties of PUD films can be improved by incorporation of imide groups into the hard segment of PU matrix. Jung et al. developed waterborne PU-imide/silica hybrids using vinyltrimethoxysilylated (VTMS) silica particles through UV-curing mechanism [50]. Using the PMDA, imide groups were introduced into the prepolymer and the vinyl terminated PU-imide prepolymer was obtained using HEA as an end-capping agent. The UV cure was made between vinyl terminated PU-imide prepolymer and vinyltrimethoxysilylated-silica particles in the presence of a photoinitiator (Darocur 1173). The extended siloxane cross-linking networks were obtained by different concentration of fumed silica (0.4, 0.8 and 1.2 wt%) and it was observed that waterborne PU-imide/silica hybrids exhibited remarkably high hardness, mechanical properties, dynamic mechanical properties and high thermal stability at 1.2 wt% silica loadings. These enhanced properties are combinatorial properties of the siloxane cross-linking networks and imide functionalities in PU matrix.

Two-component waterborne PUD (2K-PUD) system offers attractive applications over normal, one-package PUDs, due to their high cross-linking density [51, 52]. These 2K-PUDs have high performance in coating industry because of their excellent adhesion, chemical resistance, abrasion resistance and easy curing at room temperature [53]. Generally, 2K-PUDs are developed by mixing of the hydroxyl-functional waterborne PU dispersed polyol (PUDp) acts as binder and water dispersible polyisocyanate hardener acts as external cross-linking agent [52, 53]. During the mixing process, the NCO groups of polyisocyanate react with OH groups of PUDp in aqueous medium and develops the 2K-PUD with high cross-linking network structure. Yue et al. developed a series 2K-PUD-silane composites by mixing of PUDp having TMP and APTES as blocking agents and commercially available hydrophilically modified polyisocyanate (Bayhydur 304) [54]. The group has studied the effect of APTES content in different concentrations (1, 2, 3 and 4 wt%) on the properties of 2K-PUD-silane composites and they found enhanced thermal stability, tensile strength and solvent resistance with an increase in APTES content.

8.2.1.2 Titania (TiO$_2$) based PUD thermosetting composites

Titanium dioxide (TiO$_2$) has emerged as a promising nanomaterial to enhance the properties of polymer composites because of its high photostability, chemical inertness and eco-friendly nature [55, 56]. Nowadays, nano-TiO$_2$ as inorganic filler has generated great interest for a variety of applications, including coatings and membranes with improved thermal [57], mechanical [58], electrical [59], UV resistance [60], self-cleaning [61, 62], and antimicrobial properties [63, 64].

The effect of different crystalline forms of nano-TiO$_2$ on the structure and properties of waterborne PUDs were reported by Li et al., where the nanoparticles were prepared by a modified chemical method [65]. The prepared anatase and rutile TiO$_2$ were characterized by XRD, SEM, TEM and UV-visible spectrophotometer and were introduced into the commercially available waterborne PUD in different concentrations (0.25, 0.50, 0.75 and 1.0 wt% respectively). The results are indicating that the incorporation of anatase TiO$_2$ enhanced the UV protection ability, photocatalytic ability and antifungal capacity of resulting PUD films and the values are highest at 1.0 wt% loading of TiO$_2$. On the other side, rutile TiO$_2$ incorporated PUD hybrid films exhibited high tensile strength at 0.75 wt% and this may be due to the special asymmetric acicular sharp of rutile TiO$_2.$

However, due to their small size and high specific surface area, nano-TiO_2 particles tend to segregate when dispersed in the polymer matrix and the advantages of TiO_2 do not exhibit to the full [66]. To intensify the dispersion of polymers, TiO_2 need some chemical modifications to develop surface functionalized nano-TiO_2 which bond covalently to the polymer end groups. Different strategies like grafting approach and sol-gel method have been developed to reduce the nano-TiO_2 agglomeration in polymer composites [67]. Deng et al. fabricated dopamine modified TiO_2 (DA-TiO_2) using grafting approach (Fig. 8.3) and introduced as filler into the waterborne PUD system with different concentrations (0.2, 0.5, 1.0 and 2.0 w%) [68]. The grafting of dopamine on the surface of TiO_2 was confirmed by FTIR and UV-vis spectrometers. In contrast with pure PUD, the PUD/ DA-TiO_2 films showed good thermal stability and water resistance at 1.0 wt% DA-TiO_2 loading.

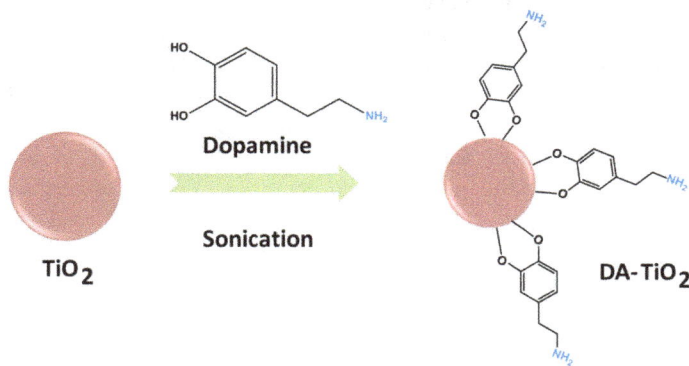

Figure 8.3 Surface modification of TiO_2 with DA through grafting approach.

Wu et al. synthesized waterborne PUD/TiO_2 hybrid composites using a sol-gel process on the basis of tetrabutyl titanate (TBT) and 3-glycidyloxypropyl trimethoxysilane (GLYMO) as coupling agent [69]. A homogeneous solution was obtained from the GLYMO and PUD which was synthesized from IPDI, polyether polyol, DMPA, and EDA. An inorganic hydrolyzed TBT solution was carefully added to the aforementioned solution in different concentrations (0.3, 0.5, 1.0, 1.5 and 3.0 wt%). The Ti–O bonds were developed among TiO_2 particles and silanol groups of GLYMO through the sol-gel process. The homogeneous dispersion of TiO_2 particles in PUD system was confirmed by SEM, TEM and AFM results. As compared to the pure PUD, the PUD/TiO_2 hybrid composites exhibited improved thermo-mechanical properties.

8.2.1.3 Zinc oxide (ZnO) based PUD thermosetting composites

Zinc oxide (ZnO) nanoparticles have attracted much attention as future materials due to their promising properties including UV-shielding, antimicrobial activity, high stiffness and hardness, low coefficient of thermal expansion and high thermal conductivity [70]. All these characteristic properties made ZnO fit to develop new composite materials for a broad range of applications. ZnO nanoparticles can be prepared by the sol-gel method, precipitation, hydrothermal synthesis and spray pyrolysis.

Awad et al. studied the free volumes, glass transition temperatures (T_g) and cross-links of a PUD-ZnO system, as a function of ZnO concentration [71]. PUD-ZnO composites were prepared by mixing the commercially available PUD (Bayhydrol,) with different loadings (1.0, 2.0 and 5.0 wt%) of commercially available waterborne ZnO dispersions (NANOBYK). The free volume properties were measured using positron annihilation lifetime spectroscopy (PALS) and these free volumes were decreased with increase in ZnO loadings (Fig. 8.4a). The composite films also exhibited two T_g values for two segmental domains of PUD (Fig. 8.4b); the lower T_g (~ -52°C) is due to soft aliphatic chains and high T_g (~ 108 °C) is due to polar hard microdomains. The loading of ZnO filler improves both T_g values and this increase mainly attributed to interfacial interactions between PU matrix and ZnO nanoparticles. The experimental results indicating that the relationship between the free volume (V_h) and physical cross-link density (X_c) is found to follow an exponential function.

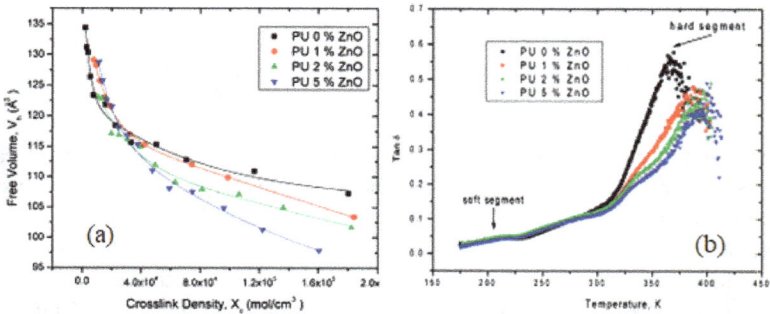

Figure 8.4 Relation between free volume and cross-link density (a) and tan δ curves (b) of PUD/ZnO nanocomposites.

Christopher et al. examined the effect of surface functionalized zin oxide with oleic acid (OA-ZnO) on properties of PUD-ZnO nanocomposites [72]. The ZnO nanoparticles were fabricated by alkaline treatment of zinc acetate and characterized by FESEM, XRD and FTIR spectroscopy. These nanoparticles were functionalized with oleic acid and the obtained OA-ZnO were loaded into a commercially available PUD (PU-687) with different dosages (0.1, 0.2 and 0.3 wt%). In contrast to the PUD composites with unmodified ZnO, the PUD with OA-ZnO exhibited improved hydrophobicity, corrosion resistance and these properties increased with increase in OA-ZnO content.

Ma et al. studied the effect of flower-like ZnO nanowhiskers (f-ZnO) on the thermo-mechanical and antibacterial properties of waterborne PUDs with different f-ZnO content (0 to 4.0 wt%) [73]. The f-ZnO was prepared via a simple hydrothermal process and functionalized with γ-aminopropyltriethoxysilane. The thermal stability, mechanical strength and water resistance properties of PUDs were strongly influenced by the loadings of functionalized f-ZnO (f-ZnO-NH$_2$). It was observed that the mechanical strength was maximum at 1.0 wt% of f-ZnO content and the water resistance and antibacterial activities were enhanced with an increase in f-ZnO content and the results were high at the loading level of 4.0 wt% f-ZnO. However, the thermal stability was initially decreased significantly and then leveled off with the increasing f-ZnO content which was completely different from other PUD composite systems reported.

Waterborne PUD/ZnAl-layered double hydroxide/ZnO composites were fabricated by Zhang et al. and the thermo-mechanical; water swelling and antibacterial properties were studied [74]. The ZnAl-layered double hydroxide nanosheets (ZnAl-LDHs) were synthesized by previously reported co-precipitation method and then ZnO nanoparticles grew onto the prepared sheets. The hydroxyl groups on the surface of ZnAl-LDHs/ZnO were reacted with NCO groups of IPDI and the obtained functionalized ZnAl-LDHs/ZnO (ZnAl-LDHs/ZnO-g-NCO) was incorporated into the PUD system with different content (0.3, 0.5, 1.0, 1.5, 2.0, 2.5 and 3.0 wt%). The composite film showed good thermal stability, mechanical strength and water resistance with an increase of functionalized ZnAl-LDHs/ZnO content up to the optimum content (1 wt%) and then declined. However, the composites showed excellent antimicrobial activity at all the concentrations of functionalized ZnAl-LDHs/ZnO and this activity can be attributed to the damage of bacterial cell membranes by oxyradicals produced from ZnO.

8.2.1.4 Other inorganic oxide-based PUD thermosetting composites

Some researchers also attempted to develop waterborne PUDs using different inorganic oxides as nanofillers and studied their functional effects on resulting PUD composites.

Thermoset Composite   Materials Research Forum LLC
Materials Research Foundations **38** (2018)   doi: http://dx.doi.org/10.21741/9781945291876

Recently worldwide material research mainly focused on the development of technology for environmental protection and energy conservation. In this sense, transparent heat insulating coatings have attracted great attention for their wide applications especially in construction technology [75, 76]. Antimony doped tin oxide (ATO) is a well-known optically transparent, infrared light insulating and electrically conducting oxide [77, 78]. Hence, Dai et al. developed eco friendly, transparent and heat insulating PUD-ATO coatings by blending different concentrations (1.0, 3.0, 5.0 and 7.0 wt%) of ATO into PUDs [79]. The TEM results indicating that ATO nanoparticles were homogeneously dispersed in PUD and the glass coated with PUD-ATO coating exhibited good heat insulating effect than uncoated glass. Furthermore, the glass transition temperature (T_g), mechanical strength and thermal stability of composite films were superior to conventional unmodified PUD and these properties gradually increased with increase in ATO content.

Alumina (Al_2O_3) is a harder material and used as a filler to improve the scratch and abrasion resistance of coatings. Dhoke et al. studied the effect of nanoalumina on electrochemical and mechanical properties of PUD coatings at different loading levels (0.1 and 1.0 wt%) [80]. The electrochemical performance of composite coatings was examined by exposing the coated mild steel panel to salt-spray, humidity, and accelerated UV weathering. The results showed an improvement in corrosion, UV and scratch resistance at 0.1 wt% loading of nanoalumina. The composite coatings also showed good pencil hardness at 1.0 wt% loading and all these results were higher than pure PUD coatings.

Fe_3O_4 is a well-known magnetic material used in electromagnetic screening devices, magnetic sensing, biotechnology, biomedicine and biomedical applications [81, 82]. Santos et al. synthesized PUD/Fe_3O_4 magnetic nanocomposites using different concentrations (0.5, 5.0, 10, 30 and 40 wt%) of Fe_3O_4-nano-talc gel suspension (33% solid content) [83]. The XRD and SEM results suggested that Fe_3O_4-talc-nanofillers were well-dispersed in PUD matrix event at high filler content and this can be attributed to hydrogen bonding interactions of filler with water. All composite films exhibited ferromagnetic behaviour below Curie temperature (~ 120 K) and a super-paramagnetic behaviour above this temperature. The surface roughness, mechanical strength, T_g, and storage modulus values of composite films increased with increase in filler content and the values are greater than pure PUD.

Oleic acid (OA) functionalized nano-Fe_3O_4 particles synthesized by Zhang et al. and the PUD-Fe_3O_4 nanocomposite films were developed by changing the concentration of OA-Fe_3O_4 nanoparticles (0.5, 1.0, 1.5, 2.0, 2.5, 3.0, 3.5 and 4.0 wt%) [84]. The SEM and XRD results indicated that OA-Fe_3O_4 nanoparticles homogeneously dispersed in PUD matrix and the electrical performance (volume and surface resistivity) of composite films increased up to 1.5 wt% loading and then declined. The thermal stability of composite films increased with increase in OA-Fe_3O_4 content and the storage modulus and T_g values are higher at 2.0 wt% loading of nanofiller. The mechanical properties were gradually decreased with increase in OA-Fe_3O_4 loadings and the composite films exhibited super-paramagnetic properties. However all these thermo-mechanical, electrical and magnetic properties are higher than pure PUDs.

Chen et al. synthesized UV−curable hyperbranched PUD/Fe_3O_4 nanocomposites by incorporating different amounts (0.5, 2.0, 4.0 and 6.0 wt%) of vinyl functionalized Fe_3O_4 nanoparticles (Fe_3O_4-VTEO) as fillers into the vinyl terminated PUD [85]. The UV-curing process was initiated by Irgacure 2959 through the formation of cross-links between the filler and PUD matrix. The thermal degradation rate decreases as filler content increases and the storage modulus, T_g values and pencil hardness values are higher at 2.0 wt% loading of nanofiller and show a downward trend at higher loadings. The water and solvent (toluene) resistance of composite films gradually increased with increase in Fe_3O_4-VTEO content and the mechanical properties are in reverse order. The composites exhibited good magnetic properties and the maximum saturation magnetization was increased with the increasing of Fe_3O_4-VTEO content.

8.2.2 PUD thermosetting composites with metal (Ag and Au) nanoparticles

Several nanomaterials introduced into the polymer matrix to develop antimicrobial coatings and it was observed that silver (Ag) nanoparticles had considerably higher antibacterial activity [86]. The nanoAg coated PU foams serve as a water purifier to remove bacteria from water [87] and bone cement loaded with 1 % nano-Ag exhibited high antibacterial activity [88]. The antimicrobial activity of silver mainly because of silver cation (Ag^+) which binds strongly to the electron donor groups of proteins, nucleic acids and cell membranes of microorganisms and thus interrupts many essential metabolic processes of microorganisms which kill them [89].

Hsu et al. studied the biocompatibility and antibacterial activity of PUD/Ag nanocomposites using different concentrations (15, 30, 50 and 75 ppm) of nanosilver with good nanoparticle dispersion up to 30 ppm in PUD matrix, confirmed by TEM analysis [90]. The PUD/Ag nanocomposites with 30 ppm nano-Ag concentration exhibited superior physico-chemical properties, cellular response (confirmed by rat skin fibroblasts), biocompatibility (confirmed by rat subcutaneous and jugular vein implantation) and antibacterial properties. The *in vitro* oxidative degradation was measured using 10% H_2O_2 and 0.05 M $CoCl_2$ as the accelerated environment and the composites exhibited good inhibition in all concentrations of nano-Ag, especially at 30 ppm. The PUD/Ag 30 ppm nanocomposite also showed good scavenging ability, enhanced fibroblast attachment and endothelial cell response as well as reduced monocyte and platelet activation. Furthermore, the inserted commercial catheters in rat jugular veins, coated with PUD/Ag 30 ppm showed milder inflammation after 3 months as compared to the bare catheters and pure PUD coated catheters.

Even having superior antibacterial activity, the enthusiasm for Ag nanoparticles (Ag NPs) has been hampered by their cytotoxicity and genotoxicity [91]. However, the cytotoxicity of Ag NPs may be reduced if they are embedded in a polymer matrix and this was evaluated by Liu et al. using waterborne PUD/Ag nanocomposites with different sizes and concentrations of Ag NPs [92]. The Ag NPs with three different sizes, i.e. small (Ag-S, 3-4 nm), medium (Ag-M, 5-7 nm) and large (Ag-L, 10-40nm) sizes were fed into PUD matrix in different concentrations (15, 30, 60, 120 and 240 ppm). The cytotoxicity of PUD/Ag nanocomposites was examined using rat skin fibroblasts and the antibacterial activity was examined using *Escherichia coli* and *Staphylococcus aureus*. The nanocomposite with 60 ppm of Ag-M exhibited good antibacterial activity, biocompatibility, and endothelial nitric oxide synthase (eNOS) gene expression. The addition of Ag NPs at all sizes further increased the thermal stability of composites.

Wattonodorn et al. examined the mechanical properties of PUD/Ag nanocomposites and also their antibacterial properties by incorporating Ag NPs in the form of different amounts (100, 200, 500, 1000 and 2000 ppm) of aqueous $AgNO_3$ solution into PUD matrix [93]. The PUD/Ag nanocomposites with 2000 ppm $AgNO_3$ concentration exhibited better distribution and smaller Ag NPs was confirmed by TEM. The composites showed good antibacterial activity by Ag^+ ion release over 21 days and this activity was increased with increase in $AgNO_3$ concentration. Furthermore, in contrast with pure PUD, the composite films exhibited superior mechanical properties and the results were highest at an $AgNO_3$ concentration of 2000 ppm.

Microbiological degradation of PUD adhesives can cause deterioration of shelf-life and product quality of adhesive through loss of viscosity, discoloration, coagulation of dispersion and loss of adhesion [94]. The incorporation of antimicrobial additives into PUD adhesive formulation can protect them from a broad spectrum of microorganisms including fungi, bacteria, and algae. Previous studies revealed the high antimicrobial activity of Ag NPs by releasing Ag^+ ions which causing the killing of bacteria [95]. Perez-Liminana et al. formulated antimicrobial PUD adhesives using a different concentration of gelatine-stabilized Ag NPs (0.005 to 400 ppm) to inhibit the bacterial growth [96]. The Ag NPs were fabricated by dissolving gelatine as a reducing and stabilizing agent in aqueous $AgNO_3$ solution. The biocidal activity of Ag NPs was determined by determination of minimum inhibitory concentration (MIC) and PUD adhesives exhibited excellent antimicrobial activity at higher MIC values of Ag NPs (100 to 250 ppm). The adhesion properties of PUD/Ag NP was determined by T−peel tests on split leather/SBR joints and the PUD with 25 and 50 ppm Ag NP exhibited good adhesive strength. However, the introduction of gelatine-stabilized Ag NPs did not produce noticeable changes in thermal stability and T_g values.

Recently, PU has become an attractive material for food packaging because of its elastomeric and shape memory properties [97]. Improving the barrier properties and antimicrobial properties through the incorporation of nanoparticles in the polymer matrix is an important area of research for the food packaging industry. The non-toxic and biodegradable hydroxylapatite (HA) possess superior barrier properties and restricts the passage of gases, vapors and organic liquids through the packaging material. Hence, the combination of HA and antimicrobial Ag NPs offers good barrier properties, high mechanical strength, and antimicrobial properties to the waterborne PUDs and make them as a good alternative to the packaging industry. Rahman et al. improved the barrier and antimicrobial properties of PUDs with HA-Ag NPs and the experiments were carried out using different contents of HA (0.56, 0.95 and 1.33 mol%) and DMPA (17.03, 22.05, and 24.02 mol%) [98]. Ag NP dispersion was prepared by the reduction of $AgNO_3$ solution with $NaBH_4$ and this silver colloid was mixed with a defined content of PUD-HA dispersion. The hydrogen bonds in the PUD nanocomposites increased with increase in HA content (1.33 mol%), which enhanced the mechanical properties and water vapor permeability (WVP) resistance of nanocomposites up to the optimum HA content. The highest tensile strength, Young's modulus, and WVP resistance were observed at PUD with 24.02 mol% DMPA content and 1.33 mol% HA content. The composite films also exhibited good antimicrobial activity against *Staphylococcus aureus*.

Halloysite nanotubes (HNTs) are often used as reinforcing filler in polymer nanocomposites because they can be easily dispersed in the polymer matrix [99]. HNTs also act as carriers for Ag NPs to prevent their agglomeration in the polymer matrix. Therefore, Fu et al. developed castor oil based antibacterial PUD-Ag nanocomposites using HNTs as a carrier for Ag NPs [100]. The HNTs were first modified with a silane coupling agent (AEAPTMS) and chitosan and then mixed with $AgNO_3$ for combining Ag^+ ions. Finally, the Ag NPs were loaded on the surface of modified HNTs by reducing Ag^+ ions with $NaBH_4$. The obtained Ag-HNTs with –NH and –OH groups were used to prepare PUD/Ag-HNT nanocomposites by changing the Ag-HNT contents (1.0, 2.0 and 3.0 wt%). The thermo-mechanical properties of PUD composites were enhanced with the increase of Ag-HNT loadings and the composites exhibited good antimicrobial activity against *Escherichia coli* and *Staphylococcus aureus*.

Materials Research Forum LLC

doi: http://dx.doi.org/10.21741/9781945291876

The hyperbranched polymer chemistry was introduced by Han et al. in PUD-Ag composite synthesis [101]. First, water soluble, three different generation hyperbranched polyesters (HBPE1, 2 and 3) were synthesized and mixed with the $AgNO_3$ aqueous solution in the presence of ascorbic acid. The yellowish colloidal solution of hyperbranched polyester protected Ag NPs were obtained with 2.94% Ag content. Three different PUD-Ag composites were obtained by addition of 10 mL colloidal solution of Ag NPs protected by a different generation of hyperbranched polyester to the neutralized NCO terminated PU prepolymer at the high shear field. The Ag NPs protected by HBPE3 exhibited a better distribution in PUD matrix and high T_g and storage modulus values.

Akbarian et al. studied the effect of Ag NPs on antibacterial and thermal properties of 2K-PUD coatings [102]. The 2K-PUD-Ag composite with 200 ppm nanosilver was obtained by mixing of Ag NPs containing polyacrylic resin with polyisocyanate resin. The SEM micrographs depicted uniform dispersion of Ag NPs in PU matrix and prepared 2K-PUD-Ag composite exhibited bacterial growth reduction against *Escherichia coli* and *Staphylococcus aureus*.

Gold nanoparticles (Au NPs) also had good antibacterial properties and exhibits better stability than Ag NPs with the same stabilizer. Au NPs utilized in therapeutic drugs for hyperthermia treatment of cancer cells [103]. However, there have been very few studies on PUD-Au nanocomposites in order to evaluate the effect of Au NPs on the biocidal and antimicrobial activity of PUDs.

The process of wound repair after arterial injury involves migration and proliferation of endothelial cells (ECs). The high EC migration rate was associated with increased levels of endothelial nitric oxide synthase (eNOS) which plays a critical role in vascular tissue protection [104]. The phosphoinositide 3-kinase (P13K) signaling pathway regulates the EC survival and proliferation [105]. Akt, (a serine/threonine protein kinase) is recruited to the cell membrane by its binding to P13K. The Akt subsequently phosphorylates and activates eNOS leading to the production of NO which regulates the EC growth, migration and angiogenesis [106, 107]. In order to explore the mechanism of eNOS induced EC migration and proliferation on biomaterials of different surface morphologies, Hung et al. had tried to develop a model system using PUD nanocomposites incorporated with different concentrations (17.4 to 174 ppm) of Au NPs [108]. It was found that ECs had migration rate on the nanocomposite containing 43.5 ppm Au NPs. The inductions of both eNOS and p-Akt on PUD-Au were abolished by the addition of P13K inhibitor and suggesting that P13K signaling pathway is an important regulator of EC cellular events. The effect at high Au NPs (65−174 ppm) was not clear and this mainly because of the agglomeration of nanoparticles.

The antimicrobial activity of Au NPs was examined by Han et al. by incorporating functionalized Au NPs into a cationic waterborne PUD system [109]. First, Au NPs were fabricated by the citrate-mediated reduction of $HAuCl_4.3H_2O$ and the obtained gold suspension was functionalized with dithioerythritol (DTET). The thiol-functionalized DTET−Au NPs were added to the NCO terminated PU prepolymer to develop the PUD−Au NP composite. The composite films with DTET−Au NPs have strong antibacterial activity against *Escherichia coli* and *Staphylococcus aureus*, and this activity was high as compared to the pure PUD and PUD with pure Au NPs.

8.2.3 PUD/clay thermosetting composites

The polymer nanocomposites with layered silicates (clays) have attracted much more attention in the recent decades because of their unique properties including high strength and modulus [110], better thermal stabilities [111], improved barrier properties and chemical stability [112] as well as flammability resistance [113]. To attain all aforementioned properties, the nanocomposite only requires the addition of a few percent 1-nm thickly layered silicate. Furthermore, clay as an inorganic filler also has the advantage of good processability, low cost, and transparency [113]. Hence, the addition of clay can enrich the thermo-mechanical and barrier properties of PUDs.

Lee et al. investigated the effect of ionic interactions between clay and PUD, using different amount (1.0, 2.0, 3.0, 4.0 and 5.0 wt%) of modified MMT clay with DMAB through cation-exchange reaction [114]. XRD and TEM results suggesting that the clay platelets are exfoliated by long alkyl chains of quaternary ammonium cations which facilitate the intercalation of polymer chains into the clay. The incorporation of clay doesn't influence the T_g of the soft segment domains but enhance that of the hard segment domains through the ionic interactions between clay and PUD hard segment. The ionic interactions which are evidenced by IR spectra of composite films, do not facilitate the migration of clay platelets to the film surface and lowers the volumetric and surface electrical resistance of composite films at higher clay contents (Fig. 8.5). The thermal resistance of composite films gradually increased with increase in clay content whereas tensile strength increased up to 3 wt% of clay loading.

Materials Research Forum LLC

doi: http://dx.doi.org/10.21741/9781945291876

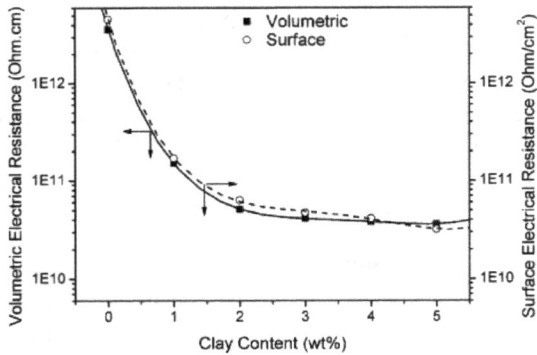

Figure 8.5   Volumetric   and   surface   electrical   resistance   of   composite   films   with different clay contents.

Rahman et al. examined the effect of clay loadings on adhesive strength of PUD/clay composites using different amount (1.0, 2.0, 3.0, 4.0 and 5.0 wt%) of organoclay (cloisite 15A) [115]. The clay with quaternary ammonium ions was exfoliated in a polyol at 80°C and the exfoliation was confirmed by XRD. The particle sizes of PUDs increased with increasing of clay loadings and the water resistance, thermal stability, mechanical strength and adhesive strength were increased up to 2 wt% clay content. Rahman et al. also examined the effect of clay loadings on water vapor permeability (WVP) of PUD/clay composites using different amount (0.5, 1.0 and 2.0, wt%) of cloisite 15A [116]. Here also, the clay was exfoliated in a polyol at 80°C and with an increase in clay content; the T_g, thermal stability and water resistance of composite films were increased. The composite films exhibited maximum tensile strength at 1 wt% loading. The WVP of PUD/clay coated nylon fabrics decreased with increase in clay content at constant temperature and at fixed clay content the WVP increased with the increase of temperature.

Subramani et al. studied the effect of silane modified clay on thermo-mechanical properties of PUD/clay composites [117]. The MMT clay was modified with APTMS through cation-exchange reaction and was incorporated in different contents (1.0, 3.0 and 5.0 wt%) into silyl terminated PUD synthesis process. The siloxane cross-linkings were developed between silylated MMT and silylated PUD through a sol-gel process. The properties of PUD with modified MMT were compared to the properties of PUD with unmodified cloisite 20A clay. The thermal stability, tensile strength, water and xylene resistance of PUD/MMT clay composites were superior to PUD/cloisite 20A composites. The thermal stability, water and xylene resistance of PUD/MMT clay composites gradually increased with increase in silylated MMT content and however high tensile strength were observed at 1 wt% loading.

Yeh et al. examined the corrosion resistance properties of PUD/clay nanocomposites using different concentrations (1.0 and 3.0 wt%) of inorganic Na^+–MMT clay and the properties were compared with neat PUD and PUD/organic-MMT (3.0 wt%) composites [118]. The cold-rolled steel (CRS) coupons coated with PUD/Na^+–MMT clay composite exhibited superior corrosion resistance in 5 wt% aqueous NaCl electrolyte and the corrosion protection performance was increased with increase in Na^+–MMT clay loading. Furthermore, PUD/Na^+–MMT clay composite coatings and films exhibited lower gas permeability, higher T_g, higher thermal decomposition and lower optical transparency as compared to the neat PUD and PUD/ organic-MMT composites.

Attapulgite (AT) is one of the natural fibrillar silicate clay mineral which can form polymer nanocomposites with remarkable exposure durability and extreme chemical stability. Peng et al. developed PUD/Attapulgite nanocomposites using different concentrations of (0.5, 1.0, 1.5, 2.0 and 2.5 wt%) functionalized AT [119]. The NCO functionalized AT (AT-NCO) was obtained by reacting IPDI with acid treated AT (AT-OH) in the presence of DBTDL catalyst. The results were indicating that AT-NCO was homogeneously dispersed in PUD matrix and the thermal stability of nanocomposites increased with increase in AT-NCO content. However, the tensile strength was increased up to 1.5% AT-NCO loading.

8.2.4 PUD/Carbohydrate thermosetting composites

Nanofillers derived from natural polysaccharides, exhibiting a reinforcing function similar to inorganic nanofillers [120]. These natural nanofillers have attracted much attention for composite synthesis due to their ready availability, no toxicity, biocompatibility, high reactivity, high rigidity and easy processability [121]. Natural nanofillers from different sources have different geometries such as rod-like nanocrystals or whiskers of cellulose and chitin, and platelet-like nanocrystals of starch; and the extent of the reinforcement depends on the dispersion of the nanofillers in the polymer matrix and the interfacial interaction between nanofillers and the polymer matrix.

8.2.4.1 Cellulose-based PUD thermosetting composites

Cellulose is the most abundant biomass carbon source consists of 1,4-β-glucopyranose units and gives the rodlike cellulose nanocrystals (CNCs) through acid treatment. Recently, CNCs have attracted increasing attention in composite materials because of their nanoscale dimensions, easy modification, abundant availability, high surface area and low thermal-expansion mechanical strength [122, 123]. These properties made CNCs as suitable reinforcing filler for polymer nanocomposites.

Cao et al. synthesized PUD/CNC nanocomposites using the different concentrations (5, 10, 15, 20, 25 and 30 wt%) of CNC dispersion prepared from acid hydrolysis of *flax fiber* [124]. The CNC fillers were well dispersed in PUD matrix due to the strong hydrogen bonding interactions and the incorporation of CNCs enhanced the phase separation between hard and soft segments which led to the decrement in T_g values of composites. The hydrogen bonding interactions between filler and matrix also enhanced the mechanical properties of composites and the values gradually increased with increasing of CNCs loading.

Gao et al. developed castor oil based biocompatible waterborne PUD/CNC nanocomposites with low-level loadings (0.2, 0.5, 1.0, 2.0, 3.0, 4.0 and 5.0 wt%) of *Eucalyptus globulus* cellulose nanocrystals (ECNCs) obtained through acid hydrolysis of *Eucalyptus* fiber [125]. The SEM results indicating that ECNCs were well dispersed in PUD matrix and the incorporation of ECNCs enhanced the hard-soft segments phase separation which led to the increased T_g values of both soft and hard segments. However, the tensile strength of composites increased up to 1.0 wt% of ECNCs loading.

Santamaria-Echart et al. were studied the reinforcing effect CNCs on thermo-mechanical and surface properties of PUD/CNC nanocomposites with different contents (0.5, 1.0, 3.0 and 5.0 wt%) of CNCs obtained by acid hydrolysis of commercially available microcrystalline cellulose (MCC) [126]. The thermal and mechanical properties of composites increased up to 3 wt% CNC loading and there is a decrease in both values at higher loadings, due to the agglomeration of CNCs. Furthermore, water contact angle and water absorption measurements suggesting the hydrophilic nature of composites, which increased with increase in CNC content.

High-quality garments and business suits are fabricated from wool because of its warmth, lightness, softness, and smoothness. However, the special scale structure in wool cuticles can cause felting shrinkage of wool fabrics under mechanical actions during laundry. Many research attempts focused on reducing or eliminating the felting behavior of wool fabrics in order to make them machine washable [127]. Unfortunately, most of these efforts were designed based on chlorination, the approach which releases the harmful halogens into water and environment. PUD can serve as an anti-felting agent at higher concentration [128], and this effect can be achieved even at lower PUD concentration through the reinforcement of PUD with nanofiller. The anti-felting effect of PUD/CNC composites for wool fabrics was examined by Zhao et al. using different concentrations

(02, 0.4, 0.6, 0.8 and 1.0 wt%) of CNCs, obtained through acid hydrolysis of MCC [129]. The wool fabrics were coated with nanocomposite through a pad-dry-cure process and the coated wool fabrics exhibited lower area-shrinking rate and higher tensile strength at 1 wt% of CNC loading and these values are comparable to the fabrics coated with high concentrated pure PUD.

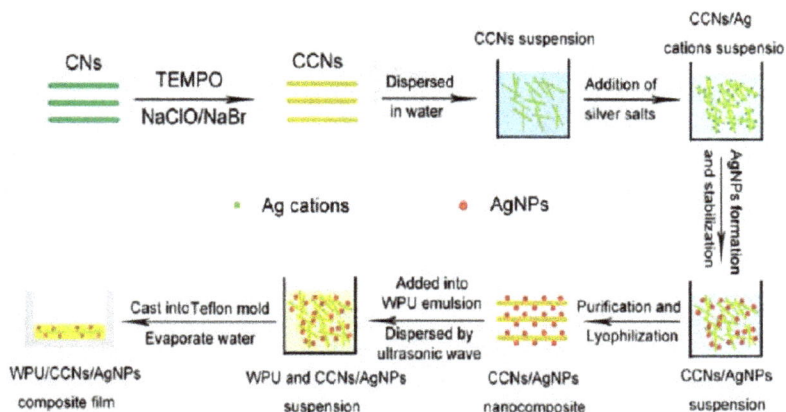

Figure 8.6 Schematic representation for CNC/Ag NP composite preparation and its incorporation into PUD.

Liu et al. observed the combinatorial reinforcing effect of Ag NPs and CNCs on properties of PUD/CNC composites with different concentrations of CNCs (5, 10, 15 and 20 wt%) and PUD/CNCs-5/Ag NP composites with different contents of Ag NPs (0.357, 1.766, 3.471 and 6.709 wt%) [130]. The CNCs were obtained from MCC through acid hydrolysis and were oxidized as carboxylated CNCs via TEMPO-mediated carboxylation approach, in which OH groups of CNC surface were converted into −COOH groups. The CNC/Ag NP suspensions were developed by mixing of different volumes of AgNO$_3$ aqueous solution into carboxylated CNC suspension at high stirring speed (Fig. 8.6). Totally four different CNC/Ag NP suspensions were obtained with different wt% of Ag NPs [0.357 wt% (1 mL AgNO$_3$), 1.766 wt% (5 mL AgNO$_3$), 3.471 wt% (10 mL AgNO$_3$) and 6.709 wt% (20 mL AgNO$_3$)]. AFM and UV-vis results suggesting that carboxylated CNCs and Ag NPs were homogeneously dispersed in PUD matrix and both nanofillers having opposite effect on mechanical properties. Up to 10 wt% of CNCs loading, the mechanical properties of composite films enhanced and then gradually decreased. On the other side, tensile properties of composite films gradually decreased with increase in Ag NP content. More importantly, PUD/CNCs-5/Ag NP composites exhibited strong antimicrobial activity against *Escherichia coli* and *Staphylococcus aureus*.

Wu et al. prepared biomass-based 2K-PUD composites using different concentrations (0.5, 1.0, 2.0, 4.0 and 5.0 wt%) of cellulose nanowhiskers (CNWs) [131]. The CNWs were developed by acid hydrolysis of MCC and mixed with a polyol (WPOL) dispersion and then the mixture was blended with hydrophilic HDI trimer. The obtained 2K-PUD/CNW nanocomposites exhibited enhanced α-relaxation temperature (T_α) and T_g values with the increase of CNWs content due to the strong interaction between CNWs and polymer matrix (Fig. 8.7). The mechanical properties of 2K-PUD/CNW composites gradually increased with increase in CNWs content and the thermal stability was slightly decreased with the increase of CNWs content.

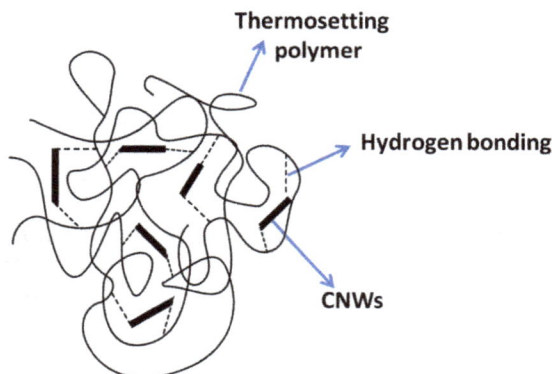

Figure 8.7 Interaction model of 2K-PUD/CNW nanocomposites.

8.2.4.2 Starch reinforced PUD thermosetting composites

Starch is a natural, renewable and biodegradable polymer produced in many plants as a source of stored energy. Starch nanocrystals (StNCs) possess platelet-like morphology similar to exfoliated layered silicates; can be prepared by acid hydrolysis of starch granules [132]. The intrinsic rigidity, strong interfacial interactions and the organization of the percolation network of StNCs contribute enhanced thermo-mechanical properties, solvent resistance and barrier properties to the composites [133]. Hence, Starch nanocomposites have attracted much attention in both academic research and practical applications.

Zou et al. synthesized PUD/starch nanocomposites through incorporating high levels (5, 10, 15, 20 and 30 wt%) of StNCs in PUD matrix [134]. The homogeneous dispersion of StNCs in PUD matrix was achieved up to 30 wt% loading. The resultant composites exhibited high tensile strength at 10 wt% loading, due to the active surface and rigidity of StNCs which facilitated the formation of an interface for stress transfer. Further increase in StNCs wt%, self-aggregation of StNCs resulted in a decrease in tensile strength; however, the rigidity of StNCs supported an increase in Young's modulus and was highest at 30 wt % StNCs loading.

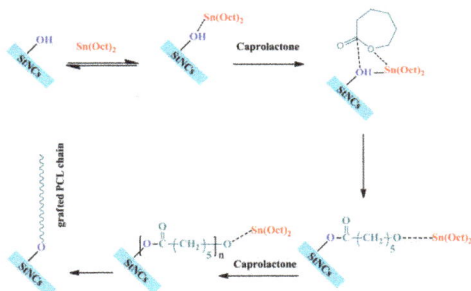

Figure 8.8 Schematic representation for grafting of polycaprolactone (PCL) chains onto the StNCs surface.

The availability of reactive OH groups on the surface of StNCs, suitable for chemical modification and grafting reactions, which facilitates the homogeneous dispersion of StNCs and inhibits their self-aggregation in composites. Chang et al. fabricated the surface functionalized StNCs by grafting polycaprolactone (PCL) chains using *"graft from"* strategy (Fig. 8.8) [135]. The obtained StNC-g-PCL nanofiller incorporated in PUD matrix with different loadings (5, 10, 15, 20, 25 and 30 wt%), to examine their agglomeration behavior. However, it was noticed that composites with 5 wt% StNC-g-PCL exhibited highest tensile strength and elongation at break values. Further increase in filler content was led to self-aggregation and inhibited improvement in tensile strength and elongation at break values, but significantly enhanced Young's modulus.

Wang et al. examined the combinatorial reinforcing effect of natural nanocrystals and whiskers on thermo-mechanical properties of PUDs by using both starch nanocrystals (StNCs) and cellulose nanowhiskers (CNWs) [136]. StNCs and CNWs were fabricated successfully from the acid hydrolysis of waxy maize starch granules and cotton linter pulp, respectively. The obtained StNCs and CNWs were loaded into PUD matrix with concentrations of 1.0 wt% and 0.4 wt%, respectively. The PUD/1% StNCs/0.4% CNWs exhibited high tensile strength, storage modulus and thermal stability and these values were higher than pure PUD, PUD/1% StNCs and PUD/0.4%CNWs. Hence, it was clear that the different polysaccharide nanocrystals and whiskers combined together to form strong hydrogen bonding interactions, which led to the synergistic reinforcement of PUD (Fig. 8.9).

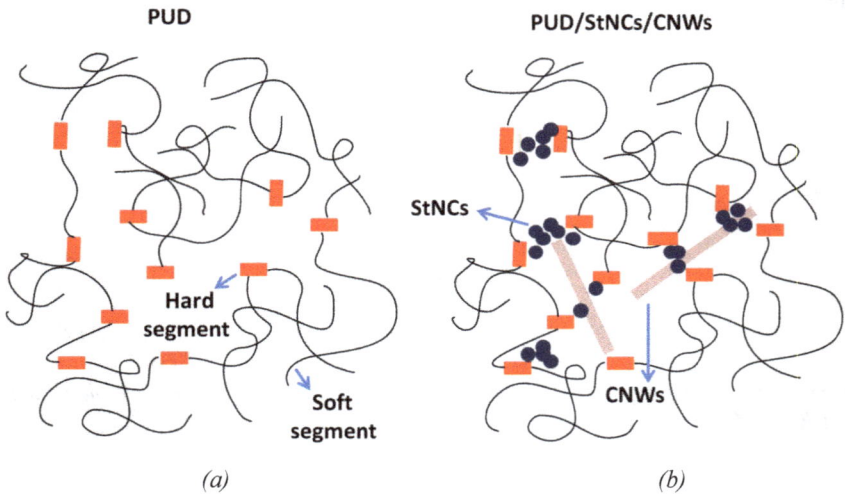

Figure 8.9 Schematic representation of pure PUD matrix (a) and reinforced networks of StNCs and CNWs in (b) PUD matrix.

8.2.5 PUD thermosetting composites reinforced with nanocarbon materials

Nanocarbon materials, especially graphene [or reduced graphene oxide (rGO)], graphite oxide (GO) and carbon nanotubes (CNTs) exhibit superior electronic and thermo-mechanical properties and have triggered tremendous interest both in fundamental research and practical applications. Polymer composites reinforced with nanocarbon materials, show fundamentally unexpected new properties because of their unique structure, low degree of filling and smaller particle sizes.

8.2.5.1 Graphene oxide (GO), and reduced graphene oxide (rGO) based PUD thermosetting composites

Graphene is a thin inorganic nanomaterial and has attracted a considerable amount of attention in the area of polymer composites because of its extraordinary mechanical properties, excellent electrical conductivity, thermal conductivity, and high surface area. The high stability of GO in aqueous media facilitates the development of robust, well-dispersed and eco-friendly waterborne polymer composites. In addition, the aqueous medium also promotes a suitable environment for the *in-situ* reduction of GO to rGO [137], which is able to disperse homogeneously in the polymer matrix and solves the problem associated with poor dispersion of graphene. Therefore, the use of graphene as a reinforcing agent has been proposed for achieving the high-performance coating applications of PUDs in various fields.

Wu et al. reported the synthesis of PUD composites using different concentrations (0.1, 0.3, 0.5 and 1.0 wt%) of amine functionalized rGO as a filler into the PUD matrix [138]. GO was first treated with TDI and gives the intermediate isocyanate functionalized GO (iGO), which further treated with hydrazine hydrate to obtain the final amine functionalized rGO (NH_2-rGO). The addition of 1 wt% of NH_2-rGO facilitated the efficient load transfer between the graphene and PUD matrix. As a result, the thermal stability, storage modulus and tensile strength of the composite with 1 wt% of NH_2-rGO loading higher than that of the neat PUD and PUD-rGO composites.

GO and reduced graphene oxide (rGO) has enhanced the flame retardant properties of polymer composites. This flame retardant property can be attributed to the formation of compact, dense and uniform char from carbon nanoadditives of the composites during combustion. Based on this phenomenon, Hu et al. reported the synthesis of flame-retardant PUD composites, using different concentrations (0.5%, 1.0%, 1.5% and 2.0 wt%) of rGO [139]. The mechanical properties of PUD-rGO nanocomposites significantly enhanced up to 2 wt% rGO loading and the thermal stability of nanocomposites slightly decreased compared to pure PUD, but carbon residue increased from 0.99 to 1.99% with 2 wt% rGO loading. The flame-retardant and smoke suppression properties of rGO-PUD nanocomposites increased up to 1 wt% rGO loading, the total smoke release and smoke factor decreased by 25 and 38% respectively, compared to the neat PUD.

Li et al. reported the fabrication of anticorrosive PUD composite coatings on electrogalvanized (EG) steel using rGO as a reinforcing agent [140]. In order to facilitate the homogeneous dispersion of rGO, graphene oxide (GO) was reduced and functionalized simultaneously with titanate coupling agent (TM-200S). The aqueous suspension of functionalized GO (TGO) was mixed with commercially available PUD in different wt% (0.2% and 0.4%) to obtain composite coatings. The PUD composite with 0.4 % TGO loading exhibited superior anticorrosion properties and this is because of the in-plane alignment of graphene layers on EG steal surface, which promotes the full utilization of high surface area of graphene layers to prevent the electrolyte from penetration.

8.2.5.2 Carbon nanotubes (CNTs) reinforced PUD thermosetting composites

Polymer-CNT nanocomposites are being considered as promising candidates for the coating industry, because of their excellent thermo-mechanical properties and superior conducting properties. Over the last decade, polymer-CNT nanocomposites have been intensively studied using different types of polymers including epoxies, polyimides and polyurethanes are reinforced with neat or functionalized CNTs. The addition of CNTs enhances the comprehensive performance, such as thermo-mechanical properties, water resistance and electrical properties of waterborne PUDs.

Zhao et al. reported the reinforcement effect of CNTs on mechanical and crystalline properties of PUD-CNT nanocomposites, with various CNTs loadings (0.5, 1.0, 2.0, 2.5, 3.0, 3.5 and 4.0 wt%) [141]. CNTs are uniformly dispersed in the PUD matrix up to 1.5% loading and further loadings of CNTs led to the aggregation of CNTs which was confirmed by the SEM and TEM analysis. The lower levels of CNTs loading ($\leq$ 1.5%) enhanced the crystallinity and tensile strength of composites with homogeneous dispersion of CNTs. On the other hand, high loading levels of CNTs ($\geq$ 2.0%) led to the agglomeration of CNTs and cause the cracks in a polymer matrix which reduced the tensile strength of composites.

Rahman et al. studied the effect of functionalized multiwalled carbon nanotubes (MWCNTs) on weather degradation and corrosion resistance performance of waterborne PUD composites [142]. The PUD- CNT composites were prepared using three defined concentrations (0.5, 1.0 and 2.0 wt%) of carboxyl functionalized MWCNTs and were coated on a mild steel panel and exposed to natural weather condition for a maximum of 1 year. Both weather degradation and corrosion protection performance of exposed coatings were characterized by potentiodynamic polarization and XPS analysis. Both properties are gradually increased with increase in loading levels of functionalized MWCNT, and the values are higher than that of pure PUD coatings. The functionalized MWCNTs acted as both a barrier (before exposure) and UV absorber (during exposed) and these combinatorial factors improved the overall protection performance of composite coatings. Furthermore, the % water swelling was decreased and contact angles were increased with the increase of CNT loadings up to 2.0 wt%.

Rahman et al. also studied the combinatorial effect of carboxyl functionalized CNT content and DMPA wt% on the adhesive strength of waterborne PUDs. Three different series of PUDs containing, 3.61% (A series), 5.16% (B series) and 5.18% (C series) of DMPA content were developed and the optimum CNT contents for these three sets were about 0.5, 1.0 and 1.5 wt% respectively [143]. Above the optimum CNT content, aggregation of CNTs takes place in all three series. The incorporation of CNTs enhanced the thermal stability of the composites and both tensile strength and adhesive strength increased up to the optimum CNT content in all series. The highest adhesive strength of PUD-CNT composites was found with the higher DMPA content (5.18%) and optimum CNT content (1.5%).

Summary

In this chapter, we reviewed some of the important literature on PUD thermosetting composites, to emphasize the role of different types of fillers on physic-chemical properties of PUD nanocomposites. The low thermal stability, storage modulus, glass transition temperature and low tensile strength of thermoplastic PUDs can considerably improve with the adding of a small amount of nanofiller. Moreover, these nanofillers give some additional properties including corrosion resistance, shape memory, flame retardancy and antimicrobial activity to the resulting PUD thermosetting composites. The reinforcement of PUD matrix with various inorganic fillers is expected to grow in the coming years for multifunctional materials and excite the researchers to solve the problems associated with incompletely explored properties of PUD nanocomposites. It is also interesting to remark that, the trend is to increase bio-based materials maintaining good overall properties with the goal for production of "green" PUD nanocomposites.

Abbreviations

AEAPTMS	3-(2-aminoethyl)aminopropyl)trimethoxysilane
AFM	atomic force microscopy
APTES	3-aminopropyltriethoxysilane
APTMS	3-aminopropyltrimethoxysilane
DETA	diethylenetriamine
DMAB	Didodecyldimethylammonium bromide
DMPA	dimethylolpropionic acid
EDA	ethylenediamine
FESEM	field emission scanning electron microscopy
FTIR	Fourier transform infrared

HEA	2-hydroxyethyl acrylate
HEMA	hydroxyethyl methacrylate
H$_{12}$MDI	4,4'-dicyclohexylmethane diisocyanate
IPDI	isophorone diisocyanate
MCC	microcrystalline cellulose
mL	milliliter
MMT	montmorillonite
OPP	oriented polypropylene
ppm	parts per million
PTMG	poly(tetramethyleneglycol)
PU	polyurethane
PUD	polyurethane dispersion
SEM	scanning electron microscopy
TDI	toluene diisocyanate
TEM	transmission electron microscopy
TEMPO	2,2,6,6-tetramethylpiperidine-1-oxyl radical
TEOS	tetraethylorthosilicate
TMP	trimethylolpropane
TMSiP-EDA	N-[3-(trimethoxysilyl)propyl]-ethylenediamine
UV	ultraviolet
VTEO	vinyltriethoxysilane
XRD	X-ray diffraction
XPS	X-ray photoelectron spectroscopy

References

[1]   J.P. Pascault, R.J.J. Williams, Overview of thermosets: structure, properties and processing for advanced applications. Thermosets, (2012) 3-27. (Elsevier). https://doi.org/10.1533/9780857097637.1.3

[2]   J. Guilleminot, S. Comas-Cardona, D. Kondo, C. Binetruy, P. Krawczak, Multiscale modelling of the composite reinforced foam core of a 3D sandwich structure. Compos. Sci.Technol. *68* (2008) 1777-1786. https://doi.org/10.1016/j.compscitech.2008.02.005

[3]   A. Yousefi, P. G. Lafleur, R. Gauvin, Kinetic studies of thermoset cure reactions: a review. Polym. Compos. *18* (1997) 157-168. https://doi.org/10.1002/pc.10270

[4]   D. Feldman, Composites, thermosetting polymers. Polymeric Materials Encyclopedia. (1996) 277-278.

[5]   S. K. Bobade, N. R. Paluvai, S. Mohanty, S. K. Nayak, Bio-based thermosetting
        resins for future generation: a review. Polym. Plast. Technol. Eng. *55* (2016) 1863-
        1896. https://doi.org/10.1080/03602559.2016.1185624

[6]   C. Zhang, S. A. Madbouly, M. R. Kessler, Biobased polyurethanes prepared from
        different vegetable oils. ACS Appl. Mater. Interfaces. *7* (2015) 1226-1233.
        https://doi.org/10.1021/am5071333

[7]   H. Bakhshi, H. Yeganeh, S. Mehdipour-Ataei, A. Solouk, S. Irani, Polyurethane
        coatings derived from 1, 2, 3-triazole-functionalized soybean oil-based polyols:
        studying their physical, mechanical, thermal, and biological
        properties. Macromolecules. *46* (2013) 7777-7788.
        https://doi.org/10.1021/ma401554c

[8]   J. M. Raquez, M. Deléglise, M. F. Lacrampe, P. Krawczak, Thermosetting (bio)
        materials derived from renewable resources: a critical review. Prog. Polym. Sci. *35*
        (2010) 487-509. https://doi.org/10.1016/j.progpolymsci.2010.01.001

[9]   K. M. Zia, H. N. Bhatti, I. A. Bhatti, Methods for polyurethane and polyurethane
        composites, recycling and recovery: A review. React. Funct. Polym. *67* (2007)
        675-692. https://doi.org/10.1016/j.reactfunctpolym.2007.05.004

[10]   J. C. Wang, Y. H. Chen, R. J. Chen, Preparation of thermosetting polyurethane
        nanocomposites by montmorillonite modified with a novel intercalation agent. J.
        Polym. Sci. Part B Polym. Phys. *45* (2007) 519-531.

[11]   D. P. Pfister, Y. Xia, R. C. Larock, Recent advances in vegetable oil-based
        polyurethanes. ChemSusChem. *4* (2011) 703-717.
        https://doi.org/10.1002/cssc.201000378

[12]   A. Zlatanić, C. Lava, W. Zhang, Z. S. Petrović, Effect of structure on properties of
        polyols and polyurethanes based on different vegetable oils. J. Polym. Sci. Part B
        Polym. Phys. *42* (2004) 809-819. https://doi.org/10.1002/polb.10737

[13]   M. Visconti, M. Cattaneo, A highly efficient photoinitiator for water-borne UV-
        curable systems. Prog.Org. Coat. *40* (2000) 243-251. https://doi.org/10.1016/S0300-
        9440(00)00147-8

[14]   D. Y. Xie, F. Song, M. Zhang, X. L. Wang, Y. Z. Wang, Roles of soft segment
        length in structure and property of soy protein isolate/waterborne polyurethane
        blend films. Ind. Eng. Chem. Res. *55* (2016) 1229-1235.
        https://doi.org/10.1021/acs.iecr.5b04185

[15]   S. A. Madbouly, Y. Xia, M. R. Kessler, Rheological behavior of environmentally
        friendly castor oil-based waterborne polyurethane
        dispersions. Macromolecules. *46* (2013) 4606-4616.
        https://doi.org/10.1021/ma400200y

[16]   H. K. Shendi, I. Omrani, A. Ahmadi, A. Farhadian, N. Babanejad, M. R. Nabid,
        Synthesis and characterization of a novel internal emulsifier derived from
        sunflower oil for the preparation of waterborne polyurethane and their application
        in coatings. Prog.Org. Coat. *105* (2017) 303-309.
        https://doi.org/10.1016/j.porgcoat.2016.11.033

[17]   S. C. Wang, P. C. Chen, J. T. Yeh, K. N. Chen, A new curing agent for self-
        curable system of aqueous-based PU dispersion. React. Funct. Polym. *67* (2007)
        299-311. https://doi.org/10.1016/j.reactfunctpolym.2007.01.002

[18]   H. Zou, S. Wu, J. Shen, Polymer/silica nanocomposites: preparation,
        characterization, properties, and applications. Chem. Rev. *108* (2008) 3893-3957.
        https://doi.org/10.1021/cr068035q

[19]   H. Sardon, L. Irusta, M. J. Fernández-Berridi, M. Lansalot, E. Bourgeat-Lami,
        Synthesis of room temperature self-curable waterborne hybrid polyurethanes
        functionalized with (3-aminopropyl) triethoxysilane (APTES). Polymer. *51* (2010)
        5051-5057. https://doi.org/10.1016/j.polymer.2010.08.035

[20]   L. Zhai, R. Liu, F. Peng, Y. Zhang, K. Zhong, J. Yuan, Y. Lan, Synthesis and
        characterization of nanosilica/waterborne polyurethane end-capped by
        alkoxysilane via a sol-gel process. J. Appl. Polym. Sci. *128* (2013) 1715-1724.

[21]   T. Gurunathan, J. S. Chung, Physicochemical properties of amino–silane-
        terminated vegetable oil-based waterborne polyurethane nanocomposites. ACS
        Sustain. Chem. Eng. *4* (2016) 4645-4653.
        https://doi.org/10.1021/acssuschemeng.6b00768

[22]   K. M. Seeni Meera, R. Murali Sankar, S. N. Jaisankar, A. B. Mandal,
        Physicochemical studies on polyurethane/siloxane cross-linked films for
        hydrophobic surfaces by the sol–gel process. J. Phys. Chem. B. *117* (2013) 2682-
        2694. https://doi.org/10.1021/jp3097346

[23]   H. T. Jeon, M. K. Jang, B. K. Kim, K. H. Kim, Synthesis and characterizations of
        waterborne polyurethane–silica hybrids using sol–gel process. Colloid Surf.
        A. *302* (2007) 559-567. https://doi.org/10.1016/j.colsurfa.2007.03.043

[24] M. M. Rahman, H. H. Chun, H. Park, Preparation and properties of waterborne polyurethane-silane: A promising antifouling coating. Macromol Res. *19* (2011) 8-13. https://doi.org/10.1007/s13233-011-0116-5

[25] A. Agirre, J. Nase, C. Creton, J. M. Asua, Adhesives for Low-Energy Surfaces. Macromol. Symp. *281* (2009) 181-190. https://doi.org/10.1002/masy.200950724

[26] A. Agirre, J. Nase, E. Degrandi, C. Creton, J. M. Asua, Improving adhesion of acrylic waterborne PSAs to low surface energy materials: introduction of stearyl acrylate. J. Polym. Sci., Part A: Polym. Chem. *48* (2010) 5030-5039. https://doi.org/10.1002/pola.24300

[27] S. K. Gaddam, S. R. Kutcherlapati, A. Palanisamy, Self-Cross-Linkable Anionic Waterborne Polyurethane–Silanol Dispersions from Cottonseed-Oil-Based Phosphorylated Polyol as Ionic Soft Segment. ACS Sustain. Chem. Eng. *5* (2017) 6447-6455. https://doi.org/10.1021/acssuschemeng.7b00327

[28] S. Pathan, S. Ahmad, Synergistic effects of linseed oil based waterborne alkyd and 3-isocynatopropyl triethoxysilane: Highly Transparent, Mechanically robust, thermally stable, hydrophobic, anticorrosive coatings. ACS Sustain. Chem. Eng. *4* (2016) 3062-3075. https://doi.org/10.1021/acssuschemeng.6b00024

[29] L. Lei, Y. Zhang, C. Ou, Z. Xia, L. Zhong, Synthesis and characterization of waterborne polyurethanes with alkoxy silane groups in the side chains for potential application in waterborne ink. Prog.Org. Coat. *92* (2016) 85-94. https://doi.org/10.1016/j.porgcoat.2015.11.019

[30] Fischer, A. M., & Frey, H. (2010). Soluble hyperbranched poly (glycolide) copolymers. Macromolecules. *43* (20), 8539-8548. https://doi.org/10.1021/ma101710t

[31] S. G. Ramkumar, K. A. Rose, S. Ramakrishnan, Direct synthesis of terminally "clickable" linear and hyperbranched polyesters. J. Polym. Sci., Part A: Polym. Chem. *48* (2010) 3200-3208. https://doi.org/10.1002/pola.24108

[32] X. H. Liu, Y. M. Bao, X. L. Tang, Y. S. Li, Synthesis of hyperbranched polymers via a facile self-condensing vinyl polymerization system–Glycidyl methacrylate/Cp2TiCl2/Zn. Polymer. *51* (2010) 2857-2863. https://doi.org/10.1016/j.polymer.2010.04.034

[33] W. Han, Synthesis and properties of networking waterborne polyurethane/silica nanocomposites by addition of poly (ester amine) dendrimer. Polym. Compos. *34* (2013) 156-163. https://doi.org/10.1002/pc.22388

[34]   K. K. Jena, S. Sahoo, R. Narayan, T. M. Aminabhavi, K. V. S. N. Raju, Novel
       hyperbranched waterborne polyurethane-urea/silica hybrid coatings and their
       characterizations. Polym. Int. *60* (2011) 1504-1513.
       https://doi.org/10.1002/pi.3109

[35]   H. C. Kolb, M. G. Finn, K. B. Sharpless, Click chemistry: diverse chemical
       function from a few good reactions. Angew. Chem. Int. Ed. *40* (2001) 2004-2021.
       https://doi.org/10.1002/1521-3773(20010601)40:11<2004::AID-
       ANIE2004>3.0.CO;2-5

[36]   C. Ornelas, J. Broichhagen, M. Weck, Strain-promoted alkyne azide cycloaddition
       for the functionalization of poly (amide)-based dendrons and dendrimers. J. Am.
       Chem. Soc. *132* (2010) 3923-3931. https://doi.org/10.1021/ja910581d

[37]   D. Sun, X. Miao, K. Zhang, H. Kim, Y. Yuan, Triazole-forming waterborne
       polyurethane composites fabricated with silane coupling agent functionalized
       nano-silica. J. Colloid Interface Sci. *361* (2011) 483-490.
       https://doi.org/10.1016/j.jcis.2011.05.062

[38]   C. E. Hoyle, C. N. Bowman, Thiol–ene click chemistry. Angew. Chem. Int. Ed. *49*
       (2010) 1540-1573. https://doi.org/10.1002/anie.200903924

[39]   M. Desroches, S. Caillol, V. Lapinte, R. Auvergne, B. Boutevin, Synthesis of
       biobased polyols by thiol– ene coupling from vegetable oils. Macromolecules. *44*
       (2011) 2489-2500. https://doi.org/10.1021/ma102884w

[40]   O. Türünç, M. A. Meier, The thiol-ene (click) reaction for the synthesis of plant oil
       derived polymers. Eur. J. Lipid Sci. Technol. *115* (2013) 41-54.
       https://doi.org/10.1002/ejlt.201200148

[41]   C. Fu, X. Hu, Z. Yang, L. Shen, Z. Zheng, Preparation and properties of
       waterborne bio-based polyurethane/siloxane cross-linked films by an in situ sol–
       gel process. Prog.Org. Coat. *84* (2015) 18-27.
       https://doi.org/10.1016/j.porgcoat.2015.02.008

[42]   K. Ishizu, N. Kobayakawa, S. Takano, Y. Tokuno, M. Ozawa, Synthesis of
       polymer particles possessing radical initiating sites on the surface by emulsion
       copolymerization and construction of core–shell structures by a photoinduced
       atom transfer radical polymerization approach. J. Polym. Sci., Part A: Polym. Chem.
       45 (2007) 1771-1777. https://doi.org/10.1002/pola.21944

[43]   L. Zhang, H. Zhang, J. Guo, Synthesis and properties of UV-curable polyester-
       based waterborne polyurethane/functionalized silica composites and morphology

of their nanostructured films. Ind. Eng. Chem. Res. *51* (2012) 8434-8441.
https://doi.org/10.1021/ie3000248

[44] Y. C. Chung, T. K. Cho, B. C. Chun, Flexible cross-linking by both pentaerythritol and polyethyleneglycol spacer and its impact on the mechanical properties and the shape memory effects of polyurethane. J. Appl. Polym. Sci. *112* (2009) 2800-2808. https://doi.org/10.1002/app.29538

[45] E. Zini, M. Scandola, P. Dobrzynski, J. Kasperczyk, M. Bero, Shape Memory Behavior of Novel (l-Lactide- Glycolide- Trimethylene Carbonate) Terpolymers. Biomacromolecules. *8* (2007) 3661-3667. https://doi.org/10.1021/bm700773s

[46] A. Lendlein, H. Jiang, O. Jünger, R. Langer, Light-induced shape-memory polymers. Nature. *434* (2005) 879. https://doi.org/10.1038/nature03496

[47] S. K. Lee, S. H. Yoon, I. Chung, A. Hartwig, B. K. Kim, Waterborne polyurethane nanocomposites having shape memory effects. J. Polym. Sci., Part A: Polym. Chem. *49* (2011) 634-641. https://doi.org/10.1002/pola.24473

[48] M. Barikani, S. Mehdipour-Ataei, H. Yeganeh, Synthesis and properties of novel optically active polyimides. J. Polym. Sci., Part A: Polym. Chem. *39* (2001) 514-518. https://doi.org/10.1002/1099-0518(20010215)39:4<514::AID-POLA1020>3.0.CO;2-4

[49] H. Yeganeh, M. A. Shamekhi, Poly (urethane-imide-imide), a new generation of thermoplastic polyurethane elastomers with enhanced thermal stability. Polymer. *45* (2004) 359-365. https://doi.org/10.1016/j.polymer.2003.11.006

[50] D. H. Jung, M. A. Jeong, H. M. Jeong, B. K. Kim, Chemical hybridization of imidized waterborne polyurethane with silica particle. Colloid. Polym. Sci. *288* (2010) 1465-1470. https://doi.org/10.1007/s00396-010-2279-6

[51] D. B. Otts, M. W. Urban, Heterogeneous crosslinking of waterborne two-component polyurethanes (WB 2K-PUR); stratification processes and the role of water. Polymer. *46* (2005) 2699-2709. https://doi.org/10.1016/j.polymer.2005.01.053

[52] M. Melchiors, M. Sonntag, C. Kobusch, E. Jürgens, Recent developments in aqueous two-component polyurethane (2K-PUR) coatings. Prog.Org. Coat. *40* (2000) 99-109. https://doi.org/10.1016/S0300-9440(00)00123-5

[53]   C. W. Chang, K. T. Lu, Natural castor oil based 2-package waterborne polyurethane wood coatings. Prog.Org. Coat. *75* (2012) 435-443. https://doi.org/10.1016/j.porgcoat.2012.06.013

[54]   S. Yue, Z. Zhang, X. Fan, P. Liu, C. Xiao, Effect of 3-aminopropyltriethoxysilane on solvent resistance, thermal stability, and mechanical properties of two-component waterborne polyurethane. Int. J. Polym. Anal. Charact. *20* (2015) 285-297. https://doi.org/10.1080/1023666X.2015.1015931

[55]   K. Gupta, R. V. Jain, A. Mittal, M. Mathur, S. Sikarwar, Photochemical degradation of the hazardous dye Safranin-T using TiO2 catalyst. J. Colloid Interface Sci. *309* (2007) 464-469. https://doi.org/10.1016/j.jcis.2006.12.010

[56]   D. S. Kim, S. J. Han, S. Y. Kwak, Synthesis and photocatalytic activity of mesoporous TiO2 with the surface area, crystallite size, and pore size. J. Colloid Interface Sci. *316* (2007) 85-91. https://doi.org/10.1016/j.jcis.2007.07.037

[57]   L. Chen, H. Shen, Z. Lu, C. Feng, S. Chen, Y. Wang, Fabrication and characterization of TiO 2–SiO 2 composite nanoparticles and polyurethane/(TiO 2–SiO 2) nanocomposite films. Colloid. Polym. Sci. *285* (2007) 1515. https://doi.org/10.1007/s00396-007-1720-y

[58]   S. C. Tjong, Structural and mechanical properties of polymer nanocomposites. Mater. Sci. Eng. R Rep. *53* (2006) 73-197. https://doi.org/10.1016/j.mser.2006.06.001

[59]   G. Polizos, E. Tuncer, A. L. Agapov, D. Stevens, A. P. Sokolov, M. K. Kidder,... I. Sauers, Effect of polymer–nanoparticle interactions on the glass transition dynamics and the conductivity mechanism in polyurethane titanium dioxide nanocomposites. Polymer. *53* (2012) 595-603. https://doi.org/10.1016/j.polymer.2011.11.050

[60]   C. Chen, Y. Wang, G. Pan, Q. Wang, Gel-sol synthesis of surface-treated TiO 2 nanoparticles and incorporation with waterborne acrylic resin systems for clear UV protective coatings. J. Coat. Technol. Res. *11* (2014) 785-791. https://doi.org/10.1007/s11998-014-9583-x

[61]   Q. F. Xu, Y. Liu, F. J. Lin, B. Mondal, A. M. Lyons, Superhydrophobic TiO2–polymer nanocomposite surface with UV-induced reversible wettability and self-cleaning properties. *ACS Appl. Mater. Interfaces. 5* (2013) 8915-8924. https://doi.org/10.1021/am401668y

[62] H. Yaghoubi, A. Dayerizadeh, S. Han, M. Mulaj, W. Gao, X. Li,... A. Takshi, The effect of surfactant-free TiO2 surface hydroxyl groups on physicochemical, optical and self-cleaning properties of developed coatings on polycarbonate. J Phys D Appl Phys. *46* (2013) 505316. https://doi.org/10.1088/0022-3727/46/50/505316

[63] F. R. Marciano, D. A. Lima-Oliveira, N. S. Da-Silva, A. V. Diniz, E. J. Corat, V. J. Trava-Airoldi, Antibacterial activity of DLC films containing TiO2 nanoparticles. J. Colloid Interface Sci. *340* (2009) 87-92. https://doi.org/10.1016/j.jcis.2009.08.024

[64] O. L. Galkina, A. Sycheva, A. Blagodatskiy, G. Kaptay, V. L. Katanaev, G. A. Seisenbaeva,... A. V. Agafonov, The sol–gel synthesis of cotton/TiO2 composites and their antibacterial properties. Surf. Coat. Technol. *253* (2014) 171-179.

[65] K. Li, J. Peng, M. Zhang, J. Heng, D. Li, C. Mu, Comparative study of the effects of anatase and rutile titanium dioxide nanoparticles on the structure and properties of waterborne polyurethane. Colloids Surf. A. *470* (2015) 92-99. https://doi.org/10.1016/j.colsurfa.2015.01.072

[66] N. Wang, W. Fu, J. Zhang, X. Li, Q. Fang, Corrosion performance of waterborne epoxy coatings containing polyethylenimine treated mesoporous-TiO2 nanoparticles on mild steel. Prog.Org. Coat. *89* (2015) 114-122. https://doi.org/10.1016/j.porgcoat.2015.07.009

[67] D. L. Reid, R. Draper, D. Richardson, A. Demko, T. Allen, E. L. Petersen, S. Seal, In situ synthesis of polyurethane–TiO 2 nanocomposite and performance in solid propellants. J. Mater. Chem. A. *2* (2014) 2313-2322. https://doi.org/10.1039/c3ta14027j

[68] F. Deng, Y. Zhang, X. Li, Y. Liu, Z. Y. Shi, Wang, Synthesis and mechanical properties of dopamine modified titanium dioxide/waterborne polyurethane composites. Polym. Compos. 2017. https://doi.org/10.1002/pc.24654

[69] D. M. Wu, F. X. Qiu, H. P. Xu, D. Y. Yang, Waterborne polyurethane/inorganic hybrid composites: preparation, morphology and properties. Plast. Rubber Compos. *40* (2011) 449-456. https://doi.org/10.1179/1743289810Y.0000000045

[70] A. M. Díez-Pascual, A. L. Díez-Vicente, Wound healing bionanocomposites based on castor oil polymeric films reinforced with chitosan-modified ZnO nanoparticles. Biomacromolecules. *16* (2015) 2631-2644. https://doi.org/10.1021/acs.biomac.5b00447

[71]   S. Awad, H. Chen, G. Chen, X. Gu, J. L. Lee, E. E. Abdel-Hady, Y. C. Jean, Free volumes, glass transitions, and cross-links in zinc oxide/waterborne polyurethane nanocomposites. Macromolecules. *44* (2010) 29-38. https://doi.org/10.1021/ma102366d

[72]   G. Christopher, M. A. Kulandainathan, G. Harichandran, Highly dispersive waterborne polyurethane/ZnO nanocomposites for corrosion protection. J. Coat. Technol. Res. *12* (2015) 657-667. https://doi.org/10.1007/s11998-015-9674-3

[73]   X. Y. Ma, W. D. Zhang, Effects of flower-like ZnO nanowhiskers on the mechanical, thermal and antibacterial properties of waterborne polyurethane. Polym. Degrad. Stab. *94* (2009) 1103-1109. https://doi.org/10.1016/j.polymdegradstab.2009.03.024

[74]   W. D. Zhang, Y. M. Zheng, Y. S. Xu, Y. X. Yu, Q. S. Shi, L. Liu,... Y. Ouyang, Preparation and Antibacterial Property of Waterborne Polyurethane/Zn–Al Layered Double Hydroxides/ZnO Nanocomposites. J. Nanosci. Nanotechnol. *13* (2013) 409-416. https://doi.org/10.1166/jnn.2013.6912

[75]   J. Ni, Q. Zhao, X. Zhao, Transparent and high infrared reflection film having sandwich structure of SiO2/Al: ZnO/SiO2. Prog.Org. Coat. *64* (2009) 317-321. https://doi.org/10.1016/j.porgcoat.2008.08.030

[76]   S. W. Kim, D. K. Lee, Y. S. Kang, Y. J. Kim, Preparation of heat insulating nanocomposite film with MPS (mercaptopropyl trimethoxysilane) coated-nanoparticles. Mol. Cryst. Liq. Cryst. *445* (2006) 81-371. https://doi.org/10.1080/15421400500367058

[77]   K. Ravichandran, P. Philominathan, Fabrication of antimony doped tin oxide (ATO) films by an inexpensive, simplified spray technique using perfume atomizer. Mater. Lett. *62* (2008) 2980-2983. https://doi.org/10.1016/j.matlet.2008.01.119

[78]   R. Outemzabet, N. Bouras, N. Kesri, Microstructure and physical properties of nanofaceted antimony doped tin oxide thin films deposited by chemical vapor deposition on different substrates. Thin Solid Films. *515* (2007) 6518-6520. https://doi.org/10.1016/j.tsf.2006.11.069

[79]   Z. Dai, Z. Li, L. Li, G. Xu, Synthesis and thermal properties of antimony doped tin oxide/waterborne polyurethane nanocomposite films as heat insulating materials. Polym. Adv. Technol. *22* (2011) 1905-1911. https://doi.org/10.1002/pat.1690

[80]   S. K. Dhoke, N. Rajgopalan, A. S. Khanna, Effect of nanoalumina on the electrochemical and mechanical properties of waterborne polyurethane composite coatings. J. Nanopart. Res. *2013*. https://doi.org/10.1155/2013/527432

[81]   A. Mohammadi, M. Barikani, M. M. Lakouraj, Biocompatible polyurethane/thiacalix[4]arenes functionalized Fe_3O_4 magnetic nanocomposites: Synthesis and properties. Mater. Sci. Eng. C. *66* (2016) 106-118. https://doi.org/10.1016/j.msec.2016.04.064

[82]   F. Yan, J. Li, J. Zhang, F. Liu, W. Yang, Preparation of Fe_3O_4/polystyrene composite particles from monolayer oleic acid modified Fe_3O_4 nanoparticles via miniemulsion polymerization. J. Nanopart. Res. *11* (2009) 289-296. https://doi.org/10.1007/s11051-008-9382-3

[83]   L. M. dos Santos, R. Ligabue, A. Dumas, C. Le Roux, P. Micoud, J. F. Meunier,... S. Einloft, Waterborne polyurethane/Fe_3O_4-synthetic talc composites: synthesis, characterization, and magnetic properties. Polym. Bull. (2017) 1-16.

[84]   S. Zhang, Y. Li, L. Peng, Q. Li, S. Chen, K. Hou, Synthesis and characterization of novel waterborne polyurethane nanocomposites with magnetic and electrical properties. Compos Part A: Appl. Sci. Manuf. *55* (2013) 94-101. https://doi.org/10.1016/j.compositesa.2013.05.018

[85]   S. Chen, S. Zhang, Y. Li, G. Zhao, Synthesis and properties of novel UV–curable hyperbranched waterborne polyurethane/Fe_3O_4 nanocomposite films with excellent magnetic properties. RSC Adv. *5* (2015) 4355-4363. https://doi.org/10.1039/C4RA13683G

[86]   M. Y. Mamaghani, M. Pishvaei, B. Kaffashi, Synthesis of latex based antibacterial acrylate polymer/nanosilver via in situ miniemulsion polymerization. Macromol Res. *19* (2011) 243-249. https://doi.org/10.1007/s13233-011-0307-0

[87]   P. Jain, T. Pradeep, Potential of silver nanoparticle-coated polyurethane foam as an antibacterial water filter. Biotechnol. Bioeng. *90* (2005) 59-63. https://doi.org/10.1002/bit.20368

[88]   V. Alt, T. Bechert, P. Steinrücke, M. Wagener, P. Seidel, E. Dingeldein,... R. Schnettler, An in vitro assessment of the antibacterial properties and cytotoxicity of nanoparticulate silver bone cement. Biomaterials. *25* (2004) 4383-4391. https://doi.org/10.1016/j.biomaterials.2003.10.078

[89] R. Kumar, H. Münstedt, Silver ion release from antimicrobial polyamide/silver composites. Biomaterials. *26* (2005) 2081-2088. https://doi.org/10.1016/j.biomaterials.2004.05.030

[90] S. H. Hsu, H. J. Tseng, Y. C. Lin, The biocompatibility and antibacterial properties of waterborne polyurethane-silver nanocomposites. Biomaterials. *31* (2010) 6796-6808. https://doi.org/10.1016/j.biomaterials.2010.05.015

[91] R. Foldbjerg, D. A. Dang, H. Autrup, Cytotoxicity and genotoxicity of silver nanoparticles in the human lung cancer cell line, A549. Arch. Toxicol. *85* (2011) 743-750. https://doi.org/10.1007/s00204-010-0545-5

[92] H. L. Liu, S. A. Dai, K. Y. Fu, S. H. Hsu, Antibacterial properties of silver nanoparticles in three different sizes and their nanocomposites with a new waterborne polyurethane. Int. J. Nanomed. *5* (2010) 1017. https://doi.org/10.2147/IJN.S14572

[93] Y. Wattanodorn, R. Jenkan, P. Atorngitjawat, S. Wirasate, Antibacterial anionic waterborne polyurethanes/Ag nanocomposites with enhanced mechanical properties. Polym. Test. *40* (2014) 163-169. https://doi.org/10.1016/j.polymertesting.2014.09.004

[94] G. T. Howard, Biodegradation of polyurethane: a review. Int. Biodeterior. Biodegrad. *49* (2002) 245-252. https://doi.org/10.1016/S0964-8305(02)00051-3

[95] S. Egger, R. P. Lehmann, M. J. Height, M. J. Loessner, M. Schuppler, Antimicrobial properties of a novel silver-silica nanocomposite material. Appl. Environ. Microbiol. *75* (2009) 2973-2976. https://doi.org/10.1128/AEM.01658-08

[96] M. A. Pérez-Limiñana, F. Arán-Aís, C. Orgilés-Barceló, Waterborne Polyurethane Adhesives Based on Gelatine-Stabilized AgNPs with Improved Antimicrobial Properties. J. Adhes. *90* (2014) 860-876. https://doi.org/10.1080/00218464.2014.884462

[97] D. Turan, G. Gunes, F. Seniha Güner, Synthesis, characterization and O2 permeability of shape memory polyurethane films for fresh produce packaging. Packag. Technol. Sci. *29* (2016) 415-427. https://doi.org/10.1002/pts.2222

[98] M. M. Rahman, Improvements of antimicrobial and barrier properties of waterborne polyurethane containing hydroxyapatite-silver nanoparticles. J. Adhes. Sci. Technol. *31* (2017) 613-626. https://doi.org/10.1080/01694243.2016.1228744

[99] M. Liu, Z. Jia, D. Jia, C. Zhou, Recent advance in research on halloysite nanotubes-polymer nanocomposite. Prog. Polym. Sci. *39* (2014) 1498-1525. https://doi.org/10.1016/j.progpolymsci.2014.04.004

[100] H. Fu, Y. Wang, X. Li, W. Chen, Synthesis of vegetable oil-based waterborne polyurethane/silver-halloysite antibacterial nanocomposites. Compos. Sci. Technol. *126* (2016) 86-93. https://doi.org/10.1016/j.compscitech.2016.02.018

[101] W. S. Han, Synthesis and characterization of hyperbranched waterborne polyurethane/Ag nanoparticle composites. Polym. Compos. 2016.

[102] M. Akbarian, M. E. Olya, M. Ataeefard, M. Mahdavian, The influence of nanosilver on thermal and antibacterial properties of a 2 K waterborne polyurethane coating. Prog.Org. Coat. *75* (2012) 344-348. https://doi.org/10.1016/j.porgcoat.2012.07.017

[103] P. K. Jain, X. Huang, I. H. El-Sayed, M. A. El-Sayed, Noble metals on the nanoscale: optical and photothermal properties and some applications in imaging, sensing, biology, and medicine. Acc. Chem. Res. *41* (2008) 1578-1586. https://doi.org/10.1021/ar7002804

[104] N. Lindenblatt, M. D. Menger, E. Klar, B. Vollmar, Darbepoetin-alpha does not promote microvascular thrombus formation in mice: role of eNOS-dependent protection through platelet and endothelial cell deactivation. Arterioscler. Thromb. Vasc. Biol. *27* (2007) 1191-1198. https://doi.org/10.1161/ATVBAHA.107.141580

[105] M. R. Abid, S. Guo, T. Minami, K. C. Spokes, K. Ueki, C. Skurk,... W. C. Aird, Vascular endothelial growth factor activates PI3K/Akt/forkhead signaling in endothelial cells. Arterioscler. Thromb. Vasc. Biol. *24* (2004) 294-300. https://doi.org/10.1161/01.ATV.0000110502.10593.06

[106] J. Doukas, W. Wrasidlo, G. Noronha, E. Dneprovskaia, R. Fine, S. Weis,... D. Cheresh, Phosphoinositide 3-kinase γ/δ inhibition limits infarct size after myocardial ischemia/reperfusion injury. Proc. Natl. Acad. Sci. *103* (2006) 19866-19871. https://doi.org/10.1073/pnas.0606956103

[107] Y. Feng, V. J. Venema, R. C. Venema, N. Tsai, R. B. Caldwell, VEGF induces nuclear translocation of Flk-1/KDR, endothelial nitric oxide synthase, and caveolin-1 in vascular endothelial cells. Biochem. Biophys. Res. Commun. *256* (1999) 192-197. https://doi.org/10.1006/bbrc.1998.9790

[108] H. S. Hung, C. C. Wu, S. Chien, S. H. Hsu, The behavior of endothelial cells on polyurethane nanocomposites and the associated signaling

pathways. Biomaterials. *30* (2009) 1502-1511.
https://doi.org/10.1016/j.biomaterials.2008.12.003

[109] J. G. Han, Y. Q. Xiang, Y. Zhu, New Antibacterial Composites: Waterborne Polyurethane/Gold Nanocomposites Synthesized Via Self-Emulsifying Method. J. Inorg. Organomet. Polym. Mater. *24* (2014) 283-290.
https://doi.org/10.1007/s10904-013-9965-z

[110] P. B. Messersmith, E. P. Giannelis, Synthesis and characterization of layered silicate-epoxy nanocomposites. Chem. Mater. *6* (1994) 1719-1725.
https://doi.org/10.1021/cm00046a026

[111] Z. Wang, T. J. Pinnavaia, Nanolayer reinforcement of elastomeric polyurethane. Chem. Mater. *10* (1998) 3769-3771.
https://doi.org/10.1021/cm980448n

[112] J. Massam, T. J. Pinnavaia, Clay nanolayer reinforcement of a glassy epoxy polymer. MRS Online Proceedings Library Archive. *520* (1998).

[113] J. W. Gilman, Flammability and thermal stability studies of polymer layered-silicate (clay) nanocomposites1. Appl. Clay Sci. *15* (1999) 31-49.
https://doi.org/10.1016/S0169-1317(99)00019-8

[114] H. T. Lee, L. H. Lin, Waterborne polyurethane/clay nanocomposites: novel effects of the clay and its interlayer ions on the morphology and physical and electrical properties. Macromolecules. *39* (2006) 6133-6141.
https://doi.org/10.1021/ma060621y

[115] M. M. Rahman, H. J. Yoo, C. J. Mi, H. D. Kim, Synthesis and characterization of waterborne polyurethane/clay nanocomposite–effect on adhesive strength. Macromol. Symp. 249 (2007) 251-258. https://doi.org/10.1002/masy.200750341

[116] M. M. Rahman, H. D. Kim, W. K. Lee, Preparation and characterization of waterborne polyurethane/clay nanocomposite: effect on water vapor permeability. J. Appl. Polym. Sci. *110* (2008) 3697-3705.
https://doi.org/10.1002/app.28985

[117] S. Subramani, J. Y. Lee, S. W. Choi, J. H. Kim, Waterborne trifunctionalsilane-terminated polyurethane nanocomposite with silane-modified clay. J. Polym. Sci., Part B: Polym. Phys. *45* (2007) 2747-2761. https://doi.org/10.1002/polb.21285

[118] J. M. Yeh, C. T. Yao, C. F. Hsieh, L. H. Lin, P. L. Chen, J. C. Wu, C. P. Wu, Preparation, characterization and electrochemical corrosion studies on environmentally friendly waterborne polyurethane/Na+-MMT clay nanocomposite

coatings. Eur. Polym. J. *44* (2008) 3046-3056.
https://doi.org/10.1016/j.eurpolymj.2008.05.037

[119] L. Peng, L. Zhou, Y. Li, F. Pan, S. Zhang, Synthesis and properties of waterborne
polyurethane/attapulgite nanocomposites. Compos. Sci. Technol. *71* (2011) 1280-
1285. https://doi.org/10.1016/j.compscitech.2011.04.012

[120] M. Paillet, A. Dufresne, Chitin whisker reinforced thermoplastic
nanocomposites. Macromolecules. *34* (2001) 6527-6530.
https://doi.org/10.1021/ma002049v

[121] M. A. S. Azizi Samir, F. Alloin, A. Dufresne, Review of recent research into
cellulosic whiskers, their properties and their application in nanocomposite
field. Biomacromolecules. *6* (2005) 612-626. https://doi.org/10.1021/bm0493685

[122] M. Matos Ruiz, J. Y. Cavaille, A. Dufresne, J. F. Gerard, C. Graillat, Processing
and characterization of new thermoset nanocomposites based on cellulose
whiskers. Compos. Interface. *7* (2000) 117-131.
https://doi.org/10.1163/156855400300184271

[123] Q. Zhao, S. Wang, X. Cheng, R. C. Yam, D. Kong, R. K. Li, Surface Modification
of Cellulose Fiber via Supramolecular Assembly of Biodegradable Polyesters by
the Aid of Host− Guest Inclusion Complexation. Biomacromolecules. *11* (2010)
1364-1369. https://doi.org/10.1021/bm100140n

[124] X. Cao, H. Dong, C. M. Li, New nanocomposite materials reinforced with flax
cellulose nanocrystals in waterborne polyurethane. Biomacromolecules. *8* (2007)
899-904. https://doi.org/10.1021/bm0610368

[125] Z. Gao, J. Peng, T. Zhong, J. Sun, X. Wang, C. Yue, Biocompatible elastomer of
waterborne polyurethane based on castor oil and polyethylene glycol with
cellulose nanocrystals. Carbohydr Polym. *87* (2012) 2068-2075.
https://doi.org/10.1016/j.carbpol.2011.10.027

[126] A. Santamaria-Echart, L. Ugarte, C. García-Astrain, A. Arbelaiz, M. A. Corcuera,
A. Eceiza, Cellulose nanocrystals reinforced environmentally-friendly waterborne
polyurethane nanocomposites. Carbohydr Polym. *151* (2016) 1203-1209.
https://doi.org/10.1016/j.carbpol.2016.06.069

[127] C. J. Silva, Q. Zhang, J. Shen, A. Cavaco-Paulo, Immobilization of proteases with
a water soluble–insoluble reversible polymer for treatment of wool. Enzyme
Microb. Technol. *39* (2006) 634-640.
https://doi.org/10.1016/j.enzmictec.2005.11.016

[128] J. Zhang, S. B. Zhao, H. H. Luo, Analysis and estimate of shrink-proof finish methods of wool? Journal ofWuhan Textile S.h.t. Institute, *10* (1997) 80–83.

[129] Q. Zhao, G. Sun, K. Yan, A. Zhou, Y. Chen, Novel bio-antifelting agent based on waterborne polyurethane and cellulose nanocrystals. Carbohydr Polym. *91* (2013) 169-174. https://doi.org/10.1016/j.carbpol.2012.08.020

[130] H. Liu, J. Song, S. Shang, Z. Song, D. Wang, Cellulose nanocrystal/silver nanoparticle composites as bifunctional nanofillers within waterborne polyurethane. ACS Appl. Mater. Interfaces. *4* (2012) 2413-2419. https://doi.org/10.1021/am3000209

[131] G. M. Wu, J. Chen, S. P. Huo, G. F. Liu, Z. W. Kong, Thermoset nanocomposites from two-component waterborne polyurethanes and cellulose whiskers. Carbohydr Polym. *105* (2014) 207-213. https://doi.org/10.1016/j.carbpol.2014.01.095

[132] L. Jiang, J. Zhang, M. P. Wolcott, Comparison of polylactide/nano-sized calcium carbonate and polylactide/montmorillonite composites: reinforcing effects and toughening mechanisms. Polymer. *48* (2007) 7632-7644. https://doi.org/10.1016/j.polymer.2007.11.001

[133] N. Lin, J. Huang, P. R. Chang, D. P. Anderson, J. Yu, Preparation, modification, and application of starch nanocrystals in nanomaterials: a review. J. Nanomater. (2011) 20. https://doi.org/10.1155/2011/573687

[134] J. Zou, F. Zhang, J. Huang, P. R. Chang, Z. Su, J. Yu, Effects of starch nanocrystals on structure and properties of waterborne polyurethane-based composites. Carbohydr Polym. *85* (2011) 824-831. https://doi.org/10.1016/j.carbpol.2011.04.006

[135] P. R. Chang, F. Ai, Y. Chen, A. Dufresne, J. Huang, Effects of starch nanocrystal-graft-polycaprolactone on mechanical properties of waterborne polyurethane-based nanocomposites. J. Appl. Polym. Sci. *111* (2009) 619-627.

[136] Y. Wang, H. Tian, L. Zhang, Role of starch nanocrystals and cellulose whiskers in synergistic reinforcement of waterborne polyurethane. Carbohydr Polym. *80* (2010) 665-671. https://doi.org/10.1016/j.carbpol.2009.10.043

[137] N. Yousefi, M. M. Gudarzi, Q. Zheng, X. Lin, X. Shen, J. Jia, ... J. K. Kim, Highly aligned, ultralarge-size reduced graphene oxide/polyurethane nanocomposites: mechanical properties and moisture permeability. Compos Part A: Appl. Sci. Manuf. *49* (2013) 42-50. https://doi.org/10.1016/j.compositesa.2013.02.005

[138] S. Wu, T. Shi, L. Zhang, Preparation and properties of amine-functionalized reduced graphene oxide/waterborne polyurethane nanocomposites. High Perform. Polym. *28* (2016) 453-465. https://doi.org/10.1177/0954008315587124

[139] J. Hu, F. Zhang, Self-assembled fabrication and flame-retardant properties of reduced graphene oxide/waterborne polyurethane nanocomposites. J. Therm. Anal. Calorim. *118* (2014) 1561-1568. https://doi.org/10.1007/s10973-014-4078-7

[140] Y. Li, Z. Yang, H. Qiu, Y. Dai, Q. Zheng, J. Li, J. Yang, Self-aligned graphene as anticorrosive barrier in waterborne polyurethane composite coatings. J. Mater. Chem. A. *2* (2014) 14139-14145. https://doi.org/10.1039/C4TA02262A

[141] C. X. Zhao, W. D. Zhang, D. C. Sun, Preparation and mechanical properties of waterborne polyurethane/carbon nanotube composites. Polym. Compos. *30* (2009) 649-654. https://doi.org/10.1002/pc.20609

[142] M. M. Rahman, R. Suleiman, H. Do Kim, Effect of functionalized multiwalled carbon nanotubes on weather degradation and corrosion of waterborne polyurethane coatings. Korean J. Chem. Eng. *34* (2017) 2480-2487. https://doi.org/10.1007/s11814-017-0145-7

[143] M. M. Rahman, E. Y. Kim, K. T. Lim, W. K. Lee, Morphology and properties of waterborne polyurethane/CNT nanocomposite adhesives with various carboxyl acid salt groups. J. Adhes. Sci. Technol. *23* (2009) 839-850. https://doi.org/10.1163/156856109X411210

Chapter 9

Classical Thermoset Epoxy Composites for Structural Purposes: Designing, Preparation, Properties and Applications

A.E. Kolosov[1*], E.P. Kolosoval, V.V. Vanin[1], Anish Khan[2]

[1]National Technical University of Ukraine «Igor Sikorsky Kyiv Polytechnic Institute», 19 Build., 37 Prospect Peremohy, 03056, Kyiv, Ukraine

[2]King Abdulaziz University, Jeddah, 21589, Saudi Arabia

a-kolosov@ukr.net

Abstract

Classical thermosetting epoxy composites for structural purpose, along with nanocomposites, are now widely used in various industries. An epoxy matrix is considered as a dominant polymer matrix in the design of such composites due to its study, high performance and wide commercial use. The optimization of processes and design and technological parameters of the equipment for their molding and processing of the polymer composite materials (PCMs), as well as the creation of PCMs with a predetermined set of properties, remains an urgent task nowadays. Equally important problems are the production of defect-free and monolithic structures of such composites while increasing the productivity of their molding. Particular attention is paid to low-frequency ultrasonic as a basic method of physical modification of the liquid epoxy media and intensification of the processes of capillary impregnation and "wet" winding.

Keywords

Thermoset, Epoxy, Composite, Prepreg, Modeling, Design, Technology, Ultrasonic

Contents

9.1 Introduction

Classical thermosetting composites of structural applicatione on the basis of reinforced fibers and epoxy matrix are widely used in many spheres of modern industry. The latter include aviation, rocket and space industries, engineering, energy, communications, gas, chemical, shipbuilding, electrical engineering and a lot of other industries. Such materials also find increasing application in the municipal economy, in particular, in technologies for connecting and repairing of polymeric pipelines. Today, there is a huge amount of polymer composite materials (PCM), which differ not only in compositions and properties, but also in production technology. However, the optimization of processes and design and technological parameters of the equipment for their molding and processing, as well as the creation of a PCM with a predetermined set of properties, remains an urgent task nowadays [1].

Manufacturing products from PCM is a relatively complex technological process, which is based on the use of certain physical and chemical patterns. Depending on the molding conditions of PCM, its physical and mechanical properties change. Therefore, the choice and justification of the regime parameters of molding, as well as the parameters of the forming equipment, are of fundamental importance. Of particular importance is the development of the theoretical foundations of molding and processing of PCM in connection with the growth of PCM production – both thermoplastic and reactoplastic. This puts high demands on existing molding technologies and equipment that implements them [2].

We will understand the modification of PCM (both on the basis of thermoplastic and on the reactoplastic matrices), as it is customary for scientific and technical literature [1, 2], as directed regulation of their structure and properties. Such modification can be carried out by both chemical and physical (or physico-chemical) methods. In addition to purely chemical methods of modification (such as copolymerization, grafting, cross-linking, etc.), the polymer processing technology operates with physico-chemical methods. Among the latter one can distinguish such methods as plasticization, filling, fusion of two or more polymers, treatment with high frequency currents, ultrasonic (US), laser and radiation [1, 2]. These methods can change the chemical structure of the polymer, its physical (supramolecular) structure, composition and phase structure of both the oligomer and the polymer binder (PB) on its basis. All this leads to a directed change in the properties of the final solidified polymer.

It should be noted that prediction and creation of a PCM with the necessary complex of properties is an extremely complex scientific and technical task for a number of reasons. First, there are still not enough clear theoretical concepts that allow to synthesize new

PCM with specific properties, and also to predict the regime parameters of their formation. Especially it concerns modeling of technological processes of impregnation, "wet" winding, and also forecasting of constructive and technological parameters of forming equipment [3]. An effective direction of modeling and forecasting of parameters of the abovementioned objects is the use of adequate structural (structural-parametric) models of the media (objects) under consideration.

If we investigate the history of the creation of polymer systems, we can see that in most cases the theoretical ideas about the properties of polymers (in particular, their adhesion) appeared after the development of the corresponding (specific) materials [4]. These theories and concepts are certainly important, as they expand our understanding of the mechanism of emerging processes. However, none of the existing theories is universal. Therefore, it seems advisable to talk only about the creation of a scientifically based system of representations, which covers a wide range of issues related to the physico-chemical modification in the molding of PCM [1 – 4]. Secondly, since the polymer forms the basis of the composition, the choice of a particular polymer is the first and decisive step in the development of efficient construction-oriented PCM. Third, PCM includes, in addition to polymer, fillers, including fiber fillers (FF), stabilizers, plasticizers, thickeners, thixotropic additives and other components. Each of these additives in the polymer system performs certain functions.

When creating a specific PCM, it is necessary to understand clearly how these components (that is, chemical modification) will affect the properties of the final PCM. Equally important is the use of various methods of physical modification. Such directions of physical modification include, for example, hardening using US, treatment with high-frequency currents, and others. At the same time, the use of mechanical oscillations of the US range, or US vibrations (USV), is one of the most promising means of physical impact on liquid or solid components. Such an effect is widely used in chemical technology to intensify a number of technological processes, in particular, the formation of reinforced plastics. For practical use, the inexpensive, high-strength, usually hard-cured epoxy polymers (EP) obtained on the basis of epoxy oligomers (EO) are of greatest interest as the reactoplastic polymer compositions. They are widely used as epoxy binders (EB) for the manufacture, in particular, of fibrous PCM for structural purposes [5]. EPs are characterized by a unique combination of a set of operational properties. Among the latter – high strength characteristics, good adhesion to various materials, high resistance to corrosive media, etc.

Epoxy PCM far exceeds traditional compositions that contain mineral binders, as well as materials based on other synthetic resins (polyester, furan, carbamide, etc.). Thus, for example, the tensile strength of EP may reach 150 MPa, compression strength – 400

MPa, bending strength – 220 MPa, and the elasticity modulus – 5000 MPa [6]. Currently in the CIS countries more than 30 grades of injection molding and impregnating epoxy resins (ER) and epoxy binders (EB) are produced [7]. Epoxy-dian resins are the most widely used, and as a result their production in the total production volume is more than 90%. At the same time, more than one hundred grades of hardeners for ER have been developed. Also, there are developments in the field of creating new types of EP and their hardeners. At the same time, it should be noted that the necessary condition for the optimal technological solutions being developed is to ensure high performance properties of molded PCM at the lowest energy costs of molding.

Thus, the aforementioned brief analysis of the different aspects of molding of classical thermoset PCM structural application on the basis of reinforcing fibers and an epoxy matrix brings out the actual directions of study. The following directions can be identified:

- analysis of aspects of physico-chemical modification of EP, including dispersed and continuous FF;

- substantiation of the effectiveness of physical modification in the form of US action for liquid epoxy matrices;

- investigation of the influence of US treatment regimes on the properties of EP;

- parametrization of devices and technologies for US treatment of liquid polymer systems;

- modeling of the structure of oriented and woven fibrous materials as the base stage of modeling the technology of capillary impregnation and "wet" winding;

- modeling of the technology of US production of PCM;

- analysis of the effective application of US in the production of thermoplastics and thermosets.

The above aspects are briefly studied in this chapter.

9.2   Methods for modifying liquid epoxy compositions

At present, the main trend of molding PCM for structural purposes on an industrial scale is not so much the development of new polymers as in the modification of known materials already used for their manufacture [7]. The modification consists in the purposeful regulation of the structure and the associated properties of the polymer at various levels of the technological process.

As noted above, the purpose of the modification is to improve the process and performance characteristics of epoxy PCM:

- increase vitality, decrease viscosity, improve deformation-strength properties, heat, bio and chemical stability,

- increase dielectric properties, reduce combustibility, improve economic performance (reduce EBs consumption, reduce their cost, recycle production waste).

Modern methods of polymer modification are divided into three main groups: chemical, physical and physico-chemical. A combination of these methods is also used, which is shortly described below.

9.2.1 Chemical modification of liquid epoxy compositions

Chemical modification is realized by changing the chemical structure of the oligomer, by varying the type of hardener, by introducing into the system the reaction additives that enter into a chemical reaction with the molecular network.

The methods of chemical modification can be classified as follows [7]:

- modification based on the chemical transformation of already synthesized macromolecules;

- modification at the stage of polymer synthesis.

Chemical modification of polymers is carried out by introducing into the composition of macromolecules small fragments of a different nature. The change in the chemical nature of oligomers and hardener allows several results to be achieved at once. Namely: to increase the length of the molecular chain of the oligomer and hardener, to vary the structure of interstitial sites of the solidified system, and to modify the terminal groups of the oligomer macromolecules.

This changes the macroscopic properties of the EP. The introduction of reaction additives capable of reacting chemically with a polymer allows a wide range of adjustment of the physical and mechanical properties of EP, in particular, its heat resistance, aging resistance etc.

9.2.2 Physico-chemical modification of liquid epoxy compositions

The most widely used are physicochemical methods of modification of the structure of EP by regulating the composition of epoxy PCM. As a result of such impact, it is possible to obtain materials with a predetermined set of operational properties [7]. Such a modification is carried out by introducing solid insoluble fillers, inert plasticizers, solvents, stabilizers, alloying additives, surfactants and other modifiers into the liquid polymer matrix.

Particular mention should be made of filling, since it is the most widely used and highly effective method of directional control of EP properties. It allows increaseing the indexes of mechanical strength and rigidity, chemical resistance, heat resistance, dielectric properties, etc. In the general case, the filling of polymers is understood as the combination of polymers with solid and (or) gaseous substances.

The latter are relatively evenly distributed in the volume of the composition and have a clearly expressed interface with the continuous polymer phase (matrix). To obtain filled polymer compositions, in most cases, solid fine dispersible fillers are used. They are presented in the form of particles of spherical shape (glass microspheres, fly ash), granular form (soot, silica, wood flour, chalk, kaolin), plate form (talc, graphite, mica), needle form (oxides, salts, silicates), as well as FF (cotton, glass, organic and carbon fiber, asbestos, cellulose).

Low-filled PCM (relative degree of filling v is from 0 to 0.3) have high values of deformation and toughness but low static strength. While highly filled structural PCM (relative degree of filling v from 0.3 to 0.7) have high values of rigidity and compressive strength with respect to unfilled polymers.

However, such composites are characterized by high brittleness and low ultimate deformation failure [7].

9.2.3 Methods of physical modification of liquid epoxy compositions

Together with the above-described methods of chemical and physico-chemical modification, methods of physical modification of EP are widely used. These include: heat treatment, modification by radiation methods, modification by vacuum-compressor treatment, periodic deformation, treatment of polymers in magnetic fields [2].

Physical modification can be carried out at different stages of production, processing and use of polymers. For example, during its synthesis, at the stage of processing the polymer into an article, when processing the finished material before or during its use under certain conditions. Often physical modification is used in conjunction with chemical or physico-chemical methods of polymer modification. One of the effective methods for increasing the operational properties of EP is heat treatment. Heat treatment significantly affects the molecular mobility, structural order, speed and depth of curing. This leads to an improvement in the physical properties of the EP. Particular attention is paid to the choice of hardening modes EC. The main parameters are the temperature and duration of hardening, as well as the rate of heating and cooling. The choice of hardening temperature depends on the type of EO and the hardener used.

EP that solidified at elevated temperatures are characterized by increased strength and rigidity, which is explained by a change in the supramolecular structure of EO [5, 7]. The solidification temperature also affects the appearance and size of supramolecular formations in the polymer. With increasing temperature of hardening due to the intensification of the thermal motion of molecules, the sizes of associates and their lifetime decrease. This affects the decrease in the size of globules. The globule size depends on the density of the grid, and, consequently, on the parameters of the elastic properties of the EP. Polymers with a finely globular structure are more durable. This is due to the fact that a decrease in the size of globules increases the probability of chemical and physical interaction of molecules due to functional groups located on the surface [5].

Increasing the depth of hardening leads to a shift in the glass transition temperature of the polymers to higher temperatures. This significantly reduces the curing time, which can significantly accelerate the production of EP. It is also possible to heat-treat already solidified polymers. This reduces the number of functional groups of EO and hardener that have not reacted, due to pre-hardening of the polymer, and also increases the hardening depth. This leads to an increase in the performance characteristics of the EP [7].

An effective method of increasing the performance characteristics of epoxy PCM is vibration processing, or low-frequency processing. Under the influence of low-frequency oscillations, the viscosity of liquid compositions sharply decreases, and the conditions for their homogenization and processing improve. The vibrational action on the oligomer before its solidification leads to a less defective and more ordered structure, which contributes to the growth of the strength of the polymers. However, as a limiting factor of such an influence, an insignificant relaxation time of the vibrotreated composition appears. Under the influence of US, the conditions for homogenization of the mixture, its viscosity, relaxation time, and the kinetics of EB hardening change.

Proceeding from the principle of representing the structure of EO as a superposition of two spatially inhomogeneous grids – a thermo-fluctuation network of physical bonds and a relatively stable molecular lattice, it is possible to single out the principle of enhancing the EP by restructuring its structure. The latter ultimately leads to an increase in its heterogeneity, which ensures the strengthening of the EP in the glassy state.

The processing of polymers in a magnetic field is widely used. Under the influence of the magnetic field, the viscosity of the polymer mixture decreases, its uniformity increases.

Thermoset Composite   Materials Research Forum LLC
Materials Research Foundations **38** (2018)   doi: http://dx.doi.org/10.21741/9781945291876

As a result of the formation of a more ordered structure of solidified compositions, their strength increases, as well as a number of other performance characteristics of the EP.

Similar results are obtained by the radiation hardening of epoxides. The methods of modification by low-temperature and electrothermal processing, ultraviolet and infrared radiation, etc. are successfully applied. In practice, to give materials a number of special properties, a combination of several methods of physical modification is often used simultaneously with other modification methods.

9.3   Physico-chemical aspects of the modification of epoxy polymers by dispersed and continuous fibrous fillers

When considering the physico-chemical aspects of modification of EP by dispersed and continuous FF, a number of questions should be considered first.

Among them are the features of the formation of clusters in a polymer composite, the surface interaction of fillers with EO, the mechanism of molecular interaction between EP and filler, as well as the adhesion between EP and filler [7].

It is also necessary to analyze possible ways of using US to improve the efficiency of physico-chemical modification of these EP.

9.3.1   Features of the formation of clusters in a polymer composite

As a rule, a cluster is understood as a group of filler particles separated by thin polymer interlayers that are completely in the film phase [7]. Cluster structures from dispersed particles are formed as a result of a number of processes (diffuse, sedimentary, etc.), associated with the involuntary relative movement of the filler particles. Such structures are also formed as a result of the forced motion of the filler particles with mixing (homogenization) of the polymer matrix. The formation of clusters begins with the interaction of two separate particles.

Since the structure of the boundary layer is formed as a result of the tendency of these filler particles to lower their surface energy, it is energetically more advantageous when the limiting (feyson) layers of individual particles begin to interact with each other. This leads to an uneven distribution of particles, but at the same time contributes to the compensation of the energy excess [7]. In this case, the filler particles begin to be structured so that the polymer in the space between them completely passes into an orientation-ordered state with the formation of linear clusters. That is, clusters _l_ are formed in which particles of radius r_c are located along a curve or some conventional line (see _Fig. 9.1_).

Figure 9.1 Scheme of cluster formation in a polymer composite, including the formation of a linear cluster and transformation of an unstable linear cluster: 1 – elementary two-particle linear cluster; 2 – annular cluster.

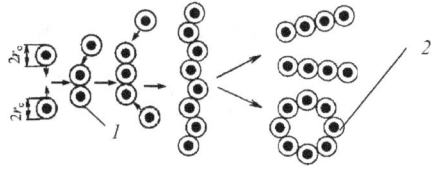

The increase in the length of linear clusters occurs up to a certain size. After that, the cluster becomes hydrodynamically unstable and decays into several small linear clusters, or ring-shaped clusters *2* are formed. The latter, in turn, are grouped together to form spatial cellular clusters in the form of irregular (distorted) spheres. The peripheral layer of the latter consists of filler particles alternating with the film phase of the polymer matrix. The inner region of the sphere is free of filler particles and contains only the polymeric matrix in the bulk state. If the filler is polydisperse, then during its polymer connection, the fine particles are "captured" by the surface of larger particles. As a result, dense ball clusters arise.

It should be noted that clusters that are freely distributed in the bulk of the polymer material and are not bonded together do not exert a strengthening effect on the solidified polymer. With the destruction of the polymer, they can serve only as a stopper for the emerging cracks, reducing the rate of their propagation. This leads to an increase in the crack resistance of solidified composites. An increase in the strength of a solidified composite filled with filler particles occurs when a spatial framework is formed in the volume of this composite from the filler particles and the film phase of the polymer matrix. The transition from individual clusters to the framework of a bulk polymeric matrix is due to the unification and enlargement of individual small clusters.

At a certain stage of filling, the entire volume of the composite is penetrated by one infinite cluster with the formation of a spatial framework. This leads to an increase in the strength of the composite. In this case, US treatment to a liquid filled polymer matrix can be considered as an intensifying factor of cluster formation.

In addition, the dispersion of the filler particles also has a significant influence on the strength of the PCM. The main reason of this is the value of the specific surface area of the filler per one particle. Thus, an increase in the size of the filler particles leads to an increase in their surface area, i.e., to a decrease in their surface energy, and hence energy of cohesion. This, in turn, leads to a decrease in the strength of the composite. And vice versa: with increasing dispersity of the filler, the strength of the PCM increases.

9.3.2 Analysis of the surface interaction of fillers with epoxy oligomers

The features of the surface interaction of fillers (both inorganic and organic) with EO are analyzed below.

9.3.2.1 Surface interaction of inorganic fillers with epoxy oligomers

When constructing PCM for structural purposes, it is necessary to strive to obtain a strong thermally and hydrolytically stable connection between the surface of the FF and the polymer matrix that ensures their joint operation. To ensure good adhesion between EP and inorganic filler, the formation of a strong and non-hydrolyzable chemical bond is necessary.That is, there must be groups on the surface of FF capable of chemical interaction with functional groups of EO. As a rule, various oxides, silicates and some other compounds are used as inorganic fillers in EP. Comparatively detailed data on the chemical structure of the surface is only for oxides and silicates. Therefore, we shall restrict ourselves to considering only these classes of inorganic fillers.

The most characteristic property of all these surfaces is the presence of a coating of hydroxyl groups. Te number of the latter can vary greatly in their behavior and chemical activity, and their number can be quite significant [5, 7]. In complex silicates, surface groups can be bound to different atoms, which lead to an even greater difference in their properties. In addition, in many cases, especially in many component polymer glasses, the surface composition differs from the composition of the mass of the material [5]. The content of active groups can vary by an order of magnitude depending on the prehistory of the filler. It is hydroxyl groups that play the most important role in the interaction of excipients with EO.

However, in addition to hydroxyl groups, water molecules are adsorbed on the surface of hydrophilic inorganic oxides and silicates, the amount of which depends on the humidity of the surrounding medium. Adsorbed water has a negative effect on the adhesion of EO. It prevents the formation of strong chemical and hydrogen bonds between the filler surface and the polymer, especially when hardening the polymer at room temperature. Coordinated-unsaturated centers turn out to be almost completely blocked by adsorbed water molecules [5]. Adsorbed molecules and atoms create new surface states or change the parameters of existing electronic surface states [5].

And the energy spectrum of the filler surface largely determines the nature of the physical and chemical interaction "polymer – filler". The effect of various impurities in ER, which are often concentrated on the surface of the filler, should also be mentioned. As impurities in ER, polar compounds are usually present. They are actively adsorbed by the filler surface and successfully compete with epoxide and hardener molecules, displacing them from the surface. However, this leads to a sharp deterioration in adhesion.

Especially large water sorption is observed in cases when the filler composition (at least in small amounts) includes soluble in water or compounds that are easily hydrolyzed (for example, alkali or alkaline earth metal oxides). In such cases, the absorption of water greatly increases as a result of the formation of a layer of an aqueous solution on the surface of the filler. Because of this, the sorption values take on values corresponding to tens and hundreds of monomolecular layers.

Thus, the surfaces of inorganic fillers are characterized by high adsorption activity [5, 7]. This leads to the presence of layers of water and various organic compounds that are adsorbed from the environment (if the filler is not subjected to special purification beforehand).They greatly impair the adhesion of EBs. At the same time, the composition of organic contaminants can vary widely. Of great importance is also the state of the filler surface, which, as mentioned above, is usually covered by adsorbed molecules of water and other compounds. This makes it difficult to wet and react the liquid polymer with the filler.

Therefore, it is important to study effective methods for cleaning the surface (fibrous glass) of the filler just before it is impregnated with EB.Such an action can be carried out, for example, by US. As a result, a reduction in the various impurities and the amount of water and organic compounds can be achieved, and adhesion is improved.

9.3.2.2 Surface interaction of organic fillers with epoxy oligomers

The interaction mechanism in the systems "polymer filler – oligomeric polymer binder (PB)" differs greatly from the mechanism described above for the interaction of oligomeric binders with mineral fillers.

This is due to the following factors:

- the elasticity moduli and thermal expansion coefficients of the filler and polymer matrix are close;

- the components of the polymer binder can penetrate the filler volume, the most active are low-molecular weight hardeners (amines and anhydrides);

- partial dissolution of the fiber in the polymer is possible;

- the interaction "polymer fiber – oligomeric binder" is not limited to the surface of the fiber, but can extend to a considerable depth of the fiber;

- polymer filler can dissolve low-molecular impurities in the binder and condensation products, which reduces the tendency to pore formation.

It can be said that the above factors generally lead to an increase in the monolithic structure of the EP. On the surface and in the volume of fibers, there are always different compounds, which are low molecular weight fractions of the fiber-forming polymer. These are textile lubricants, solvent residues and other technological impurities, as well as various impurities absorbed by the fiber during its manufacture, storage and processing. For example, depending on the chemical nature, polymer fibers can absorb up to $10 - 12\%$ water [5].

The presence of such impurities is almost inevitable and get rid of them without changing the properties of the fiber is very difficult. In the manufacture of PCM, these compounds partially pass into the EB and change its properties. In this regard, the choice of effective methods for cleaning the surface of organic fillers from unwanted impurities immediately before the impregnation of EB, for example, with US, is of scientific and practical interest.

Thus, the interaction of the oligomeric binder with organic excipients, for example, kapron, is much more complicated than with mineral fillers, and at present it is still little investigated. Therefore, it is relevant to study the results of this interaction, in particular, experimentally.

9.3.2.3 The mechanism of molecular interaction between epoxy polymer and filler

Molecular interaction between the polymer and the filler can proceed through different mechanisms [5]. Thus, between the active functional groups of the EB and the filler, chemical interaction takes place with the formation of strong chemical bonds. In addition, the existence of the whole spectrum of physical bonds is observed: from Van-der-Waals to hydrogen bonds, which cause the phenomena of wetting, adhesion and the formation of interphase layers [5, 7].

Despite the importance of the processes of interfacial molecular interaction in filled polymers, many aspects of these processes have been little studied. In addition, in the scientific literature on this subject, there are different, sometimes opposing, opinions [1, 2, 5, 7].

Thus, it has been established that the chemical interaction of EB with the substrate surface from the filler material or mineral filler can proceed according to several mechanisms:

1) the reaction of the surface OH-groups of the filler with epoxy groups:

2) the reaction of the surface OH-groups of the filler with a hardener (in particular, anhydride);

3) the reaction of the surface OH-groups of the filler with the OH- or ester groups of the EP;

4) the interaction of various surface centers with EP.

As a result of the reactions between the EB and the polymer, there are mainly strong chemical bonds of the type C-O. They largely determine the high adhesion of EP. However, such bonds are readily hydrolyzed, which is the reason for the low water resistance of the filled EP.

With chemical modification of the FF surface, the liquid epoxy matrix can interact with grafted bifunctional molecules on the surface of the filler to form nonhydrolyzable Si-O-Si and C-C bonds. This significantly increases the water resistance of the produced EP.

In the interaction of EB with other active surface centers that do not contain an OH-group, obviously, other reactions can occur. But they have practically not been investigated yet [10]. Hydroxyl OH-groups play the most important role in the interaction of FF with EB. However, in addition to hydroxyl OH-groups, molecules of water and other compounds adsorbed from the air can be located on the surface of the filler. They make it difficult to wet and react the polymer with the filler. Consideration of the ways of solving these problems lies in the framework of consideration of mechanics of adhesion between EP and FF.

9.4   Effect of ultrasonic treatment regimes on the properties of epoxy polymers

9.4.1   Technological and operational properties of epoxy polymers

To implement the optimal regime of US treatment of liquid media, it is first of all necessary to choose the effective (optimal) values of the intensity I and the frequency f of the US vibrations (USV) [8]. Most US treatment of liquid technological processes is associated with US cavitation and a sound-capillary effect [9]. Therefore, by changing the conditions of cavitation, various cavitation effects can be intensified or weakened [10, 11].

Analysis of some effective methods for obtaining prepregs (impregnated semifinished products) on the basis of EC with the use of US modification suggests the following. For each composition of EC and PCM based on them, the effective parameters of the US modification should be established, as a rule, experimentally. Thus, in particular, it was found that as a result of vibration, the dynamic viscosity decreases, as well as the modulus of elasticity of polymer melts [9]. The extreme dependence of the viscosity of the ED-20 oligomer as a function of the vibroprocessing time τ at medium US frequencies was established in [12], and the effective time of the US action was 30 min. US treatment is effected by means of a concentrator of longitudinal US at a frequency of $(17 - 44)$ kHz, amplitude $(50 - 120)$ μm, intensity $(15–30)$ W/cm^2 and temperature $(70 - 90)$ °C for $(30 - 45)$ min. The volume density of the US energy introduced into the EC in this case was $<\omega> = (1,2 - 2,5)$ W/cm^3 [13].

To assess the effectiveness of the US modification, comparative experimental studies of the composition for impregnation on the basis of EO brand ED-20 and hardener diethylenetriamine (DETA) were carried out. After US treatment, 100 w.p. ED-20 was mixed with 10 w.p. hardener DETA. The resulting mixture was cured by the 20 °C/24 hour mode +130 °C/6 hours. The effect of US treatment on the properties of solidified EP was determined. At the same time, the change in technological characteristics (dynamic viscosity η, the minimum equilibrium contact angle of the EO on the glass substrate Θ_{min} at 20 °C, the maximum height of the EO on the glass fiber h_{max} at 50 °C) and the operating characteristics (glass transition temperature T_g) were monitored. *Table 9.1* compares the results of measurements of some technological characteristics of the initial EO and performance characteristics of solidified EP based on the composition ED-20 + DETA.

Table 9.1 Influence of US processing parameters on technological and operational properties of EO and EC in ED-20 + DETA composition [13].

Parameters of US treatment					EO properties			
f, kHz	A, μm	I, W/cm^2	T, °C	τ, min	η, Pa·sec	Θ_{min}, °	h_{max}, mm	T_g, °C
The starting epoxy resin ED-20								
–	–	–	–	–	0,78	30	0,90	104
US processing of epoxy resin ED-20 according to the developed energy-saving technology								
16	10	4	50	25	0,83	18,6	2,40	118,5
16	30	10	80	35	0,83	18,8	2,50	119,0
18	10	4	50	25	0,83	18,5	2,40	118,5
18	30	10	80	35	0,83	18,9	2,48	118,5
20	20	7	65	30	0,83	18,4	2,35	120,0
22	10	4	50	25	0,83	18,5	2,25	118,8
24	20	7	60	30	0,83	18,5	2,10	119,0

It has been established that the time dependence of the wetting contact angle Θ and the wetting power $\sigma cos\Theta$ for low US frequencies f and the temperatures of heating (treatment) is also extreme (see *Fig. 9.2*). This made it possible to find the effective time range of the sonification τ.

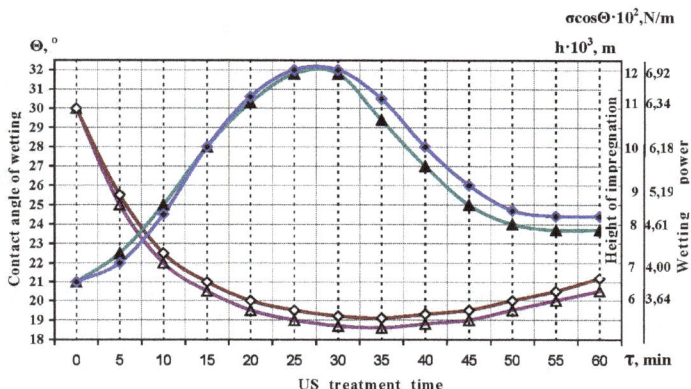

Figure 9.2 Change of the contact angle of wetting Θ on the glass substrate and the wetting capacity $\sigma cos\theta$ of EO brand ED-20 at 20 °C, depending on the time of sonification τ in the low-frequency US range $f = (16 - 20$ kHz):
frequency $f = 16$ kHz for Θ ($\triangle$) and for $\sigma cos\theta$ ($\blacktriangle$);
frequency $f = 20$ kHz for Θ ($\lozenge$) and $\sigma cos\theta$ ($\blacklozenge$);
h – height of lifting of EO brand ED-20 on glass capillary

As the amplitude of the oscillations A decreases (and, consequently, also of the intensity I), the contact angle of wetting θ of the EO with respect to the glass fiber decreases somewhat, and the value of the wetting power $\sigma cos\Theta$ increases. After mixing the components of the composition to improve its homogenization, the EC is treated at the same values of f, A, I, but at a temperature of $(20 - 25)$ °C and within $(10 - 25)$ sec. The following was established experimentally.

If the use of US treatment for mixture of resin ED-20 (both sonificated and not sonificated) and the DETA hardener at a temperature exceeding 25 °C and also for a time exceeding 25 sec are treated, a gradual uncontrolled hardening of the mixture occurs with a sharp deterioration of the final properties of the polymer. Thus, the physical modification of the epoxide-amine composition ED-20 + DETA due to the application of effective regimes of the US process allows one to solve the technical problem of obtaining a modified EC. This composition is used both for the impregnation of woven fibrous composites, and for the formation of EP, which has high performance characteristics.

In addition, an analysis of the peculiarities of the physical (namely, US) modification of liquid media indicates that the US modification of the EO is promising both in the low-frequency and mid-frequency US ranges [14]. In turn, the use of excess pressure in the implementation of US liquid polymer media is an important factor in increasing the intensity and reducing the time of US treatment [15]. After all, the intensity increases and the time of US treatment is reduced. Therefore, from this point of view, it is expedient to study the effective parameters of US modification of EO under both normal and excess pressure.

9.4.2 Physico-mechanical and technological properties of sonificated epoxy matrices

The significant efficiency of US treatment of liquid EC, which changes not only the technological characteristics of binders, but also the physical properties of the network polymers obtained after their solidification, is established as a result of a number of studies [16, 17, 18]. It has also been found that US treatment of these compositions leads to an increase in the speed and completeness of the crosslinking process, and also reduces the time of their heat treatment by $(20 - 30)\%$ compared to untreated systems. Acceleration of the solidification process is explained by more intensive mixing of components and uniform distribution of hardener in the volume of the matrix.

The variance of the values (the coefficient of variation) of the strengths treated with US and solidified EC under compression can be reduced by 40%, impact strength by 80%, hardness by 20%, and for shear strength by $(1.5 - 2)$ times in depending on processing conditions [18]. The dependence of the density ρ_n of the solidified mesh EP on the time of sonification τ is also of an extreme nature with a maximum in the range $(25 - 35)$ min. It is characteristic that the glass transition temperature T_g of sonificatied and solidified EPs also increases by $15 - 17$ °C, reaching a maximum in the same time interval of sonification.

Thus, it can be concluded that the sonification of EO on effective parameters leads not only to an increase in the density of molecular packing, but also to the growth of the effective density of the macromolecular network. This leads to a corresponding strengthening of the EP. Such a change in the density ρ_n characterizes the growth of the molecular packing density of the glassy polymer, the total energy of the intermolecular interaction, and thus the cohesive strength of the EP. The dependence of the parameters of the asociative structure of isochordal EO as a function of the time of sonification τ has the same character [18]. This allows us to conclude the implementation of the principle of structural heredity in the physical modification of EO and EC.

It was found that the change in the asocyanate structure as fluctuation formations with a long viability time is of a relaxational nature and obeys the temperature-time superposition principle. And the time of structural relaxation reaches several tens of hours under normal exposure conditions.

Effective technological regimes for processing of liquid epoxy polymers using low-frequency US cavitation technology are developed. The use of the developed regimes allows to increase the technological characteristics of EC and the performance characteristics of solidified polymers based on them, in particular, adhesion to the surface of structural materials, on average by (20 – 30) %. Also, the use of US to process the EC also reduces the time of their solidification by about (2 – 4) times. In this case, the coefficient of variation of the operational characteristics is reduced by a factor of (2.5 – 3) times. Also, the wetting power of the polymer matrix increases by (30 – 50) %, which contributes to a faster and higher quality impregnation.

This creates the prerequisites for a directional optimization of the technological parameters for the manufacture of products, and in the end results in increased productivity, as well as significant energy savings.

9.5 Ultrasonic intensification of prepregs formation

9.5.1 Process of capillary impregnation

In the manufacture of PCM based on thermoplastic matrices and FF, one of the most important aspects are the technological processes of capillary impregnation (hereinafter impregnation), as well as dosed deposition. These processes are basic when forming prepregs (pre-impregnated and partially dried FF). This is due to the fact that the performance properties of products from oriented and woven polymer composites are due to the strength of adhesion between FF and the polymer matrix.

A prerequisite for achieving high bond strength is good wetting and high-quality impregnation of the filler with a liquid polymer binder. The monolithic and the density of the structure of the composite, its physico-mechanical characteristics also depend on the quality of impregnation. In particular, such materials as orthopedic implants and endoprostheses, decorative laminated plastics, synthetic plywood, getinax, electrical insulating materials (textolite, insulation tapes, etc.), artificial veneer, construction materials for rocket - car - and shipbuilding, construction, medicine, agriculture and other industries are impregnated with roll (oriented) fibrous materials with liquid EB.

Thus, the capillary structure of FF should be attributed to the most important factors that affect the speed and completeness of the impregnation process. After all, the inflow of impregnating liquid into these fillers is a process that is mainly governed by the laws of capillarity and viscosity [9]. At the same time, the optimal pulling force of the capillary impregnation process and the "wet" winding, which affects the strength of the PCM obtained, is the optimal pulling force N_{opt} of the oriented FF (OFF) during its impregnation and "wet" winding [19].

Therefore, the study of impregnation processes, as well as means of intensification of impregnation and "wet" winding, for example, by physical (namely, US) modification with simultaneous hardening of the polymer matrix, acquires particular urgency [20]. After all, this allows us to outline the optimal ways to intensify the processes of obtaining high-strength reinforced reactoplastic PCM [21].

9.5.2   Effect of ultrasonic modification regimes on the kinetics of impregnation of continuous fibrous fillers

The following regimes of the process of longitudinal US impregnation of continuous FF were studied:

1). impregnation of the filler by PB that is not sonicated and sonicated according to the optimal mode and US-contact impact of the concentrator on the impregnated filler, which is impregnated, by not sonicated and sonicated binder (first impregnation variant);

2). short-term US impact of the concentrator on the surface of the pre-impregnated filler and US treatment of the impregnated filler with an annealed binder (second impregnation variant).

In the US treatment of the filler, which is impregnated, the concentrator was positioned in the space between the mandrel and the binder mirror and the heat exchanger. This facilitated the removal of air inclusions from the interfiber space of the filler. And due to the presence of excess EB in the working area of the concentrator during the motion of the filler through the binder, a uniform distribution of the EB along the cross-section of the impregnated filler occurred. Optimal tension force of N_{opt} glass-fiber FF during its impregnation, determined in accordance with the procedure [13, 15], was 3.5 N/m.

Fig. 9.3 shows the kinetic curves of the longitudinal impregnation of fiberglass by the ED-20 oligomer at a temperature of 50 °C for the two above-mentioned impregnation variants.

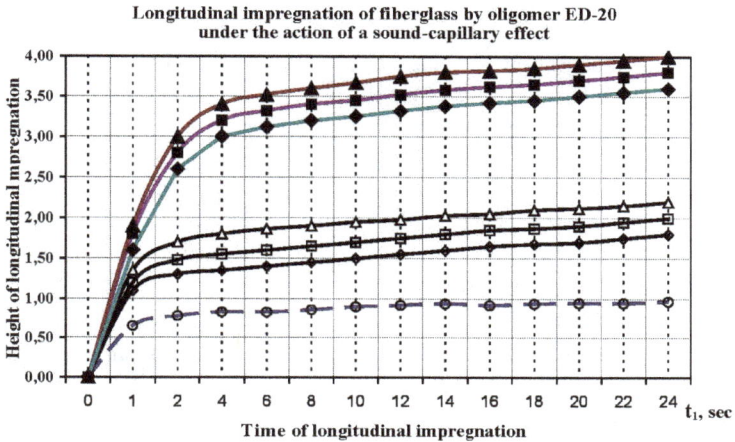

Figure 9.3 Kinetics of longitudinal impregnation of the ED-20 oligomer on glass fibers according to variants I and II of the US impregnation:

(○) – the kinetic curve of the "free" impregnation;

(□) – the theoretical kinetic curve obtained from the longitudinal impregnation equation [12] for the first (I) variant of US impregnation at a frequency f = 17 kHz;

(△) is the experimental kinetic curve obtained by the first (I) variant of US impregnation at a frequency f = 17 kHz;

(◊) is the experimental kinetic curve obtained by the first (I) variant of US impregnation at a frequency f = 22 kHz;

(■) is the theoretical kinetic curve obtained from the longitudinal impregnation equation [12] for the second (II) version of the US impregnation at a frequency f = 17 kHz;

(▲) is the experimental kinetic curve obtained by the second (II) variant of US impregnation at a frequency f = 17 kHz;

(◆) is the experimental kinetic curve obtained by the second (II) variant of US impregnation at a frequency f = 22 kHz.

Upon contact with the ED-20 oligomer, the kinetic curve of the longitudinal impregnation monotonically rises upward and after 15 sec leaves at the end of the saturation plateau (variant I of the impregnation). In the case of impregnation of the filler with a sonificated oligomer for a combined (optimal) regime, the time to reach the saturation region is reduced by a factor of two or three. And the height of the rise of the oligomer along the fiber increases two to three times (variant II of the impregnation). As a result of the contact US action, activation of the surface of pre-impregnated glass-fiber fillers takes place for a short time (0.5 – 10) sec. [14].

There is also an improvement in the wetting power of ER on glass fibers and an increase in the total specific surface area of the fillers Sss under the action of shock waves arising during the convolution of cavitation cavities [15]. US activation begins at an intensity I of US that exceeds a certain threshold value. This value depends on the state of the surface of the solid phase (filler), as well as on the nature and magnitude of the forces of interaction between individual particles of the solid phase. After switching on the US generator (USG), a sharp increase in the lifting of the binder was carried out on a glass-fiber surface pre-treated with a concentrator.

Further, the kinetic curve emerged into the saturation region (the upper kinetic curve 1 in Figure 9.3). After the USG was switched off at the point where saturation began, a certain residual increase in EB over the filler was observed. And the total height of lifting of the binder in the case of US impregnation of OFF by the impregnation variant II increased by (2.5 – 4) times [12, 13, 21].

9.6 Ultrasonic processing devices for liquid polymer systems

The effect of elastic oscillations of US frequency bands on polymer liquids makes it possible to widely use these oscillations for the intensification of a number of processes, including the production of PCM for structural purposes [13]. In connection with this, it is important to develop improved technical means for cavitation processing that generate USV necessary for specific technological processes [22]. Also, appropriate methods for calculating the parameters of these regimes and the equipment that implements them are needed [23].

For this purpose, a method was developed for calculating the US cavitation device with a radiating plate. This technique is used for US modification and intensification of the impregnation and dosing processes in the molding of PCM for structural purposes [24]. The initial data for the development of this technique is the resonance frequency frod of USV, which is characteristic for a particular technological process of US modification. For example, this can be the operating frequency of the USG. Also, the initial parameters are the intensity I of US cavitation and the width of the processed woven filler (or the volume of the US bath for sonification of PB).

The procedure is illustrated in Fig. 9.4 – Fig. 9.6. *Fig. 9.4* shows a diagram of the regular placement of US vibrators on the lower surface of the working radiating plate. *Fig. 9.5* shows the location and connection of US vibrators assembled on piezoelectric transducers, which transmit the US to the radiating plate. *Fig. 9.6* shows the design of a single-wavelength US disperser with a symmetric piezoelectric packet transducer.

Figure 9.4 The scheme of the regular placement of US vibrators on the lower surface of the radiating plate, which carries out resonance bending vibrations (the dotted line denotes the boundaries of a similar section of the plate with the elements of its attachment to the US vibrator): 1 – the radiating plate of width H_{pl}; 2 – places of attachment of US vibrators on the lower surface of the working radiating plate; 3 – waves of bending oscillations along the length L_{pl} of the plate of the plate; 4 – waves of flexural vibrations along the width B_{pl} of the plate.

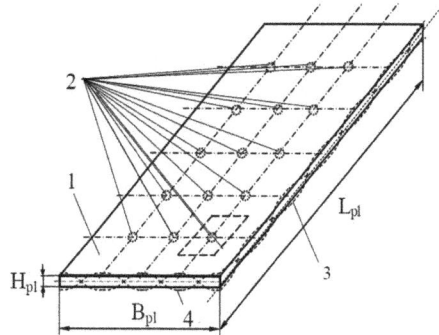

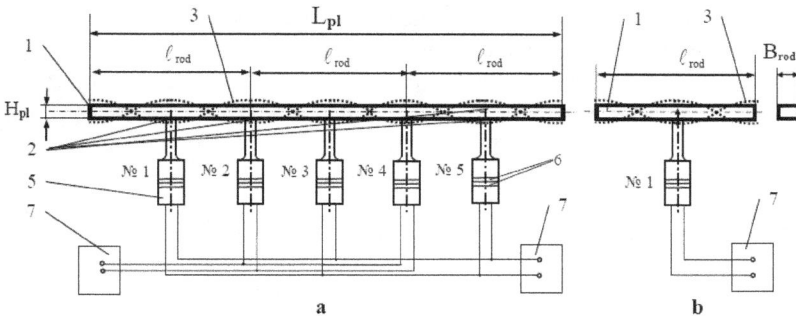

Figure 9.5 The order of placement and connection of ultrasonic vibrators assembled on piezoelectric transducers on the lower part of the radiating plate along its length (a) and on a rod (waveguide) of width B_{pl}, which performs bending vibrations (b): 1 – radiating plate (or rod); 2 – the places of attachment of US vibrators on the lower surface of the working radiating plate; 3 – waves of bending vibrations; 5 – US vibrators (№1 – №5), assembled on the basis of piezoceramic transducers 6; 7 – USG.

Figure 9.6 The design scheme of a single-wavelength US disperser with a symmetric piezoelectric packet converter:

1 – radiating cylinder of equivalent cross-section (modeling plate); 6 – piezoceramic transducers; 7 – USG; 8, 10 – the overlays reducing US frequency; 9 – electrodes of thickness 0.2 – 0.3 mm; 11 – USV concentrator (speed transformer); 12 – a wave of longitudinal vibrations of length λ.

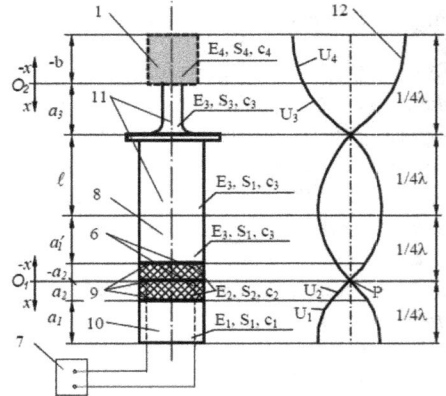

As controlled parameters of low-frequency ultrasonic vibrations (16 – 24 kHz), the frequency f, the intensity I, and the amplitude A of the elastic vibrations generated by the outer surface of the radiating plate in the environment were chosen. It was experimentally established that when contacting US treatment of both impregnated fabrics with a width of 1120 mm and the same fabric but impregnated with a PB, it is necessary to have the following values of technological parameters of cavitation processing: frequency of USV is $f_{US} = f_{rod} = (18 - 22)$ kHz; amplitude A of USV is $(3 - 5)$ microns; the intensity I of USV is $(2 - 4)$ W/cm^2.

The technique involves calculating the US cavitation device with a radiating plate, which performs bending vibrations, and consists of the following consecutive stages:

1) are set by the desired resonance frequency f_{rod} of USV, which is characteristic for a specific technological process (impregnation, dosing or wet-wound). For example, this can be the oscillation frequency of USG 7 – see Fig. 9.5), and determine the intensity I of US cavitation.

2) Select the specific material the speed of sound propagation c_{pl}, the modulus of elasticity E_{pl}, and the thickness H_{pl} of the radiating plate 1 (see Fig. 9.4), based on the structural and technological features of the realization of the technological process being studied.

3) For a selected resonant frequency USV f_{rod} are determined in accordance with [24] for the wavelength ℓ_{rod} of the stee bending vibrations of the radiating plate 1, taking into account the desired oscillation mode n_k (or of the order of the frequency λ_{rod}).

4) The overall dimensions of the radiating plate *1*, i.e., its length L_{pl} and width B_{pl}, are chosen to be multiples of the wavelength of the ℓ_{rod}, i.e., $L_{pl} = N_L \cdot \ell_{rod}$, $B_{pl} = M_B \cdot \ell_{rod}$, where N_L, M_B – integers, depending on the dimensions of the processed woven fillers.

5) Depending on the number of antinodes obtained during bending vibrations with wavelength ℓ_{rod} along and across the radiating plate *1*, determine the number of US emitters that must be installed along the length and width of plate *1*.

6) Calculate the mass of the mounting elements of the US radiator, taking into account the attached mass of a single section (section) of the radiating plate *1*, that is, the characteristics of an equivalent cylinder of density γ_4 and the cross-sectional area S_4, that is, (γ_4, c_4, E_4, S_4).

7) For the resonant frequency $f_{rod} = f_{pl}$ of the radiating plate *1*, the acoustic dimensions of the elements of the composite US radiator (a_1, a_3, ℓ) are calculated – see Fig. 9.5.

The initial data for such calculations are the geometric dimensions and physical parameters of the applied piezomaterials *6* (a_2, c_2, E_2, S_2), the speed transformer *11* (c_3, E_3, S_3), which reduce the frequency of the lining *8*, *10* (c_1, E_1, S_1), and specific acoustic power. The developed technique for calculating the US cavitator with a radiating plate was tested when several models of cavitation apparatus for the chemical industry were created. In particular, its approbation was carried out to regulate the dosed application of EB in impregnated fiberglass electrically insulating cloth with a width of 1120 mm. On the basis of the technique, experimental US cavitation devices with a power of 2.5 kW were created. They are designed for US impregnation of insulating glass fabrics and the subsequent formation of foil dielectrics. Thus, the technique allows calculating the gamma of the sizes of cavitation devices for different sizes of the radiating plate and various processing conditions of US treatment.

9.7 Modeling of the structure of oriented and woven fibrous materials

Currently, there are two main approaches to the design of technology parameters and the structural analysis for constructional elements of composites based on OFF: phenomenological and structural [13].

The first approach is based on considering the composite as a monolithic material with a certain tensor of reduced elastic (viscoelastic) characteristics. The second approach is based on the assumption of the heterogeneity of the properties of the investigated continuum with cylindrical inclusions (fibers) of arbitrary cross section.

The last approach, which was chosen as the base one, allows correctly to model the parameters of the technology and to carry out design of the technology parameters. With its help, it is also possible to calculate the stress-strain state of structures from PCM, taking into account the specific type of load. This is due to the fact that this approach takes into account the actual structure of the investigated fibrous composites.

This approach involves the analysis of some structural models of capillary-porous bodies. On the basis of these models, an adequate physical model of the capillary-porous medium on the basis of OFF is selected. The parameters of the latter, in turn, are used to predict the technological parameters for obtaining products from PCM, in particular, in the impregnation process [25] and "wet" winding process [26].

9.7.1 Physical models of a capillary-porous medium based on oriented fibrous fillers

To describe the transfer processes of impregnating liquid or moisture in impregnation technologies, new model physico-mathematical and procedural-computerized approaches are used and are being developed. They are realized, in particular, using physical (structural) models [27]. First of all, these are: 1) Structural-Network models of porous and corpuscular medium of various structures – "Pore Network Models" (random, regular and other lattices and packages); 2) methodology and mathematical apparatus of the Theory of Percolation; for our purposes, it is considering the transport conditions for a liquid or gas, depending on the statistical characteristics of the lattices that model a dispersed or porous medium (percolation in bonds, percolation through knots); 3) Cluster Analysis (cluster – bundle) – statistical identification of isolated structures.

In the transport problems, associations are studied-clusters of conducting regions, particles, molecules, etc., their dimensions, characteristics, and connection with the transport properties of the medium. The varieties of the cluster approach have long been used in problems of classification analysis. Now the term "cluster" has become fashionable and is used in a wide variety of areas, from molecular processes to economics and politics.

4) Fractal Analysis (Fractal Theory) – modeling and studying the properties of geometric elements (lines, surfaces, volumetric formations), taking into account the change in length, area, volume when the scale of measurement is changed.

And all of the above approaches are often used together, or in an arbitrary combination. In the case under study, the structural-network model of impregnation of OFF as a capillary-porous body is used to determine the refined mathematical model of the technological process for impregnating OFF with liquid EB. This also uses the classical theory of filtration for the laminar flow of a viscous non-Newtonian fluid that is non-compressible [25]. The basic question in determining the parameters of the kinetic equation of the impregnation process is the correct determination of the structural characteristics of the physical model of an OFF, on the basis of which this equation was obtained.

First of all, these characteristics include porosity ε, specific internal surface S_{sp} and effective (hydraulic) capillary radius r_{ef} OFF. In constructing of structural (physical) OFF model four options to determine its parameters have been developed, in particular, the effective (hydraulic) capillary radius r_{ef} [13]. The first of these options is based on microstructural analysis of the cross section of the composite on the basis of OFF. The second is the determination of the r_{ef} on the basis of the analysis for the characteristic kinetic curve of the process of capillary impregnation. The third one is the calculation of the r_{ef} with calculated value for a highly compacted OFF of circular cross-section (for hexagonal packing of OFF fibers). The fourth is the determination of the r_{ef} rate, depending on the force of its tension N during impregnation and "wet" winding.

In this first embodiment are useful for determining the structural characteristics of both optically transparent and optically opaque (e.g., metalofibrous) compositionally-oriented fibrous media. This variant is based on an experimental study of the microstructure of these media and the construction of the corresponding distribution curves (histograms) of structural parameters. In this variant, an adequate model of the structure of OFF, first of all, should take into account the stochastic character of fiber distribution in the structure of the composite on the basis of OFF [28]. *Fig. 9.7* shows a computational model of a regular-structure OFF.

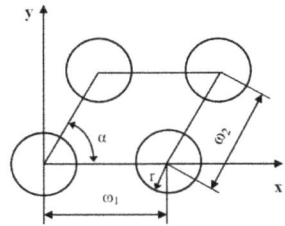

Figure 9.7 Adequate structural oriented fiber filaments' model with equal-sized nodal circles.

As the fiber packing parameters, the lengths of its sides ω_1 and ω_2 and the angle between them α are chosen. The unit of measurement of distances in the model plane is, for convenience, the base length of the side of the unit cell ω_1. A criterion for the adequacy of the structural model of the composite under study is the coincidence of the coefficients and the probability characteristics of the distribution curves of the lengths of chords of circles and the distances between neighboring fibers (the so-called ext-chords) to a certain degree of accuracy. The values of "porosity" ε and the relative binder content in the structural model and in the natural composite, respectively, are also taken into account. In practice, the coincidence of these parameters is (90 – 95)%.

On the basis of the parameters of the structural model and the type of fiber packing in the cross section of the composite on the basis of OFF, the optimal reinforcement scheme is selected in the optimum. Based on the latter, the development of rational technology at the design stage is carried out. For example, it can be an estimation of process parameters of "free" or US impregnation of OFF by means of PB [29]. The parameters of the structural model of OFF are also used to predict the stress-strain state of the structure from the PCM, taking into account the nature of the acting load [30].

9.8   Modeling of technical means for production of polymer composite materials

9.8.1   The technology of ultrasonic production of long-length epoxy composites

The conducted complex studies have shown that the use of US effects on OFF impregnated in the bath, as well as on impregnated OFF, can be considered as an effective method for automatically maintaining the required amount of EB content in impregnated OFF. *Fig. 9.8* shows the general scheme of the arrangement of technological US devices, with which the developed technology was realized [30]. These US devices were used for impregnation and dosing of EBs on impregnated OFFs on a Ukrainian MPT-3 impregnating-drying machine [31].

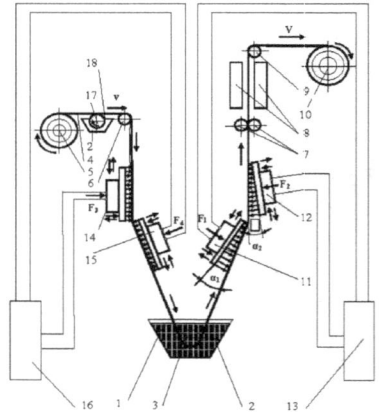

Figure 9.8   Scheme of US production of PCMs:
1 – impregnating bath;
2 – polymer binder (EB);
3 – immersion roll;
4 – impregnated material (long-length OFF);
5 – a coil reel;
6, 9 – enveloping rolls;
7 – squeezing rollers;
8 – drying chamber;
10 – take-up reel;
11, 12 and 14, 15 – pairs of working US tools with plates;
13, 16 – USG;
17 – alluvial rolls;
18 – capacity with sonificated EB.

After winding from the reel 5 OFF 4, its one-side pre-impregnation of EB 2 is carried out from one side with the ankle shaft 17, which rotates in the capacity 18 with the sonificated PB 2. After this, US activation of the surface and degassing of the structure of the pre-impregnated OFF with the intensity of USV I_3 and I_4 and a dosed force of pressing F_3 and F_4 are carried out. This is done by means of a pair of working tools 14 and 15 having individual drives from USG 16 and in contact with pre-impregnated material 4 over the entire surface of the radiating plates.

By means of the action of the second pair of working tools 11 and 12 inclined to the material being dosed at angles α_1 and α_2, and also the pressing forces F_1 and F_2, an analogy of the physical effect is obtained in the form of peristaltic movement of liquid and pasty media. The radiating plates $11 - 14$ must be made of a material with a high bending strength and corrosion resistance.

Due to both contact and non-contact impact of US testing on pre-impregnated and pre-impregnated OFF, the following results are achieved in the developed device: improvement of homogenization of EB by US; US activation of the surface of the OFF to improve its wettability of EB; degassing the structure of the OFF immediately before its impregnation; increasing the productivity of impregnation processes and dosing of EB by increasing the speed of broaching the OFF while improving the properties of the final PCM [32]; reduction of the total time for obtaining prepregs (at least $2 - 3$ times).

Strengthening of classical PCM, obtained according to this scheme with using of combined US treatment, is no less than $(15 - 18)$ %. The obtained results of the developed US technology can be used in the development of energy-efficient regimes and design-technological parameters of impregnating-drying machines [33 – 36].

Further modeling of the technology for obtaining PCM should be carried out based on the principle of the system approach [37]. This approach consists in the development of an appropriate structural-parametric model of the technology under investigation. In this case, it involves analyzing the object of US technology of obtaining PCM simultaneously as a set of certain interrelated elements in the form of technological operations, and as a potential component of the highest hierarchical level. In particular, it seems expedient to separate the technological scheme shown in Fig. 9.8 into such separate structured blocks I – III (they are not marked in Fig. 9.8), which are further investigated:

I – block of US treatment of EB and preparation of impregnating composition;

II – block of "free" impregnation of OFF by liquid EB;

III – block for dosing the application of liquid EB on impregnated OFF of woven type.

Later in the synthesis, only the above-mentioned enlarged (base) blocks I – III and their constituent structural elements are analyzed, and also the interrelations between them are investigated [38].

9.8.2 Modeling of technical means for thermoplastic production

It should also be noted briefly on some results of modeling the design and technological parameters of thermoplastic molding, which are based on a number of approaches common to thermosets. The main basic difference in the rheological properties of these two classes of polymers is, first of all, in the values of the dynamic viscosities of the processed melts of thermoplastics and solutions of thermosets, which differ substantially. Despite this, some results of modeling the design and technological parameters of molding thermoplastics deserve special attention.

Thus, in study [39] constructive-technological parameters of the equipment intended for forming corrugated polymer pipes are considered. The effects of the volume variation as well as the wall thickness of the pipe billet are analyzed from the point of view of the productivity of the forming equipment. In study [40], the developed approach is described, which establishes the relationship a priori of the design parameters and the selected corrugation scheme with the geometric dimensions of the corrugated product of different shapes and sizes. At the same time, the obtained results are generalized to any types of corrugated tubular products from thermoplastics.

The results of profiling the corrugated pipe in the case of forming both low-profile and high-strength corrugations with vertical and inclined walls, as well as with a flat or circular arc are analyzed in [41]. The physical and technological conditions for welding corrugated layers have been formulated, and a numerical solution of the temperature distribution in the weld zone of the corrugated layers has been obtained.

Features of extrusion molding of tubular polymeric billets for manufacturing corrugated pipes are described in the article [42]. The study [43] describes a non-isothermal process of mixing in a barrier mixer of a worm machine by numerical simulation. As a result of numerical studies, the distribution of the velocities of the polymer melt flow is established, which flows in the working channels of the barrier mixer, under different temperature and deformation processing conditions, as well as the design parameters of the mixer.

The results of modeling the viscoelastic flow of the thermoplastic polymer at the exit of the extruder tool using the Phan-Thien-Tanner model are presented in investigation [44]. The results obtained can be applied to project modeling of geometric parameters of extrusion heads.

In study [45], improved physical and mathematical models of the melting of thermoplastic polymer in the channels of screw extruders are proposed. It is assumed that in the melting zone, a solid porous plug, a melt mixture, solid pellet residues and gas, and a fully melted thermoplastic material cooperate together.

9.9   Other applications of ultrasonic in the production of thermosets and thermoplastic

9.9.1   The effectiveness of ultrasonic treatment for the production of epoxy nanocomposites

The use of US in the production of thermosets and thermoplastics is extremely wide. So, for example, it concerns first of all epoxy nanocomposites. It should be noted that the discovery of carbon nanotubes (CNTs) as an independent material with unique physico-chemical characteristics is undoubtedly one of the most significant achievements of modern science in the last three decades. At the same time, the study of the structure and properties, as well as the applications of CNTs, is of considerable fundamental and applied interest [46].

A number of technological and operational characteristics of both elastomers and rigid polymers can be significantly, sometimes even times, improved by modifying them with small additions of nanoparticles – fullerenes, CNTs, nanofibres, inorganic nanoparticles, etc. [47]. It has been experimentally established that the necessary conditions for the creation of effective nanomodified (NM) PCM are the small size and maximally uniform distribution of nanoparticles in the polymer matrix [48].

In addition, the tendency of nanoparticles to aggregate prevents the formation of stable dispersions in water and in organic media, including solutions of liquid polymers. That is why an intensive search is being made for effective methods that facilitate the disaggregation (disintegration) of CNT in organic solvents and liquid polymer media [49]. Such methods include: chemical modification of nanoparticles by low-molecular compounds and polymers with the formation of covalent bonds between the molecules of the modifier and nanoparticles, as well as non-covalent modification of nanoparticles with the help of surfactants of both low-molecular and polymeric nature.

It was found that the most stable and uniform dispersions are also formed when a submerged US disperser is used in low-frequency cavitation regimes. These regimes are especially important in the production of reactoplastic NM PCM using EO modifications with CNT [50].

One of the promising types of structural OFF is carbon fiber, so it is given special attention. Despite the fact that carbon plastics are used to manufacture structural elements in modern military aircraft, space vehicles, the wide use of carbon composite materials in civil aircraft designs is definitely limited. This is due primarily to the fact that, having a high dimensional stability, heat resistance and resistance to various external influences of the environment, carbon plastics belong to brittle materials. The latter have low values of transversal and shear strength.

One of the ways to solve this problem is to modify the surface of carbon macrofibers and EB to improve the mechanical properties when creating carbon plastics combined filling. In the latter, a continuous carbon fiber is combined with a PB, in the volume of which the ultradisperse carbon particles are uniformly distributed [51]. The prospects of using such NM carbon plastics as load-bearing elements of power structures are described in [52].

Modeling the electrostatic control over depth of the introduction of intelligent sensors into a NM PCMs are studied in [53]. In addition, a promising area of research is the use of low-frequency US in the creation of so-called intelligent (smart) PCMs. This direction is an advanced trend in modern polymeric material science [54].

9.9.2 Repair technologies for the maintenance and restoration of polyethylene pipelines

There are other areas of combined use of US for both thermoplastics and thermosets. So, in particular, among the class of thermoplastic polymers, polyethylene (PE) takes a leading position both in terms of production volume and volume of use. For example, more than 50% of PE pipelines (PEPL) are used in the structure of the world gas supply, water supply, sewerage network, etc., and this share is constantly growing [55]. In particular, PE gas pipelines of small diameter (up to 220 mm) are widely used, the life of which is up to 50 years.

However, the low surface energy of PE, in turn, causes a low adhesion of its surface to most structural materials. An effective direction of solving this problem is directional modification of the surface of the PEPL to increase the surface energy of PE. This can be achieved, in particular, due to contact US-action [56]. One of the most important issues in the operation of PEPL is the restoration (repair) of damaged areas. For this purpose, along with the traditional method of welding, various glue repair technologies based on the use of epoxy adhesive compositions (EAC) [57], as well as the use of coupling compounds [58] and banding joints [57], are widely used in the world practice. Also in a number of cases, the use of thermoresistive couplings with shape memory is effective for connecting damaged PEPL sites in the field [59].

Analysis of the properties of known adhesives used for bonding metals, shows that for technological and physical and mechanical performance for these purposes are more suitable EAC. EAC have good technological characteristics and can be used both in liquid and solid state (powders, rods, films). When carrying out installation and repair work, especially in the field, EAC with a moderate solidification temperature (40 – 80) °C, as well as cold-setting adhesives (without heat supply) were most used. As the basis of the resin part for the preparation of such glues, low-molecular epoxy-dian resins of the brands ED-16, ED-20 and ED-22 have found application. Hardening agents such as DETA, triethylenetetraamine (TETA), polyethylene polyamine (PEPA), as well as low molecular weight polyaminoamide oligomers are widely used for their hardening.

Fig. 9.9 shows the scheme for obtaining products with heat shrinkage molded from filled and unfilled EP.

Figure 9.9 The scheme for obtaining
 thermosetting products from filled and unfilled EP:
1 – mandrel;
2 – the bushing;
3 – base plate;
4 – EP preparation

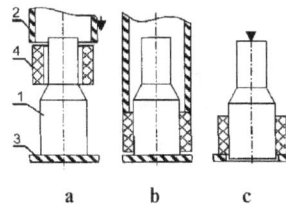

Manufacture of couplings with EP on the basis of EB (both filled and unfilled), intended for the implementation of the technology of the muft-glued joint during the repair of PEPL, is carried out in the form of a sequence of such operations [57 – 59]:

1. The filler in the mixture with the rigid and elastic EC components prior to the preparation of the initial blank was subjected to the bulk effect of low-frequency USV for several tens of minutes.

2. Mix the sonificated resin part of the composition with the filler with a mixture of accelerator and hardener, obtaining the initial liquid casting EC.

3. Tubular billets are prepared by pouring the resulting liquid EC into specially prepared metal molds and then solidifying them in a stepwise temperature-time regime. This mode is selected based on measurements of the electrical resistivity of the material of the preforms.This operation ensures the formation of an extremely hardened structure with 95–97% sol-fraction content.

4. The finished tubular billets are transferred to a highly elastic state by heating to a certain temperature. After which they are placed on the mandrel, which provides the necessary increase in the internal diameter of the tubular billets (see Fig. 9.9, a).

5. After the workpiece has been deformed along the internal diameter to the specified size, it is cooled to a temperature lower than the glass transition temperature Tg. This allows further to realize practically the 100% degree of shrinkage of the coupling (Fig. 9.9, b).

6. After some holding at this temperature, the workpieces are removed from the mandrel (Fig. 9.9, c) and their visual inspection is carried out. After successful passing of the control, the blanks are recognized ready for use as heat shrinkable couplings when connecting PEPL and for other purposes.

It should also be noted another application of US, concerning with the formation of thermal facade insulation from PCM for exterior walls of buildings and structures during their thermomodernization [60]. Engineering analysis of thermal-load components in the process of heating of PET performs, in which as a variant can be used US influence of the polymer melt, is done in [61]. It is also possible to use US treatment for producing polymer products by extrusion blow molding [62].

Conclusions

Despite the creation of new types of PCM (nanocomposites, intelligent PCM), classical thermoset composites of structural design based on reinforced fibers and epoxy matrix take an important place in the creation of a wide range of products and structures. The use of physical modification in the form of bulk US, along with chemical modification, opens up new possibilities for the directed regulation of the structure and properties of EC. This is due to a number of positive factors from effective US treatment.

In particular, the result of US treatment is an increase in the deformation-strength and adhesion characteristics of PCM. In addition, such an impact allows, to reduce the level of residual stresses, increase the durability, and also significantly reduce the total hardening time of PCM. The experimentally found optimal parameters for US modification of liquid EC (frequency, amplitude, intensity, temperature, static pressure) are used in determining the design and technological parameters of the corresponding shaping US equipment (such as volumetric sonification, impregnation, dosing and activation equipment based on US cavitation).

As for nanotechnology, the use of US is the most effective method, which facilitates the disaggregation (disintegration) of CNTs in organic solvents and liquid polymer media, and also improves the performance characteristics of NM PCM. Also, the use of US is effective in muff-adhesive and bandage technologies connecting PEPL during their repair. At the same time, an increase in the performance characteristics of repair connections is achieved by establishing an effective composition of EC and EAC, as well as optimizing the operating parameters of their US modification.

References

[1]   Yu. S. Lipatov, Structure and Strength of Polymers, Chemistry, Moscow, 1980.

[2]   V. N. Kuleznev, V. K. Gusev, Fundamentals of plastics processing technology, Khimiya, Moscow, 1995.

[3]   O. G. Tsyplakov, Scientific Bases of Fiber-Composite Material Technology, Part 1 [in Russian], Perm, 1974.

[4]   A. S. Freidin, R. A. Turusov, Properties and Design of Adhesive Compounds [in Russian], Khimiya, Moscow, 1990.

[5]   I. Z. Chernin, F. M. Smekhov, Yu.V. Zherdev, Epoxy polymers and compositions [in Russian], Khimiya, Moscow, 1982.

[6]   Yu. S. Zaitsev, Yu. S. Kochergin, M. K. Pakter, R. V. Kucher, Epoxy Oligomers and Adhesive Composition, Naukova Dumka, Kiev, 1990.

[7]   A. G. Voronkov, V. P. Yartsev, Epoxy Polymer Solutions for Repair and Protection of Building Objects and Structures: Textbook [in Russian], TGTU, Tambov, 2006.

[8]   Ultrasonic, Small Encyclopedia [in Russian], Moscow, 1979.

[9]   B. G. Novitskii, Use of Acoustic Vibrations in Chemical Technology Processes, Khimiya, Moscow, 1983.

[10]   M. A. Margulis, Sound-Chemistry Reactions and Sound-Luminescence [in Russian], Khimiya, Moscow, 1986.

[11]   I. M. Fedotkin, I. S. Gulyi, Cavitation, Cavitation Engineering and Technology, Their Use in Industry. Part II, OKO, Kiev, 2000.

[12]   O. E. Kolosov, Disturbances of Processes and Equipment Used in Fabrication of Articles from Epoxy Polymer Compositions by Means of Ultrasonic Modification [in Ukrainian], Author's Abstract of Dissertation for the Degree of Doctor of Technical Sciences, Specialization 05.17.08: Processes and Equipment of Chemical Technology, Sichkar, Kiev, 2010.

[13] O. E. Kolosov, V. I. Sivetskii, O. P. Kolosova, Preparation of Fiber-Filled Reactoplastic Polymer Composite Materials with Ultrasonic Treatment [in Ukrainian], VPK Politekhnika, Kiev, 2015.

[14] A. E. Kolosov, Efficiency of liquid reactoplastic composite heterofrequency ultrasonic treatment, Chem. and Petrol. Eng. 50 (3–4) (2014) 268–272, https://doi.org/10.1007/s10556-014-9893-y.

[15] A. E. Kolosov, Low-Frequency Ultrasonic Treatment of Liquid Reactoplastic Media with Pressure Variation, Chem. and Petrol. Eng. 50 (5–6) (2014) 339–342, https://doi.org/10.1007/s10556-014-9904-z

[16] A. E. Kolosov, Low-Frequency Ultrasonic Treatment as an Effective Method for Modifying Liquid Reactoplastic Media, Chem. and Petrol. Eng. 50 (1–2) (2014) 79–83, https://doi.org/10.1007/s10556-014-9859-0.

[17] A. E. Kolosov, Effect of low-frequency ultrasonic treatment regimes on reactoplastic polymer composite material operating properties, Chem. and Petrol. Eng. 50 (3–4) (2014) 150–155, https://doi.org/10.1007/s10556-014-9871-4.

[18] A. A. Karimov, A. E. Kolosov, V. G. Khozin, V. V. Klyavlin, Impregnation of fibrous fillers with polymer binders. 4. Effect of the parameters of ultrasound treatment on the strength characteristics of epoxy binders, Mech. Compos. Mater. 25 (1) (1989) 82–88, https://doi.org/10.1007/bf00608456.

[19] A. E. Kolosov, I. A. Repelis, V. G. Khozin, V. V. Klyavlin, Impregnation of fibrous fillers with polymer binders. 2. Effect of the impregnation regimes on the strength of the impregnated fillers, Mech. Compos. Mater. 24 (3) (1988) 373–380, https://doi.org/10.1007/bf00606611.

[20] A. E. Kolosov, Prerequisites for using ultrasonic treatment for intensifying production of polymer composite materials, Chem. and Petrol. Eng. 50 (1–2) (2014) 11–17, https://doi.org/10.1007/s10556-014-9846-5.

[21] A. E. Kolosov, A. A. Karimov, V. G. Khozin, V. V. Klyavlin, Impregnation of fibrous fillers with polymer binders. 3. Ultrasonic intensification of impregnation, Mech. Compos. Mater. 24 (4) (1989) 494–502, https://doi.org/10.1007/bf00608132.

[22] A. V. Donskoi, O. K. Keller, G. S. Kratysh, Ultrasonic Electrotechnical Devices [in Russian], Energoizdat, Leningrad, 1982.

[23] A. E. Kolosov, A. S. Sakharov, V. I. Sivetskii, D. E. Sidorov, A. L. Sokolskii, Method of selecting efficient design and operating parameters for equipment used

for the ultrasonic modification of liquid-polymer composites and fibrous fillers, Chem. and Petrol. Eng. 48 (7–8) (2012) 459–466, https://doi.org/10.1007/s10556-012-9640-1.

[24] A. E. Kolosov, V. I. Sivetskii, E. P. Kolosova, E. A. Lugovskaya, Procedure for analysis of ultrasonic cavitator with radiative plate, Chem. and Petrol. Eng. 48 (11–12) (2013) 662–672, https://doi.org/10.1007/s10556-013-9677-9.

[25] A. E. Kolosov, Impregnation of fibrous fillers with polymer binders. 1. Kinetic equations of longitudinal and transverse impregnation. Mech. Compos. Mater. 23, I. 5 (1988) 625–633, https://doi.org/10.1007/bf00605688.

[26] A. E. Kolosov, I. A. Repelis, Saturation of fibrous fillers with polymer binders 5. Optimization of parameters of the winding conditions, Mech. Compos. Mater. 25 (3) (1989) 407–415, https://doi.org/10.1007/bf00614811.

[27] A. F. Kichigin, A. E. Kolosov, V. V. Klyavlin, V. G. Sidyachenko, Probabilistic-geometric model of structurally inhomogeneous materials, Soviet Mining Sci. 24, I. 2 (1988) 87–94, https://doi.org/10.1007/bf02497828.

[28] A. E. Kolosov, V. V. Klyavlin, Determination of the parameters of a geometric model of the structure of directionally reinforced fiber composites, Mech. Compos. Mater. 23 (6) (1988) 699–706, https://doi.org/10.1007/bf00616790.

[29] A. E. Kolosov, V. V. Klyavlin, Several aspects of determination of the adequate model of the structure of oriented fiber-reinforced composites, Mech. Compos. Mater. 24 (6) (1989) 751–757, https://doi.org/10.1007/bf00610779.

[30] E. P. Kolosova, V. V. Vanin, A. E. Kolosov, V. I. Sivetskii, Modeling of processes and equipment for the manufacturing of thermosetting materials [in Ukrainian], Igor Sikorsky KPI, Kiev, 2017.

[31] A. E. Kolosov, A. S. Sakharov, V. I. Sivetskii, D. E. Sidorov, A. L. Sokolskii, Substantiation of the efficiency of using ultrasonic modification as a basis of a production cycle for preparing reinforced objects of epoxy polymer composition, Chem. and Petrol. Eng. 48 (5–6) (2012) 391–397, https://doi.org/10.1007/s10556-012-9629-9.

[32] A. E. Kolosov, A. A. Karimov, I. A. Repelis, V. G. Khozin, V. V. Klyavlin, Impregnation of fibrous fillers with polymeric binders. 6. Effect of parameters of ultrasound treatment on strength properties of wound fibrous composites, Mech. Compos. Mater. 25 (4) (1990) 548–555, https://doi.org/10.1007/bf00610711.

[33] O. E. Kolosov, V. I. Sivets'kii, E. M. Panov, I. O. Mikulyonok, V. V. Klyavlin, D. E. Sidorov, Mathematical Modeling of Basic Processes in the Fabrication of Polymer Composite Materials with Use of Ultrasonic Modification [in Ukrainian], VD Edelveis, Kiev, 2012.

[34] O. E. Kolosov, Molding of Polymer Composites with the Use of Physicochemical Modifications. Part 1. Studies of the Prerequisites for Directed Realization of Physicochemical Modifications [in Ukrainian], NTUU KPI, Kiev, 2005.

[35] O. E. Kolosov, V. I. Sivetskii, Molding of Polymer Composites with the Use of Physicochemical Modifications. Part 2. Effective Regimes and Equipment for Realization of Physicochemical Modifications [in Ukrainian], NTUU KPI, Kiev, 2006.

[36] O. E. Kolosov, V. I. Sivets'kii, Ye. M. Panov, Technology for Production of Multicomponent Epoxy Polymers with Use of Intentional Physicochemical Modification [in Ukrainian], NTUU KPI, 2010.

[37] A. E. Kolosov, G. A. Virchenko, E. P. Kolosova, G. I. Virchenko, Structural and technological design of ways for preparing reactoplastic composite fiber materials based on structural parametric modeling, Chem. and Petrol. Eng. 51 (7–8) (2015) 493–500, https://doi.org/10.1007/s10556-015-0075-3.

[38] E. P. Kolosova, V. V. Vanin, G. A.Virchenko, A. E. Kolosov, Modeling of manufacturing processes of reactoplastic composite-fibrous materials [in Ukrainian], VPI VPK Politekhnika, Kiev, 2016.

[39] D. E. Sidorov, V. I. Sivetskii, A. E. Kolosov, A. S. Sakharov, Shaping of corrugation profiles during production of corrugated tubular articles, Chem. and Petrol. Eng. 48 (5–6) (2012) 384–390, https://doi.org/10.1007/s10556-012-9628-x.

[40] A. E. Kolosov, A. S. Sakharov, D. E. Sidorov, V. I. Sivetskii, Aspects of profile shaping of corrugated tubular components. Part 1. Modeling of parameters of different profiles of corrugations, and also their shaping equipment, Chem. and Petrol. Eng. 48 (1–2) (2012) 60–67, https://doi.org/10.1007/s10556-012-9575-6.

[41] A. E. Kolosov, A. S. Sakharov, D. E. Sidorov, V. I.Sivetskii, Manufacturing Technology: Aspects of profile shaping of corrugated tubular components. Part 2. Modeling the extrusion welding of layers of corrugated tubular articles, Chem. and Petrol. Eng. 48 (1–2) (2012) 131–138, https://doi.org/10.1007/s10556-012-9588-1.

[42] A. E. Kolosov, A. S. Sakharov, D. E. Sidorov, V. I. Sivetskii, Aspects of profile shaping of corrugated tubular components. Part 3. Extrusion shaping of tubular

polymeric blanks for manufacture of corrugated pipes, Chem. and Petrol. Eng. 48 (3–4) (2012) 199–206, https://doi.org/10.1007/s10556-012-9598-z.

[43]   A. S. Sakharov, A. E. Kolosov, A. L. Sokolskii, V. I. Sivetskii, Modeling the mixing of polymeric composites in an extrusion drum mixer, Chem. And Petrol. Eng. 47 (11–12) (2012) 799–805, https://doi.org/10.1007/s10556-012-9553-z.

[44]   K. G. Kovalenko, A. E. Kolosov, V. I. Sivetskii, A. L. Sokolskii, Modeling Polymer Melt Flow at the Outlet from an Extruder Molding Tool, Chem. and Petrol. Eng. 49 (11) (2014) 792–797, https://doi.org/10.1007/s10556-014-9837-6.

[45]   A. S. Sakharov, A. E. Kolosov, V. I. Sivetskii, A. L. Sokolskii, Modeling of Polymer Melting Processes in Screw Extruder Channels, Chem. and Petrol. Eng. 49 (5–6) (2013) 357–363, https://doi.org/10.1007/s10556-013-9755-z.

[46]   A. E. Kolosov, Preparation of Nano-Modified Reactoplast Polymer Composites. Part 1. Features of used nanotechnologies and potential alication areas of nanocomposites (a review), Chem. and Petrol. Eng. 51 (7–8) (2015) 569–573, https://doi.org/10.1007/s10556-015-0088-y.

[47]   O. E. Kolosov, Preparation of Traditional and Nanomodified Reactoplastic Polymer Composite Materials [in Ukrainian], Izd. VPI VPK Politekhnika, Kiev, 2015.

[48]   A. E. Kolosov, Preparation of Reactoplastic Nanomodified Polymer Composites. Part 2. Analysis of means of forming nanocomposites (patent review), Chem. and Petrol. Eng. 51 (9–10) (2016) 640–645, https://doi.org/10.1007/s10556-016-0100-1.

[49]   A. E. Kolosov, Preparation of Reactoplastic Nanomodified Polymer Composites. Part 3. Methods for dispersing carbon nanotubes in organic solvents and liquid polymeric media (review), Chem. and Petrol. Eng. 52 (1–2) (2016) 71–76, https://doi.org/10.1007/s10556-016-0151-3.

[50]   A. E. Kolosov, Preparation of Reactoplastic Nanomodified Polymer Composites. Part 4. Effectiveness of modifying epoxide oligomers with carbon nanotubes (review), Chem. and Petrol. Eng. 52 (7–8) (2016) 573–577, https://doi.org/10.1007/s10556-016-0235-0.

[51]   A. E. Kolosov, Preparation of Reactoplastic Nano-Modified Polymer Composites. Part 5. Advantages of using nano-modified structural carbon-fiber composites (a review), Chem. and Petrol. Eng. 52 (9–10) (2017) 721–725, https://doi.org/10.1007/s10556-017-0259-0.

[52]   A. E. Kolosov, E. P. Kolosova, Functional Materials for Construction Application
       Based on Classical and Nano Composites: Production and Properties, in: Rita
       Khanna, Romina Cayumil (Eds.), Recent Developments in the field of Carbon
       Fibers, InTechOpen, 2018, ISBN: 978-953-51-6055-7.

[53]   I. Ivitskiy, V. Sivetskiy, V. Bazhenov, D. Ivitska, Modeling the electrostatic
       control over depth of the introduction of intelligent sensors into a polymer
       composite material, East.-Europ. J. of Enterprise Technol., 1 (5 (85)) (2017) 4–9,
       https://doi.org/10.15587/1729-4061.2017.91659

[54]   V. I. Sivetskii, A. E. Kolosov, A. L. Sokolskii, I. I. Ivitskiy, Technologies and
       equipment for the molding of products from traditional and intelligent polymeric
       composite materials [in Ukrainian], VPI VPK Politekhnika, Kiev, 2017.

[55]   A.E. Kolosov, O.S. Sakharov, V.I. Sivetskii, D.E. Sidorov, S.O.
       Pristailov, Effective hardware for connection and repair of polyethylene pipelines
       using ultrasonic modification and heat shrinkage. Part 1. Aspects of connection
       and restoration of polymeric pipelines for gas transport, Chem. and Petrol. Eng. 47
       (2011) 204–209, https://doi.org/10.1007/s10556-011-9447-5.

[56]   A. E. Kolosov, O. S. Sakharov, V.I. Sivetskii, D. E. Sidorov, S.O.
       Pristailov, Effective hardware for connection and repair of polyethylene pipelines
       using ultrasonic modification and heat shrinkage. Part 3. Analysis of surface-
       treatment methods for polyethylene pipes connected by banding, Chem. and
       Petrol. Eng. 47 (2011) 216, https://doi.org/10.1007/s10556-011-9449-3.

[57]   A. E. Kolosov, O.S. Sakharov, V.I. Sivetskii, D.E. Sidorov, S.O.
       Pristailov, Effective hardware for connection and repair of polyethylene pipelines
       using ultrasonic modification and heat shrinkage. Part 4. Characteristics of
       practical implementation of production bases developed using epoxy-glue
       compositions and banding, Chem. and Petrol. Eng. 47 (2011) 280,
       https://doi.org/10.1007/s10556-011-9460-8.

[58]   A.E. Kolosov, O.S.Sakharov, V.I.Sivetskii, D.E. Sidorov, S.O. Pristailov,
       Effective hardware for connection and repair of polyethylene pipelines using
       ultrasonic modification and heat shrinkage. Part 2. Production bases for molding
       of epoxy repair couplings with shape memory, Chem. and Petrol. Eng. 47 (2011)
       210, https://doi.org/10.1007/s10556-011-9448-4.

[59]   A. E. Kolosov, O. S. Sakharov, V. I. Sivetskii, D. E. Sidorov, S. O. Pristailov,
       Effective hardware for connection and repair of polyethylene pipelines using
       ultrasound modification and heat shrinking. Part 5. Aspects of thermistor

couplings and components used in gas-pipeline repair, Chem. and Petrol. Eng. 47 (2011) 285, https://doi.org/10.1007/s10556-011-9461-7.

[60] A. V. Yeromin, A. E. Kolosov, Modeling of energy effective solutions regarding the heating system and facade heat insulation during implementation of thermomodernization, Technol. Audit and Product. Reserves. 1/8 (91) (2018) 49–58, https://doi.org/10.15587/1729-4061.2018.123021.

[61] D.É. Sidorov, A.E. Kolosov, O.V. Pogorelyi, I.A. Kazak, Engineering Analysis of Thermal-Load Components in the Process of Heating of Pet Preforms, Journ. of Eng. Phys. and Thermophys. 2 (2) (2018) 1–5, https://doi.org/10.1007/s10891-018-1768-1.

[62] D. E. Sidorov, E. P. Kolosova, A. E. Kolosov, T. A. Shabliy, Analysis of blown process for producing polymer products by extrusion blow molding, East.-Eur. J. of Enterpr. Technol. 2/1 (92) (2018) 14–21, https://doi.org/10.15587/1729-4061.2018.126015.

Thermoset Composite
Materials Research Foundations **38** (2018)

Materials Research Forum LLC
doi: http://dx.doi.org/10.21741/9781945291876

Chapter 10

A Review on Tribological Performance of Polymeric Composites Based on Natural Fibres

Umar Nirmal[1,*], Alvin Devadas[1], M.M.H. Megat Ahmad[2,a], M.Y. Yuhazri[3,b]

[1]Center of Advance Materials and Green Technology, Faculty of Engineering and Technology, Multimedia University, Jalan Ayer Keroh Lama, 75450, Melaka, Malaysia

[2]Faculty of Engineering, Department of Mechanical Engineering, National Defence University of Malaysia, Sungai Besi Camp, 57000 Kuala Lumpur, Malaysia

[3]Faculty of Engineering Technology, Universiti Teknikal Malaysia Melaka, Hang Tuah Jaya, 76100, Durian Tunggal Melaka, Malaysia

*nirmal@mmu.edu.my, *nirmal288@zoho.com, [a]megat@upnm.edu.my, [b]yuhazri@utem.edu.my

Abstract

The need for stronger, more lightweight materials have led the way for the development of synthetic fibre reinforced polymer composites. Composites have unique properties which can only be obtained by the combination of specific fibres and matrix materials. Today these composites are used in many areas such as in the automotive and construction industries which require materials with good mechanical and tribological properties. However, shortcomings in the usage of these fibres, primarily environmental concerns, have motivated researchers to explore the possibility of using natural fibres as an alternative. The current work attempts to provide a review of recent works (year 2007 to 2017) done on tribological performance of natural fibre reinforced polymer composites. Topics related to natural fibre, polymer, composites, and tribology will be discussed, along with critical findings from these published works.

Keywords

Natural Fibres, Polymer, Composite, Tribology, Wear

Contents

10.1 Introduction

Composite materials have been in use for hundreds of years, and dates as far back as the times of the Egyptian empire [1]. More recently, polymer composites based on synthetic fibres or fillers have been used to produce parts such as bushes, seals, and gears for the automotive and aerospace industries, and are continuously finding new market in other areas [2,3]. The combination of a polymer matrix which distributes loading onto fibres as well as the fibres themselves which acts as the load carrying members, can give unique material properties. Examples of common synthetic fibres include glass, graphite, carbon, and calcium carbonate [4,5]. However, the environmental issues associated with the production and usage of these types of composites is of concern. Crude oil, the raw material used to make these composites, must be extracted from oil wells around the globe, including from the sea bed. During extraction, accidental spills do occur, and oil is constantly discharged into the sea as part of vessels and oil tankers operations. It is estimated that 21% of oil spills in the ocean is contributed by sea transport operations [6]. These oil spills cause long lasting degradation of marine and human life as well as ecological and environmental damage to the surrounding areas [7]. The steps required to process crude oil into polymers itself uses up a considerable amount of energy. These processes, which relies on the burning of fossil fuels, release carbon dioxide to the environment and contributes to global warming [8]. Apart from this, energy is also required to produce the polymer matrix and the synthetic fibres [9]. The issues associated with crude oil together with the expected rise in global demand for crude oil [10] has resulted in the growing number of research being done to reduce dependency on crude oil-based materials. As a consequence, there is growing interest in using natural fibres as reinforcement in polymer composites [11]. The current work aims to discuss various topics such as natural fibres, polymers, composites, as well as tribological and wear mechanisms. Also, published works done on tribological performance of polymeric composites based on natural fibres will be reviewed, and critical findings presented.

10.2 Natural fibres

Natural fibres are fibres which grow naturally, or harvested. Table 10.1 summarizes the main types of natural fibres which can be found.

Table 10.1 Types of Natural Fibres [12].

Bast Fibre	Leaf Fibre	Seed Fibre	Core Fibres	Reed Fibres	Others
Jute	Sisal	Coir	Kenaf	Wheat	Wood
Hemp	Abaca	Cotton	Hemp	Corn	Roots
Ramie	Pineapple	Kapok	Jute	Rice	
Kenaf		Betelnut			
Flax					

The chemical compositions of natural fibres are primarily cellulose, hemicelluloses, pectin and lignin [13], [14]. These four constituents contribute to the overall property of the natural fibre. Proportions of the constituents can vary irrespective of fibre type. For example, hemi-cellulose contributes to biodegradation, thermal degradation and moisture absorption, whereas lignin is responsible for UV degradation [15]. Compared with synthetic fibres, natural fibres offer advantages such as non-toxicity - which allows for safer processing - flexible usage, high specific strength, low density, are less abrasive to processing equipment, and are more environmentally friendly [15–17]. Also, as these fibres grow in abundance, they are cheap and relatively easy to obtain [18]. When used to form composite materials, these fibres offer substantial environmental benefits. S.V. Joshi and associates [19] have done a Life Cycle Assessment (LCA) study on natural fibre and glass fibre composites. They identified four primary benefits in the LCA of natural fibre composites, to the environment. Firstly, there is less impact to the environment in the production of natural fibre, as cultivating these fibres mainly rely on solar energy with little non-renewable fossil fuel used during extraction. Glass fibre production, on the other hand, is a heavily fossil fuel-dependent process. Heavier use of fossil fuel in the production of glass fibres also results in significantly higher polluting emissions. Secondly, the volume of fibre in natural fibre composites is higher compared to glass fibre composites, which results in lower usage of base polymers that require more energy and releases more polluting emissions. Thirdly, natural fibres have lower densities than glass fibres, which gives natural fibre composites a lower weight. Comparatively, there is approximately a 20-30% weight reduction. When used in applications such as automobiles, this improves fuel efficiency. Finally, when incinerated during its end of life, energy can be recovered and, at the same time, there is theoretically no addition to carbon dioxide emissions. This is because the plants which natural fibres are obtained absorb carbon dioxide during their growth, which is released during the combustion of natural fibres. There are however, some drawbacks associated with the usage of natural fibres. High levels of moisture absorption, poor wettability and poor adhesion of fibres to polymer matrix are among them. Strength and stiffness of natural fibre composites are also much lower than synthetic fibres [20,21]. There is also a high degree of variation in the characteristics of natural fibres, which affects its mechanical properties. In a previous work done [14], researchers compiled a list of mechanical properties of kenaf fiber from various sources, as shown in Table 10.2.

Table 10.2 Variations in Kenaf Fibre Mechanical Properties[14]

Density (g/cm^3)	Tensile Strength (MPa)	Tensile Modulus (GPa)	Elongation (%)
-	692	10.94	4.3
-	930	53	1.6
1.45	930	53	1.6
1.4	284-800	21-60	1.6
-	295-1191	2.86	3.5
1.5	350-600	40	2.5-3.5
0.75	400-550	-	-
0.6	-	-	-
0.749	223-624	11-14.5	2.7-5.7
1.2	295	-	3-10

Based on this finding, there is a high variation in tensile strength and tensile modulus reported by different authors. This is because variations in kenaf fibre exist in terms of place of origin and retting method. Also, the cross section of a strand of natural fibre varies along the length of the fibre [22]. This high variation in mechanical and physical properties makes it challenging to standardize the performance of natural fibres.

10.3 Polymer

Polymers are substances which contain repeated units of monomers, which are molecules. These molecules are joined together in a process called polymerization, which forms a long chain of hundreds or sometimes thousands of molecules. These long chains of monomers are reason why polymers are sometimes referred to as macro-molecules [23]. Polymers can either be natural (cellulose, protein, DNA, wood, etc.) or synthetic (nylon, polyethylene, etc.). Table 10.3 shows some common synthetic polymers and the monomer molecules which are used to create them. Synthetic polymers are derived from petroleum, which comprises of crude oil, natural gas and solid hydrocarbons, as shown in

Figure 10.1. Petroleum is extracted from various oil fields around the world. Once extracted, it is brought to a refinery to be processed. During processing, the petroleum is first heated to a gas state in a process called 'distillation'. As the gasses condense, they form separate groups of hydrocarbons called fractions. These fractions are then broken down into smaller hydrocarbon molecules, such as ethylene, propylene, and styrene in a process called 'cracking'. After that, the small hydrocarbon molecules are combined in a process called 'polymerization', to produce polymers. These polymers are then used as raw materials for various products, including plastics and composites. Although there are many ways to classify the various types of synthetic polymers, they are generally grouped into two main categories; non-cross linked (thermoplastics) and cross-liked (thermosets) polymers [24]. In non-cross linked polymers, the bonds between molecules can be broken and reattached. This implies that the polymer can be melted and remolded. In cross-linked polymers, the bond between molecules cannot be reattached once broken.

Table 10.4 summarizes the difference between thermoplastics and thermosets. Polymers are extensively used in sliding components such as gears and cams due to their favorable lubricating properties [25]. However, compared to metals, unfilled polymers have low wear resistance and low thermal conductivity. Therefore, to improve their tribological and thermal properties, fillers are added to form composites [26]. Apart from this, the addition of fillers can also result in a material with enhanced mechanical properties.

Table 10.3 Common Polymers and Their Respective Monomer Molecules.

Monomers	Polymers
Ethylene	Polyethylene
H H C = C H H	H H H H - C - C - C - C - H H H H $\quad n$
Propylene	Polypropylene
H CH3 C = C H H	H CH3 H CH3 - C = C - C = C - H H H H $\quad n$

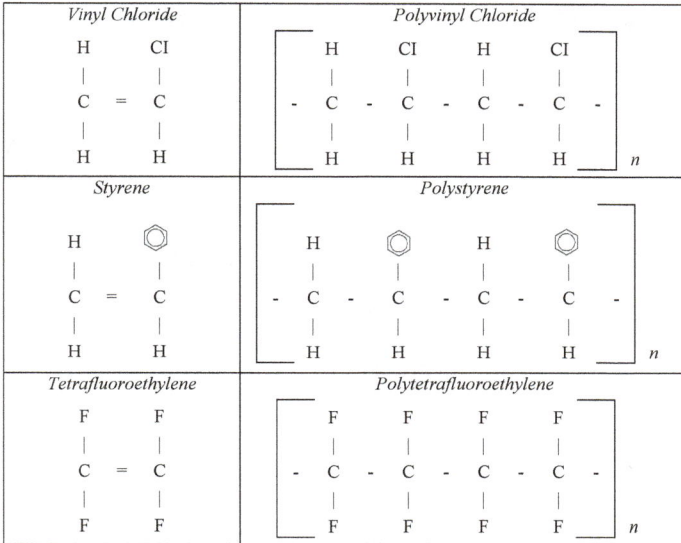

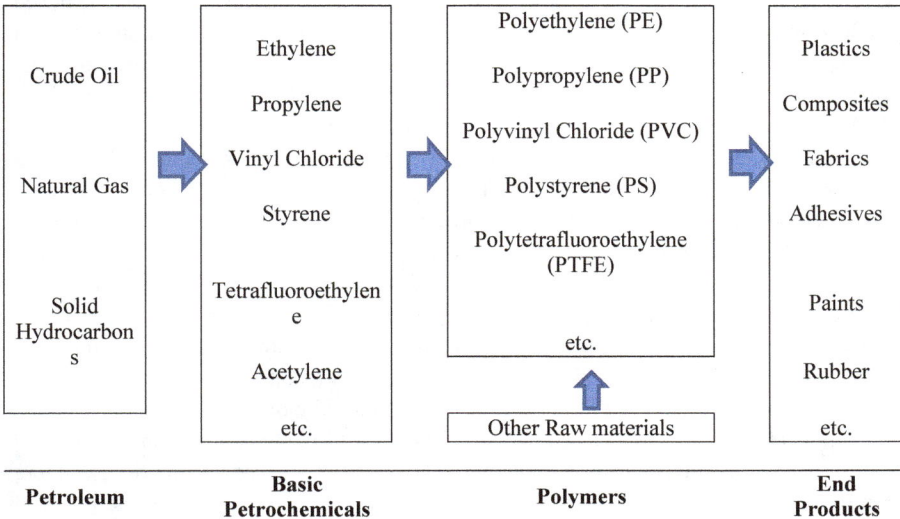

Figure 10.1 Process of Making Polymers, from Raw Petroleum into End Products [27].

Table 10.4 Differences between Thermoplastics and Thermosets [23,24].

	Thermoplastics	**Thermosets**
Feature	Curing process is reversible	Curing process is non-reversible.
Advantages	Recyclable Chemical Resistant High Impact Resistance Superior Finishing	Resistant to high temperatures Can be molded thin High dimensional stability Low cost
Disadvantages	Expensive compared to Thermosets	Cannot be remolded or recycled
Examples	Polystyrene, Polypropylene, poly vinyl chloride (PVC), Polyether ether ketone (PEEK)	Bakelite, Epoxy, Polyester

10.4 Composite

Composites are a class of materials which are made by combining two or more materials to create a new one. This combination of materials is done to achieve unique characteristics such as light weight, high stiffness, high strength, high wear resistance, high thermal conductivity etc. Composites have been used for centuries and can be traced as far back as the ancient Egyptian empire which used plywood- and straw-strengthened mudbricks for construction. In the early nineteenth century, discovery of polymers spearheaded the development of composite materials based on a polymer matrix [28]. Today, composite materials are used in buildings, airplanes, automobiles, boats, etc. Composites can be classified primarily into two categories; particle reinforced and fibre reinforced, as shown in Figure 10.2. Particle reinforced composites can further be classified per the size of the particle reinforcement; small particle (up to 100 nm in size) or large particle. The structure of this composite consists of reinforcements embedded in a matrix. The reinforcement material provides strength and stiffness, and is responsible for the overall mechanical properties of the composite. It can occupy up to 70 % of the volume in a composite, depending on the application, cost, etc. Common examples are glass, metal, and organic materials. The matrix supports and protects the reinforcement, maintains the overall shape of the composite, and is usually of a lower density material. Examples are polymers, metal, and ceramics [28–30]. The structure of fibre-reinforced composites is similar to the structure of particle-reinforced composites. Fibre reinforcements can either be long or short, with different diameters to suit the intended application. It can be randomly distributed, or in a mat form. Other types of composites with different structures include laminar, honeycomb, and sandwich structures.

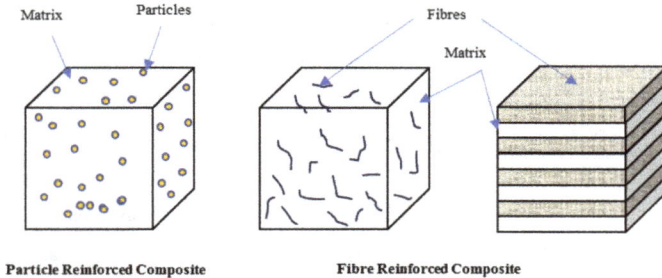

Particle Reinforced Composite   Fibre Reinforced Composite

Figure 10.2 Types of Composites [28,29].

Depending on the applications, fillers are sometimes added to improve the performance of the composite in a way that can't be achieved with the matrix and reinforcement alone [31]. Examples of filler additives are graphite, silicon carbide (SiC) and aluminum oxide (Al$_2$O$_3$) which are usually added to improve wear performance of the composites. The usage of composite materials are varied, ranging from bridges and structures, gears and boats to race car bodies and airplane bodies and parts.

10.5 Tribology

Tribology is the science that deals with design, friction, wear and lubrication of interacting surfaces in relative motion [32]. Although related studies and published work have been done from as early as the 1900s, the word 'Tribology' was only coined in 1966. A few years prior to this, there was an increase in the reported failures of plants and machinery due to wear and associated failures. This led the British government to set up a Working Group to investigate the present state of awareness on tribology-related matters and give opinions on its needs. Their findings led to the establishment of the Committee on Tribology which had duties including advising government departments and other bodies on matters associated to tribology. It has been reported that 90% of mechanical parts failures are a result of tribological loadings [33]. As such, a proper understanding of tribology and its principles can save a significant amount of money in repair costs. Table 10.5 summarizes the common testing methods for tribology tests of polymeric composites based on natural fibres. Figure 10.3 shows a schematic view of various composite orientation for tribology testing.

Table 10.5. Summary of Various Types of Tribological Tests.

Test Method	Features
1. Pin on disk/Block on disk [34–43]	• Follows ASTM G99 procedure. • Specimen is held vertically or horizontally under loading against a rotating counterface. • Typical operating parameters includes sliding distance, sliding velocity, applied load, wet or dry sliding, abrasive or adhesive contact condition. • Contact area of specimen is constant with respect to sliding time. • Recommend samples size: 10 x 10 x 20 mm^3
2. Pin on drum [39,44–46]	• Follows ASTM A514 procedure. • Specimen travels horizontally (linearly) against a rotating drum. • Testing can be either abrasive or adhesive. • Simulates wear in applications such as conveyor belts.
3. Block on ring [39–41,47–49]	• Follows ASTM G77 procedure. • Specimen is held against a rotating ring or wheel at 90° to the ring or wheels axis of rotation. • Samples can be of any kind, including metals. • Typical operating parameters include sliding distance, sliding velocity, applied load, temperature, and wet or dry sliding. • Contact are of specimen varies with respect to sliding time. • Recommended sample size: 10 x 20 x 50 mm^3. • Simulates wear in applications such as pulleys and camshafts.

Table 10.5 (cont'd): Summary of Various Types of Tribological Tests.

Test Method	Features
4. Linear tribo machine [39.50]	• Specimen is held in a container filled with abrasive particles while sliding counterface moves horizontally (linearly). • Can also be used for adhesive testing. • Simulates wear in applications such as sliding window panels and drawers.
5. Dry sand rubber wheel [38–40,45,51,52]	• Follows ASTM G65 procedure. • Specimen is held against a rubber wheel, while sand is introduced to the rubber interface to simulate abrasive test. • Can also be used for adhesive testing. • Simulates wear in applications such as tyres, bearings, and rollers.
6. Scratch test [53] (a) Schematic diagram showing the scratch test machine: 1, specimen; 2, shaft; 3, stylus; 4, cantilever; 5, frame; 6, dead weight; 7, balancing weight; (b) Close up view of the scratching mechanism.	• Specimen (1) is fixed on a servo motor driven shaft (2) which is capable of rotating at very low speed range. • A conical stylus (3) with a specified apex angle and tip radius is attached to its holder at the cantilever arm (4). • The cantilever arm is supported on a frame (5). The specimen is loaded normally by placing dead weights (6) onto the stylus holder. • The cantilever is kept in equilibrium by a balancing weight (7) to ensure no initial normal load is present. • The setup is equipped with a force gage of 0.001 N sensitivity and 1 ms^{-1} sampling speed, which allowed the tangential force to be monitored and recorded during scratching.

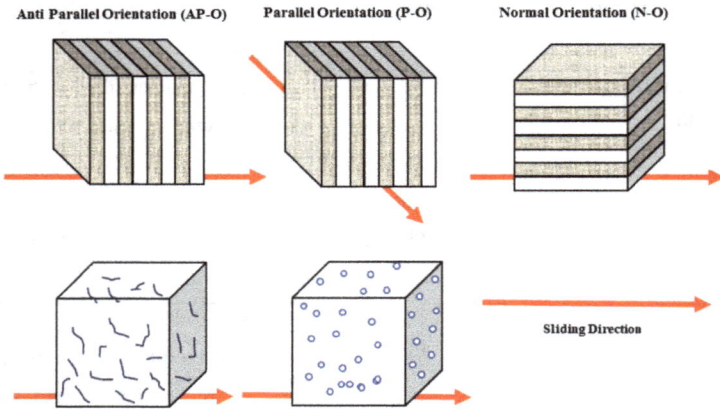

Figure 10.3 Different composite orientations during sliding.

10.6 Friction and wear

Friction is an object's surface resistance to motion. In other words, it is the opposing force which is generated when a surface is slid across another surface. Wear, on the other hand, is the progressive loss of material on a surface caused by rubbing by another surface, fluid, or gas. In tribology, friction and wear depends on factors such as rubbing surface roughness, relative motion, type of material, temperature, normal force, stick slip, relative humidity, lubrication, and vibration [54]. Friction often leads to various types of wear such as adhesive wear and abrasive wear [55]. Other types of wear include fretting wear, erosion wear, and fatigue wear.

a) Adhesive wear

Wear occurs when two smooth surfaces contact or slide against each other. Figure 10.4 shows a microscopic view of the wear mechanism of a smooth steel surface sliding against a stationary polymer surface. When stationary and under loading (F_n), surface forces of both materials form a bond at localized junctions. As the surfaces begin to slide, friction force is generated (F_f) to overcome these bonds. The bonds eventually break and cause either fractures at the interface or polymer transfer from the weaker material [56].

Thermoset Composite

Materials Research Forum LLC

Materials Research Foundations **38** (2018)

doi: http://dx.doi.org/10.21741/9781945291876

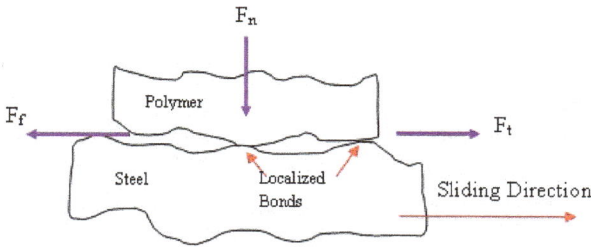

Figure 10.4 Mechanism of Adhesive Wear

b) Abrasive wear

Material loss occurs when one of the sliding surfaces is harder or more abrasive than the other. In this case, material loss will be seen on the softer or smoother material. Figure 10.5 shows a macroscopic view of the mechanisms of an abrasive surface sliding against a stationary polymer surface. The sliding motion causes the abrasive surface to plough into the softer polymer, chipping material and causing material loss [57].

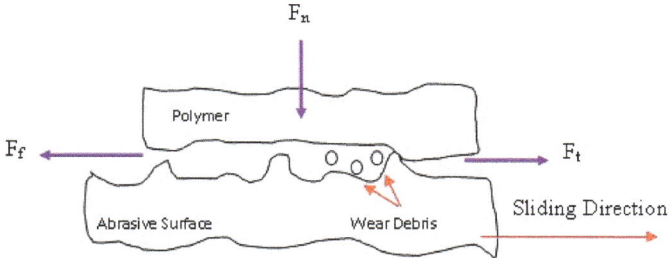

Figure 10.5 Mechanism of Abrasive Wear

For both cases, the described wear mode is a two-body wear. At times, the wear debris or other external debris would collect at the rubbing interface, changing the mode to a three-body wear.

10.7 Tribology of natural fibre polymeric composites

a) Oil palm (Elaeis) fibre composites

The oil palm tree originates from West Africa and is now grown in over 42 countries worldwide. Malaysia is the world's largest oil palm producer, accounting for 60 % of the world's supply. Its primary use is to produce of palm oil [58].

Yousif and associates investigated the wear and frictional characteristics of polyester composites made using oil palm fibres (OPRP). Results were compared against neat polyester (NP) [59]. Dry contact condition tests were conducted using the Pin-on-Disk (POD) Machine; the test specimen was slid against a polished stainless steel counterface at different sliding distances (0 - 5 km), different sliding velocities (1.7 - 3.9m/s) and different applied loads (30 – 7 N). It was reported that the specific wear rate (W_S) was higher for NP compared to the OPRP composite at various sliding distances. At higher applied loads, the W_S was lower in OPRP by three or four times. Meanwhile, as sliding velocity increases, there is a significant reduction in W_S at higher loads for the OPRP composite, whereas for NP, higher sliding velocity shows higher W_S at all applied loads. This was due to the strengthening of the polyester by oil palm fibres. Also, when observed under the Scanning Electron Microscope (SEM), the oil palm fibres assisted in the protection of exposed rubbing layer of the composite during sliding, especially at longer sliding distance and higher applied loads. For NP, high deformation was observed due to high temperatures generated by friction. The absence of oil palm fibres worsened the surface characteristics of NP, and at higher loads, the adhesive wear mode changed to abrasive wear. The effects on friction coefficient using oil palm fibres as reinforcement are noticeable at higher applied loads, where there was a 5 – 23 % reduction. The wear mechanisms of OPRP composite was mostly the debonding and bending of fibres, splitting of fibre bundles, and polyester deformation.

In another work, Yousif and associates also investigated the effects that aged treatment oil palm fibres had on wear and frictional characteristics of polyester composites [60]. As part of the fibre preparation, the fibres were treated with 6 % Sodium Hydroxide (NaOH) solution for 24 hours. SEM images showed that this treatment created hollow spots in the fibre, which could help interfacial adhesion strength between the fibre and matrix. Upon composite fabrication, samples were immersed in water, salt water, diesel, petrol and engine oil for a period of three years before testing. Testing was conducted on a POD machine. Dry sliding of treated OPRP composite was tested against a stainless steel counterface at different sliding distance (0-6.72 km) and at a fixed sliding velocity of 2.8 m/s and a fixed normal load of 25 N. W_S for treated OPRP immersed in water and salt water showed gradual increment at 4 km and 2.5 km respectively. However, no sign of a steady state was observed for both. Overall, treated OPRP immersed in water showed approximately 18 % lower W_S. This could be attributed to the fibres, which were immersed in 50 % salt water, experiencing cell wall densification which made them weak in the resin. On the other hand, treated OPRP immersed in Diesel, petrol, and engine oil showed similar a trend whereby steady state of W_S was seen at 5 km. Petrol-immersed treated OPRP had much higher W_S compared to those immersed in diesel and engine oil. This could be due to the lower absorption rate of diesel-treated OPRP, at only 0.2 % compared to 0.5 % when immersed in petrol. Furthermore, the presence of engine oil at the interface during sliding would lower the thermo-mechanical loading for this composite. In the case of friction coefficient, treated OPRP composite immersed in diesel showed the lowest friction coefficient, which could be due to its low absorption rate, among all tested immersion solutions, at only approximately 0.2 %. Frictional performance for all the five immersion solutions followed the order of diesel > engine oil > water > petrol > salt water, where diesel showed superior frictional performance. Finally, comparing results from the current work against previous, treated OPRP aged in various solutions showed improved wear performance compared to un-aged treated OPRP. Also, friction performance improved by approximately 43.2 % when immersed in diesel. This finding demonstrates that aged treated OPRP could be used in applications where ageing of the composite would take place, such as diesel tanks, oil tank containers, non-structural bearings, and sliding materials subjected to tribology loading conditions.

Fazillah and associates investigated the effects of temperature and fibre composition on the friction and wear of oil palm as well as kenaf fibre reinforced epoxy composites (OPRE and KFRE) [61]. 30 %, 50 % and 70 %wt. fibre content were considered, and dry sliding tests were conducted on a POD machine at a 49.05 N applied load and a sliding speed of 1000 rpm. The temperature range for testing was 23°C, 40°C, 100°C and 150°C. It was reported that friction showed a decreasing trend as temperature increased for both types of composite. This occurred due to the existence of a thin lubricating layer consisting of molten resin that develops at high temperature, which reduced friction. Also, the 70 % wt. fibres OPRE composite showed the highest friction, whereas in the case of the KFRE composite, a composition of 30% wt. fibre content showed higher friction. This difference was said to be due to the different in hardness of both composites. The wear rate, on the other hand, showed an increasing trend as temperatures increased, for both composites. This was due to the reduction in hardness at higher temperatures for both composites. However, at higher fibre contents, KFRE showed better wear resistance than OPFE. This was said to be due to the transition from mild to severe wear, which occurred at a much lower %wt. of fibre in the case of KFRE.

b) Kenaf (Hibiscus) fibre composites

The word Kenaf originates from the Persian language in the late nineteenth century [62]. Kenaf fibre is produced from the bast of a plant called *Hibiscus cannabinus*. This is a plant from the Malvaceae family. It consists of 35 % to 40 % bast fibre and 60 % to 65 % core fibre by weight of the stalk [63]. The plant can grow more than 3 m tall with a stem diameter of 25 – 51 mm within three months. The fibre is harvested in countries such as Malaysia, Thailand, India, Bangladesh, Africa, and Southeast Europe and used mainly for making paper, rope, twine, and coarse cloth. Modern applications include textiles, building materials, and absorbents However, nowadays there is demand for this fibre to be used as a reinforcement for polymers [64]. The sustainable properties of kenaf are that the plant absorbs carbon dioxide at a significantly high rate, absorbs nitrogen and phosphorus from the soil, whereas the fibre is easily recyclable and lightweight [19][65,66].

Narish and associates have published two papers on adhesive wear performance of natural fibre composite using kenaf fibres as a reinforcement. In their first paper [67], testing was done using treated kenaf fibre reinforced polyurethane (KFRP) composite on a Block On Disk (BOD) machine. Three types of fibre orientations were considered: Parallel (P-O), Anti-Parallel (AP-O) and Normal (N-O). Results were also compared against neat polyurethane (N-PU). Results showed that there was a significant improvement in Ws when N-PU was reinforced with kenaf fibres, averaging approximately 78 %. Also, for treated KFRP, the fibre mat orientation highly influenced Ws. Wear performance followed the order of AP-O > N-O > P-O, whereby AP-O orientation showed the best wear performance. Predominant wear mechanisms for AP-O were detachment of fibres and plastic deformation, whereas P-O showed micro- and macro-cracking, detachment of fibres, micro-delamination, and plastic deformation.

In their second paper, a similar work was done with comparisons were made between findings from dry and wet contact conditions [68]. They reported that in general Ws showed a decreasing trend as load increased. Also, Ws was lower in dry contact conditions, for both P-O and AP-O for loads up to 60 N. Under higher applied loads in wet contact, Ws in AP-O and N-O were lower overall. Friction coefficient under wet sliding showed a reduction of more than 90 % for all fibre orientations. This showed that the presence of water cooled the interface temperature, reducing friction and at higher loads, as well as reducing Ws.

In another work, Chin and associates investigated the potential of using kenaf fibres as a reinforcement in epoxy (KFRE)-based composites [69]. In the study, sliding was done against a stainless steel counterface using a BOD machine. Various applied loads (30 –

100 N), sliding distance (0 – 5 km), and sliding velocities (1.1 - 3.9 m/s) were considered, along with neat epoxy (NE), P-O, AP-O and N-O fibre orientations. They reported that under all tested conditions, wear resistance was higher in KFRE compared to NE. Moreover, KFRE oriented in N-O showed an improvement of approximately 85%. Among the three tested KFRE composite orientations, N-O showed better wear performance. With regards to surface morphology, the wear mechanism of KFRE in N-O was predominantly micro-cracking, with no fibre pull out seen. This indicates high interfacial adhesion between the kenaf fibre and the matrix. Friction was high for NE and KFRE orientated in N-O, compared to other orientations. This was due to the higher surface temperatures generated during sliding. Comparing the current results from previous work done, KFRE composites display better wear and frictional performance than composites based on cotton, oil palm, jute, sugar cane, and even glass fibre.

Nirmal and associates has previously investigated the effects of using kenaf fibres as particle fillers in epoxy composite (KPafRE) [70]. Test was conducted under dry contact on a POD machine. The composite was subjected to 5-30 N applied loads, at a velocity of 2.83 m/s. Various weight percentages (%wt.) of kenaf particle (5-20%) composites were used alongside with NE for comparison. It reported that all KPafRE composites showed lower wear rate and friction as compared to NE. Also, 15 %wt. KPafRE composite showed the best performance, whereby wear rate and friction improved by 67 % and 56 % respectively. The superior performance of 15 %wt. KPafRE composite was attributed to large amounts of back transfer film on the composite surface, especially at higher loads. This generated film protects the rubbing surface from further wear. This is further proven by roughness measurements of the test samples, whereby 15 % wt. KPafRE composite showed highest surface roughness; an increment of 48.2 %, as compared to the virgin composite before testing.

c) Bamboo (Phyllostachys pubescens) fibre composites

Bamboo can be found in many areas especially in the Asia-Pacific region. Countries like China, India, Bangladesh, Malaysia, and Vietnam have an abundance in supply [71]. Traditionally used in various living facilities and tools, bamboo has a high strength to weight ratio. This makes it potentially suitable to be used in polymer composites [72].

Nirmal and associates have investigated the wear and frictional performance of bamboo fibre reinforced epoxy composites (BFRE) [73]. Dry sliding tests were conducted on a POD machine under normal load of 30 N, at different sliding distances (1 - 4 km) and sliding velocities (1.7 - 3.96 m/s). Three types of fibre orientations were considered in this study; Random (R-O), P-O and AP-O. NE samples were also prepared as a benchmark. Results showed that in general, the wear rate of NE tends to improve when using bamboo fibres for all fibre orientations. Wear performance of the composite following the order AP-O > P-O > R-O. AP-O had superior wear performance compared to the other orientations. This could be due to the fibres experiencing low shear resistance in the rubbing zone, thus lowering material removal. Microscopic images of the worn surface show the fibres were still in good shape, even at higher speeds. Back transfer film was also evident, which acts as a protective layer and prevents fibres from being ruined or detached. In the case of R-O, fibres receive less support in this orientation, and there was a gradual increase in W_S at about 3 km sliding distance at higher sliding velocity. Material removal rate was high from both resin and fibre regions during the test, indicating thermo-mechanical loading between interacting surfaces which increased interface temperature. Higher temperatures degrade the wear performance of the composite. Predominant wear mechanisms here were broken fibres, back film transfer, and plastic deformation associated with fractures on the generated back film transfer. In terms of its frictional performance, results indicate that there is an improvement in friction coefficient of all three fibre composite orientation as compared to NE. In the case of P-O and AP-O, there is a rise in friction values as sliding velocity increases. This is mainly due to the higher material removal rate at higher velocities, which could have caused a third body interaction between the interfaces. For R-O, at low sliding speeds there was a rise in frictional coefficient through the test until 4 km. However, at higher sliding speeds, friction coefficient showed a drop. This could be because at low speeds, the contact mechanism was predominantly from the resinous region. At higher speeds, there was the absence of mechanical interlocking, and improvement in the intimating contact between specimen and counterface.

d) Sugarcane (Saccharum Officinarum) fibre composites

Sugarcane is a major crop cultivated in tropical countries like Brazil, China, India, Thailand, and Australia. Brazil is the largest cane producer, contributing to 25% of the world's production. The production of sugar is the main commercial use for sugarcane. Sugarcane fibres are among the lower-cost natural fibres which can be considered for use as a reinforcement in polymer composites [74,75].

El-Tayeb reported at least two papers on the usage of sugarcane fibres. In his first paper, wear and frictional performance of sugarcane fibre reinforced polyester (SCRP) and glass fibre reinforced polyester (GRP) composite were compared under dry sliding conditions [76]. Two types of fibres were prepared; randomly distributed (1, 5, 100 mm fibre lengths for C-SCRP and C-GRP) and unidirectional mat fibres (U-SCRP and CSM-GRP). Both P-O and AP-O sliding conditions were considered for unidirectional mat fibres. Operating parameters were loads (20 – 80 N), 2.5 m/s speed and 2.25 km distance. Results showed that for C-SCRP, the friction coefficient showed similar trend for all fibre lengths where there was an increase of friction coefficient up to 40 N, then reduction with increasing load. The initial increase could be due to mechanical interlocking between interacting surfaces. As load increases, wear debris formed during testing is compressed and acts as a protective layer on the composite, thus reducing friction. This formation of protective layer, or back transfer film, on the composite also increased the wear resistance of all three tested length composites at loads of 40 N and higher. In terms of fibre length, 5 mm length fibres showed much better wear performance than 1 mm or 10 mm. In the case of C-GRP, frictional coefficient values were comparable to those of C-SCRP. Wear resistance however, showed a decrease at 40 N, and stabilized at 60 N and 80 N. This decrease is evident because unlike sugarcane, glass fibres, which are hard and brittle, do not deform and inhibits the formation of a protective layer on the composite surface. Looking at the study on fibre orientation, U-SCRP composite showed lower friction values when tested in AP-O and higher in P-O when compared to CSM-GRP. Wear resistance was higher for U-SCRP in AP-O compared to P-O, and even higher than CSM-GRP in AP-O and same for CSM-GRP in P-O at 40 N and 60 N. It was concluded that overall it could be said that sugarcane fibre composites have a strong potential and is comparable to glass fibre composites.

In his second paper, a similar test was done to consider the effects of both adhesive and abrasive wear on C-SCRP and U-SCRP [77]. In adhesive testing, wear for C-SCRP reduced as applied load increased. Also 5 mm fibre length showed better wear performance compared to 1 mm or 10 mm fibre lengths. The decrease in wear as load increased was due to the presence of transfer polymer film on the wear track, as mentioned previously. At higher loads, the formation of this transfer polymer film protects the composite interface, reducing wear. For U-SCRP, wear was higher when the composite was oriented in P-O, compared to AP-O, due to the support the fibres received in AP-O from the matrix. In abrasive tests, on the other hand, wear increased with load for C-SCRP. This was because at lower loads, there were less abrasive particles which penetrated the composite surface. At higher loads, more particles penetrated, resulting in more material removal by severe plastic deformation. Again, a 5 mm length fibre showed lower wear compared to 1 mm or 10 mm. This was because the short 1 mm fibres cannot be embedded as well into the matrix, and hence, was easier to remove by abrasive particles. In the case of U-SCRP in P-O, higher wear was observed compared to AP-O. This was because in P-O wear, debris generated were easily removed due to the free path ahead on the wear track. Also, the abrasive particles were abrading the fibre and matrix at the same time, which could have increased the extent of micro cutting and ploughing in the matrix and shearing or tearing of the fibre. In AP-O, abrasive particles were moving across different layers of fibre and matrix alternately, which limited micro cutting action due to phase continuity. Also, the alternating layers gave more resistance to abrading particles, resulting in lower wear. Additionally, U-SCRP gave better abrasive wear performance than C-SCRP. Finally, it was concluded that SCRP composite has a lot of potential as a low-cost polymeric composite. However more work needs to be done.

e) Betelnut (Areca catechu) fibre composites

Betelnut, or *Areca catechu* is a palm species plant grown in Southeast Asia, particularly in India. It is grown commercially for oral consumption and interior decorations [78].

Yousif and associates published at least three papers on using betelnut fibres as a reinforcement in polymer composites. In their first paper, they investigated the adhesive wear and frictional performance of polyester composites using betelnut fibres (BFRP) [79]. Experiments were conducted on a BOD machine, under dry and wet sliding conditions. At 2.8 m/s sliding speed, 5 - 30 N loads were applied. For wet sliding, due to low weight loss, applied loads were 30 - 200 N. It was reported that under dry contact condition, the W_S showed a decreasing trend (also known as running in) until about 3.4 km, after which steady state was reached. This could be due to the presence of fibres at the interacting surface, which reduce the softening of the resin due to heat. Also, high interfacial adhesion between betelnut fibres and resin prevented fibre pull out. For wet contact condition, a similar trend was seen whereby W_S showed a running in region and reaching a steady state at about 4.2 km. However, the presence of water acted as a cleaner and cooling agent at the interface, reducing thermo-mechanical loading. Under wet sliding, there was a 49 % reduction on W_S at 30 N applied load alone. Friction coefficient values under dry sliding were in the range of 0.22 - 0.65. No steady state was observed. The presence of water at the sliding interface greatly reduced the friction coefficient values by approximately 94 %. SEM micrographs of the worn specimens showed that the predominant wear mechanisms overall were macro and micro cracks, debonding, and fibre pull-out. However, under wet contact conditions, there was less damage to the fibres even at higher loads. Compared with previous work done on glass fibre and oil palm composites under wet sliding, BFRP composites showed better wear performance. This was due to the higher interfacial adhesion of betelnut fibres with polyester resin. Also, the abrasive nature of glass fibres causes the wear debris formed at the interface to act as a third body, which damages the counterface and composite during sliding.

In their second paper, a similar study was done whereby the adhesive wear and frictional behavior of BFRP composites were studied under wet and dry sliding contact condition. However, two variations of fibre mats were considered [80]; P-O and N-O. Sliding was done against stainless steel counterface on a BOD machine with sliding velocity of 2.8 m/s, sliding distance of 0 - 6.72 km, and 20 – 200 N applied loads. Under wet sliding, the

W_S for BFRP composite in both P-O and N-O orientations showed similar trends as described earlier i.e. decrease as sliding distance increases, until a steady state is reached at 2.52 km. Also, 30 N exhibited highest W_S, followed by 70 N, 130 N and finally 200 N. N-O showed lower W_S compared to P-O. Under dry sliding at 30 N, the W_S was much higher for both fibre orientations. Also, under N-O, the W_S showed an increasing trend. Comparing both wet and dry contact conditions, the reductions in W_S were 93-100 % for N-O and 85 – 90 % for P-O under wet contact. The friction coefficient value was also reduced under wet sliding, by 91 %. Under wet sliding, friction coefficient was higher and fluctuated for N-O. This could be due to the constantly changing composite surface during sliding, i.e. either fibre or polyester was exposed to the counterface. Wear mechanism was mostly debonding of fibre and ploughing. At higher loads, micro and macro-cracking in the resinous region were seen.

In their third paper [50], the effects of three body abrasion on wear and frictional performance of treated betelnut fibre reinforced epoxy (BFRE) composite were studied. Treated BFRE composites were slid against particles of varying sizes; 500 μm (fine), 714 μm (grain) and 1430 μm (coarse) on a Linear Tribo Machine (LTM). Work was done for six different sliding velocities (0.026 - 0.115 ms^{-1}) under 5 N applied load. It was reported that the W_s was highest when the composite was subjected to coarse particles, and followed the order or coarse>grain>fine. This was due to the more evenly distributed stress on the particles, which caused low particle penetration into the composite. Larger particles caused higher stress concentration, where the predominant wear mechanisms were fibre detachment, pitting, macro-cracks, and particle penetration. Friction coefficient value was also highest when the composite was subjected to coarse particles, and lowest when subjected to fine particles. All three particles showed reducing friction coefficient as velocity increased. This could be due to the more gradual wear mechanism at higher sliding velocities, as well as the reduced abrasion effects.

f) Sisal (Agave sisalana) fibre composites

Sisal is a type of hard fibre, which is extracted from the leaves of the sisal plant *Agave sisalana*. Mainly produced in Tanzania and Brazil, sisal is the most widely used natural fibre in yarns, ropes, twines, cords, rugs, carpets, mattresses, mats, and handcrafted articles. Nearly 4.5 million tons of sisal fibres are produced every year [81,82].

Xin and associates investigated the friction and wear behavior of sisal fibre reinforced phenolic resin brake composites (SFRPR) [83]. Tribological tests were conducted on a constant speed tester at a constant speed of 2 m/s, using a cast iron friction disk, at various temperatures (100 – 350°C) and fibre volume (10 – 30 %wt.). They reported that the friction and wear properties of SFRPR composites were at its best when the resin and fibre proportion was 3:4. Lower proportions of fibre would result in more resin on the contacting surface, thus resulting in degrading wear properties. On the other hand, if the fibre content was higher, there would not be enough resin to properly cover the fibres. At lower temperatures (< 250°C), the wear mechanisms were cutting abrasion on the fibre and fatigue cracking on the resin. At higher temperatures (> 250°C), fibre decomposition can be seen. Overall it was reported that sisal fibres have a good potential as a substitute for asbestos for brake composites.

In another work [84], Ashok and associates investigated the friction performance of treated and untreated sisal-glass-fibre reinforced epoxy composite (SGFRE). Tests were conducted on a POD machine under different sliding velocities (0.2 - 4 mm/s) and fibre lengths. It was reported that there was a decreasing trend of friction as sliding speed increased, up to 2 cm fibre length, after which, the friction coefficient increased. Treating the fibres with 5 % NaOH improved the friction properties of the composite. Also, the optimum fibre length which gave the lowest friction coefficient was 2 mm.

Navin and associates [85] investigated the effects of treated and untreated sisal fibres reinforced polyester (SFRP) composites under tribological loading. Wear and friction tests were done in a POD machine, against a stainless steel counterface. Operating parameters were 6330 m sliding distance, constant sliding velocity of 1.75 m/s, and 10 – 100 N applied load. They found that in general, all samples showed an increase in weight loss as load increased. NP had the poorest P-V limit amongst all the tested samples, with the specimen failing at 30 N. The SFRP (10, 27, and 42 %wt.) all showed better structural integrity, with both 27 %wt. and 42 %wt. samples reaching up to 80 N. However, weight loss increased on addition of these fibres. This is due to the poor bonding between sisal fibres and the polyester matrix. Also, there was more energy dissipation caused by higher friction. The treated 27 %wt. SFRP composite showed significant improvement in wear performance, compared to untreated 27% wt. SFRP composite. Moreover, it was the only sample that did not fail right up to 100 N applied load. The authors also reported that the addition of sisal fibre as a reinforcement in polyesters shows an increase in friction coefficient values, which further increased with adding Silane as coupling agent. SEM micrographs revealed that at 30 N applied load, NP shows sheared destruction of material that may be attributed to thermo-mechanical loading beyond the elastic limit of the material. The 10 %wt. SFRP composite showed poor bonding and the polyester did not allow the sisal fibre to be aligned. 27 %wt. SFRP composite showed debonding of fibres after 60 N and deforming of the polyester. The treated 27 %wt. SFRP composite showed better bonding even after 80 N of applied load. Overall it was concluded that incorporating sisal fibres extended the PV limit of NP, with further improvement on the addition of Silane treatment. Also, Silane treatment significantly reduced the wear of SFRP composite, due to increased adhesion between fibre and matrix.

g) Jute (Corchorus capsularis) fibre composites

Jute fibres are extracted from the bark of the jute plant *Corchorus capsularis*. They are mainly grown in South Asia, particularly India and Bangladesh, where 95% of the world's supply is grown. These high strength, low cost, and durable fibres are used to make shopping bags, carpets, and rugs [86].

Temesgen and associates [87] investigated the effects of using jute fibres as a reinforcement on the tribological performance of polypropylene (JFRPP). Results were compared with polypropylene (PP), using a computerized POD test machine. Various speeds (1 - 3 m/s), applied load (10 – 30 N) and a 3 km sliding distance were considered. It was reported that there was a reducing trend in the friction coefficient for both PP and JFRPP composite as applied load increased, for all tested speeds. This was because, at the start of test with lower loads, mechanical interlocking at the sliding interface occurred, which increased friction and temperature. At higher loads, with the increase in temperature, the polymer begins to soften and degrade, forming a thin protective layer to cover the fibre cross section. This effect reduces friction. Overall there was little difference in the friction coefficient values. JFRPP composite however, showed much lower W_S, a reduction of 65%.

In another work [88], the effects of adding ceramic fillers, silicon carbide (SiC) and Aluminum Oxide (Al_2O_3), into jute fibre reinforced epoxy (JFRE) composites were investigated. Dry sliding tests were conducted on a POD machine, at different sliding velocities (3 - 6 m/s), and applied loads (30 – 50 N). It was found that the incorporation of ceramic fillers improved the wear properties of jute epoxy significantly. The lowest coefficient of friction was recorded in 15 %wt. Al_2O_3 filled JFRE composite. Also, Al_2O_3 filled JFRE composites showed lower friction and wear loss compared to SiC filled composites. Wear mechanism of unfilled JFRE was predominated by fibre breakage and plastic deformation, whereas in filled composites, microcracking, pit and debris formation were predominant.

h) Cotton (Gossypium) fibre composites

Among natural fibres, cotton is the most popularly used in several applications. Cotton waste are used as reinforcement in composites mainly due to their cost effectiveness as well as excellent biodegradability [89].

Hashmi and associates investigated the effects of using graphite on friction and wear of cotton-polyester (CFRP) composites at different applied pressures [90]. Dry sliding was conducted on a POD machine, against a polished stainless steel disk. Test samples were prepared for NP, CFRP, and graphite modified CFRP in different compositions. They reported that the W_S for NP was the highest, whereas the sample with highest amounts of graphite was lowest. In fact, graphite-modified samples showed nearly constant W_S under all tested loads. Friction coefficient was also lowered with the addition of graphite, whereby the sample which had the highest amount of graphite displayed the lowest friction coefficient. This demonstrates the lubricating effect of graphite. CFRP composite however, showed the highest coefficient of friction. This could be due to the area of resin in contact which was replaced by the less heat of friction-sensitive cotton fibres, as compared to NP, which undergoes thermal softening. The result is resistance to movement. Contact surface temperature was also lowered by the addition of graphite in CFRP.

In a different study, Zhang and associates investigated the tribological performance of hybrid PTFE/cotton resin composites filled with multi-walled carbon nanotubes (MWCNTs) [91]. Testing was conducted on a POD machine at loads between 156.80 N to 250.88 N, with temperature ranging from 25°C to 120°C and a sliding speed of 0.26 m/s. Comparisons were made between three types of samples: unfilled PTFE/cotton, MWCNTs + PTFE/cotton (untreated) and covalent modified or MWCNTs-g-MA + PTFE/cotton (treated) composites. Using various percentages of MWCNTs, it was found that the composite with 1% MWCNTs gave the most optimum wear and frictional behavior. At increasing applied load, it was found that wear rate increased on all three tested samples. Unfilled PTFE/cotton composites showed a much higher wear rate, especially at higher loads. MWCNTs-g-MA + PTFE/cotton composites, on the other hand, gave the lowest overall wear rate. It is believed that the covalent modification done improved the adhesion strength between MWCNT and the polymer matrix. As temperatures increased under load of 156.80 N, it was found that friction coefficient showed a decreasing trend for unfilled PTFE/cotton and MWCNTs-g-MA + PTFE/cotton composites, whereas the wear rate showed an increasing trend. This was due to the softening of resin at higher temperatures. MWCNTs-g-MA + PTFE/cotton composites showed better wear performance at elevated temperatures due to the good thermal conductivity of MWCNTs. Also, the nanoparticles could have acted as ball bearings, and rolled rather than slid during testing, thus reducing shear stress and contact temperature.

i) Coconut fibre

Coconut, or *Cocos nucifera linn* is a type of palm plant belonging to the Arecaceae plant family. They are grown in more than ninety countries, with India, Sri Lanka, Philippines, and Indonesia being the largest producing countries in the world. These countries contribute to 78% of the world's coconut supply. Industrially, coconut oil is used to produce soap, hair oil, and cosmetics, while the outer thick fibrous fruit, known as the husk, is used as a source of fibre [92].

Ibrahem [93] investigated the effects on friction and wear on polyester (CORP) and (CRP) composites reinforced with coconut and carbon fibres and powders. Polyester resin was used as the matrix, and tests were carried out under dry sliding condition on a POD machine. The sliding velocity used was 2.35 ms^{-1}, under 4 - 8 N applied loads, with a sliding time of five minutes. It was reported that reinforcing polyester with either carbon or coconut helped to reduce both friction and wear. Also, powder reinforcements were better at reducing friction, while fibres were better at reducing wear. Under 8 N applied load, there was a 17 % reduction in friction for CRP composite, whereas for CORP composite there was a 27 % reduction. On the other hand, wear rate reduced by 86 % with carbon fibre reinforcement, and 95 % when reinforced with coconut fibres.

In another work, Ravikantha and associates [94] investigated the tribological properties of treated coconut fibre reinforced epoxy (CFRE) composites. Various fibre percentages were used (0 - 15 %), and dry sliding tests were performed on a POD machine at different sliding velocities (1.5 - 2.5 m/s) under different applied loads (50 – 70 N). It was found that CFRE composite with a 15 % fibre content showed the lowest wear results in the adhesive and abrasive wear tests, while NE showed the highest. There was a 55 % reduction in wear for abrasive tests, while in the adhesive wear tests at constant load of 70 N and sliding velocity of 2.5 m/s, the reduction was approximately 55 %. Friction was found to have also reduced upon the addition of coconut fibres, with 15 % fibre content sample showing the lowest friction among all tested samples.

j) Banana (Musaceae) fibre

Banana plants are of the Musaceae family and can be found mostly in Africa, Asia, and Australia. It is the fourth largest fruit crop, and the most popular fruit in the world. Economically, they are grown as a food source, and the fibres are used to make ropes and coarse textiles [95].

Very few works have been reported on the tribological properties of banana fibre reinforced polymeric composites. On such work studied the effects of fibre size, resin type, and curing agent on the tribological properties of musaceae reinforced polymeric composites (MFRP) [96]. Dry sliding tests were conducted on a pin on disk machine under 4.9 N applied loading, at a constant speed of 200 $m.min^{-1}$, for 3 km. It was reported that in the case of friction, there was not much variation with the different hardeners used, i.e. hardeners had little effect on friction of this type of composite. Friction value recorded was in the range of 0.25 to 0.5. The wear recorded was lower for the composites as compared to NP. Also, the S901 type resin showed the best wear performance with an

85 % improvement compared to NP. The primary wear mechanisms were surface fatigue and crazing. Finally, the authors also compared the results of the current work with other similar works done using different natural fibres. Here it can be concluded that the current friction and wear results were comparable, indicating that this type of fibre performs as well as other natural fibres.

k) Rice husk fibre

Rice is the seed of the grass plant species *Oryza*, and is consumed in many parts of the world as a food source, especially in Asia. Rice is among the world's highest produced agricultural crop, with 650 million tons produced annually [97]. During rice production, the by-products are mostly rice bran and rice husks. Rice husks, which contain silica, are generally regarded as waste and are difficult to dispose of, but has been used in applications such as electricity generation and well as producing light weight concrete and insulation materials [98–100].

Navin and associates studied the effects of different fibre weights and surface treatments on wear properties of rice husk reinforced PVC (RFRPVC) composites [101]. 10, 20 and 30 %wt. percentage fibres were used, and results were compared with neat PVC. A POD machine was used to conduct the abrasive wear test, using three grades of abrasive papers, under 10 N applied load, and up to a sliding distance of 31.4 m. Foaming agents were used in the composite to reduce density. They concluded that wear was highest for the PVC sample, whereas lowest for the RFRPVC composite with 30% wt. percentage treated fibre. Also, in the case of the RFRPVC composites, higher fibre weight content (up to 30% wt.) gave better wear resistance. The worn surface study revealed that huge microchips were evident in the PVC sample. The treated rice husk composites showed much less damage.

Rice husks can also be carbonized through mixing with phenolic resin in an inert gas environment, forming Rice Husk ceramics, or 'RH ceramics'. In 2009, Tuvshin and associates investigated the friction and wear properties of RH ceramics under dry and wet contact conditions [102]. A BOD machine was used, under various loads (0.98 - 9.8 N) and sliding velocities (0.02 - 1 m/s). They reported that friction was higher in wet contact condition, compared to dry. A similar observation was seen for the wear rate as well. Photo micrographs showed that a much thicker transfer and silica rich transfer film developed during dry sliding. This layer is hydrophilic and induces low friction between mating surfaces. On the other hand, for wet sliding, the silica wear debris generated were washed away, resulting in a much thinner transfer layer consisting mostly of carbon, which induces more friction. Under both conditions, the friction and wear values were considered low for composites; overall less than 0.2 and 1.0×10^{-8} mm^2/N respectively.

In another study [103], friction and wear properties of polyamide (PA66) filled with rice bran ceramics (PA66/RBC) and glass beads (PA66/GBs) were investigated. It was reported that pure PA66 showed higher friction and wear compared to PA66/GB and PA66/RBC, particularly at low sliding speeds. At higher sliding speeds, there was not much difference in friction between the three tested materials. Wear was also significantly higher for pure PA66, as high as 1×10^{-7} mm^2/N. For both PA66/RBC and Pa66/GB, the Ws was less than 1×10^{-8} mm^2/N. SEM observations showed that for pure PA66, large wear particles were observed at low sliding speeds, caused by friction-induced surface breakage of the material. At higher sliding speeds, plastic flow is more evident. For the PA66/RBC and PA66/GB, observations were similar. At low speeds, smaller wear particles than pure PA66 were seen and at higher sliding speeds, voids can be seen due to RBC or GB removal at the rubbing interface. This confirms that reinforcing PA66 with RBC or GB is effective, particularly at low sliding speeds.

l) Calotropis gigentea fruit fibre

Calotropis gigentea, also known as crown flower, is a plant species belonging to the *calotropis* species. It can grow up to 4 m tall, and has purple or white crown-shaped flowers with a white wooly stem. This species is native to Malaysia, Indonesia, Philippines, India, Thailand, Sri Lanka, and China. It is used primarily as traditional medication for the treatment of asthma, bronchitis, paralysis, swelling, and intermittent fevers [104].

Babu and associates investigated the wear performance of *calotropis gigentea* fruit fibres in polyester resin [105]. Various volume fraction of fibres were used, and testing was conducted using a tribometer, at a speed of 2000 rpm and at loads of 10 to 30 N. It was discovered that weight loss increased with fibre content, and the highest weight loss was recorded for 30% volume fraction composite under 30 N load. In this case, pure polyester actually had better wear resistance compared to *calotropis gigentea* reinforced polyesters. The reason for this observation is not clearly understood or reported. On the other hand, the friction coefficient showed a decreasing trend with an increase in fibre content.

Summary

The tribological performance of natural fibre reinforced polymer composites were reviewed. Important findings from these reviewed works are summarized below:

a) There was general improvement in wear and frictional performance by reinforcing polymers with natural fibres or particles. The addition of these fibres added strength and stiffness to neat polymers. All reviewed works reported improvement in wear. In some cases, almost a 100% improvement was seen by adding fibre reinforcements. Table 10.6 summarizes the improvement in wear and friction by reinforcing polymer with natural fibres.

Table 10.6 Improvement in Wear and Friction Summary.

Natural Fibre/Particle	Polymer	Improvement in Ws (%)	Improvement in Friction (%)	Reference
Oil Palm	Polyester	66 (50 N, 2.8 m/s)	8.4	[106]
Kenaf	Polyurethane	65 (AP-O, 50 N)	5.3	[107]
Kenaf	Epoxy	86 (N-O, 50 N)	16	[69]
Kenaf	Epoxy	63 (30 N, 15 %wt.)	40	[70]
Bamboo	Epoxy	39 (P-O, 2.83 m/s)	29	[73]
Jute	Polypropylene	70 (30 N, 3 m/s)	27	[87]
Cotton	Polyester	31 (20 N)	-83	[90]
Coconut	Polyester	95 (25 %wt.)	27	[93]
Coconut	Epoxy	67 (20 %wt., 50 N, 2.5	-	[94]
Banana	Polyester	m/s)	19	[96]
Rice Bran	Polyamide 66	97 (MEK, S901)	30	[103]
		94 (0.01 m/s, 5 N)		

b) Numerous works have reported on the formation of 'film transfer' and 'back film transfer' during testing. [69,70,107–110,73,85,87,90,91,96,102,106]. 'Film transfer' formation occurs when wear debris from the composite, consisting of detached fibres and polymer, collects and forms a layer on the counterface surface during wear test. This film protects the composite surface against sliding, thus reducing the wear rate. It was said that this occurs at steady state, when the wear rate is stable. 'Back film transfer' is similar to 'film transfer', except that the protective layer is formed in the composite surface instead. Synthetic fibres such as glass, which is among the most widely used, is abrasive in nature and hinders this film formation. Natural fibres are soft, and adhere well to this transfer film formation. This makes the tribological performance of some natural fibre reinforced composites comparable to that of synthetic fibres and, in some cases, even better.

c) Composites based on unidirectional or woven mats, which allows for three type of composite orientation (AP-O, P-O and N-O) during testing, have also been reviewed [69,73,80,107–109,111]. It was observed that there would be one orientation which would give good wear and friction performance compared to the rest in each test. In particular, the AP-O orientation showed superior performance in three of the reviewed work. It was reported that the fiber-polymer layers arranged in a perpendicular direction to sliding seem to be able to trap wear debris, which then form a protective layer on the composite surface.

d) Numerous reviewed works have investigated the effect of fibre length and fibre volume on tribological properties. The highlights are summarized in Table 10.7. It can be observed that for both cases, there are optimum length and volumes. In terms of length, shorter fibres are easily removed from the composite surface during sliding, whereas a longer fiber would be entangled in the polymer matrix and would not detach as easily. For fibre volume, having too little fibre results in no noticeable fibre whereas having too much fibre would cause the composite property to degrade as there is not enough polymer to effectively 'wet' or cover all the fibers completely. This is undesirable, as the polymers function is to support and maintain the overall shape of the composite.

Table 10.7 Optimum Fibre Length and Volume Summary

Fibre Length Range (mm)	Fibre Volume (%wt.)	Optimum Fibre Length (mm)	Optimum Fibre Volume (%wt.)	Reference
NA	30 - 70	NA	50	[61]
NA	5 - 20	NA	15	[70]
1 - 10	NA	5	NA	[108]
1 - 10	NA	5	NA	[109]
NA	10 – 30	NA	20	[83]
10 - 30	NA	20	NA	[112]
NA	4 - 42	NA	27	[85]
NA	5 - 25	NA	25	[93]
NA	5 – 20	NA	20	[94]
$152 – 638 \times 10^{-3}$	NA	NA	287×10^{-3}	[96]

e) Some of the reviewed works incorporated fibre treatment as part of the fibre preparation process, as shown in Table 10.8. The most common used treatment for fibres were Sodium Hydroxide (NaOH). Other treatments include Ammonia, Silane, Acetic Acid (CH_3COOH), Maleic Anhydride (MAH) and Benzyl Peroxide (BP). It was reported that fibre treatment creates micro holes on the fibre surface. This increases the surface wettability between fibre and polymer matrix, therefore improving the interfacial adhesion between the two. It was reported in some cases that fibre treatments improve wear resistance by 70%, as compared to untreated fibres.

Table 10.8 Fibre Treatment Summary

Treatment	Improvement in Ws (%)	Improvement in Friction (%)	Reference
6 % NaOH	NA	NA	[60]
6 % NaOH	NA	NA	[61]
6 % NaOH	NA	NA	[107]
6 % NaOH	NA	NA	[111]
6 % NaOH	NA	NA	[50]
10 % NaOH + Ammonia	NA	NA	[83]
5 % NaOH	-	15 (2mm, 0.2 mm/s)	[112]
1 % Silane	70 (27 %wt., 40N)	-	[85]
NaOH + CH$_3$COOH	NA	-	[94]
MAH + BP	25 (30 %wt., 600 SiC)	NA	[101]
		-	

f) Several of the reviewed works investigated the effects of reinforcing natural fibre composites by adding filler materials, as shown in Table 10.9. It was reported that these fillers would further enhance the wear properties and reduce the friction of natural fibre reinforced composites. Ceramic fillers, like Silicon Carbide (SiC) and Aluminum Oxide (Al$_2$O$_3$), graphite, and multiwalled carbon nanotubes (MWCNTs) have all been considered in this review. There is significant improvement in wear rate, and friction is generally reduced.

Table 10.9 Filler Treatment Summary

Additional Fillers	Improvement in Ws (%)	Improvement in Friction (%)	Reference
SiC and Al$_2$O$_3$	87 (Al$_2$O$_3$, 50 N, 3 m/s)	19	[88]
Graphite	75	50	[90]
MWCNTs	58 (188.1 N, 25°C)	-44%	[91]

g) There were four works reviewed which tested the performance of natural fibre composites under wet contact conditions. It was reported that the use of water during sliding reduces wear and friction, as water acts as a coolant, which keeps the rubbing interface temperature low and reduces the effects of thermos-mechanical loading of the composite. Water also washes away wear debris, which keeps the rubbing interface clean. The improvement in wear and friction under wet sliding compared to dry is summarized in Table 10.10. One of the four reviewed work contradicted this [102]. In this work, it was reported that in dry contact, the formation of thick transfer film consisting of amorphous silica with carbon particles (which was the wear debris) helped in lowering wear and friction.

Table 10.10 Wet Sliding Tests Summary

Improvement in Ws (%)	Improvement in Friction (%)	Reference
33 (AP-O, Steady state)	99	[111]
50	94	[110]
85	85	[80]
-	-36	[102]

h) There were three papers reviewed in which abrasive testing was conducted [50,101,109]. It was observed that this type of wear results in high wear rate, especially when compared to adhesive wear. This is because of the constant ploughing of abrasive particles against the composite surface. A lower grit size (bigger abrasive particles) results in higher wear rate compared to higher grit size.

Future Developments

The current review is comprehensive but not exhaustive. There are still areas which could be investigated further to expand this current review:

1) There are still many other natural fibres, such as flax fibre, which were not considered in this work.

2) The different types of composite fabrication methods, such as hand lay-up and compression molding, were not discussed. These and other composite fabrication methods can be investigated and compared.

3) The effects of test parameters such as load and sliding distance and test findings on sliding interface temperature as well as roughness were not explicitly discussed. These parameters and findings can be investigated in detail to study their effects on wear and friction performances.

4) It has been reported that variation do exist in the physical and mechanical properties natural of fibres due to the different cultivation methods and geographical origin of these fibres. Therefore, the effects of this type of fibre variation on tribological performance can be considered. A proposal to standardize these fibre properties can also be included.

5) A comprehensive comparison of properties between natural and synthetic fibre reinforced polyester composites can be done. This would give an indication of whether the performance of natural fibre composites are up to par with synthetic fibre composites, which are currently more widely used.

6) The usage of bio-based plastics such as PLA (Polylactide Acid) and PHB (Polyhydroxy butyrate) has been growing in recent years [17]. The usage of these plastics, together with natural fibres, would form a 100% bio-composite. This would completely negate the reliance on fossil fuel-based materials. However, the tribological performance of these composites have not been fully investigated. It has been reported that the mechanical properties of these bio-plastics are comparable to polystyrene and polypropylene [113].

7) Modelling or soft-computing is a collection of computational methods which can be used to predict the performance of composites. However, very few works have been reported on predicting the wear for natural fibre composites. Methods such as Artificial Neural Networks (ANN), Fuzzy Inference Systems (FIS), and Adaptive Neuro-Fuzzy Inference Systems (ANFIS) can be employed to predict desired responses, based on the highly non-linear relationship that exist between the input parameters [114]. These methods give accurate responses in a timely manner at low cost, which allows for exponential growth in this field.

8) Natural fibres can be mixed with synthetic fibres to form a composite. This type of material combines the strengths of both types of fibres, and has been reported to have good mechanical and physical properties [115,116]. More works can be done here, to study the effects of various combination of natural and synthetic fibre and its effects on both mechanical and tribological properties.

References

[1] H. Pihtili, An experimental investigation of wear of glass fibre-epoxy resin and glass fibre-polyester resin composite materials, European Polymer Journal. 45 (2009) 149–154.

[2] A. Shalwan, B.F. Yousif, In State of Art: Mechanical and tribological behaviour of polymeric composites based on natural fibres, Materials & Design. 48 (2013) 14–24. https://doi.org/http://dx.doi.org/10.1016/j.matdes.2012.07.014.

[3] N.S.M. El-Tayeb, B.F. Yousif, Evaluation of glass fiber reinforced polyester composite for multi-pass abrasive wear applications, Wear. 262 (2007) 1140–1151. https://doi.org/HTTPS://DOI.ORG/ 10.1016/j.wear.2006.11.015.

[4]   U. Nirmal, K.O. Low, J. Hashim, On the effect of abrasiveness to process equipment using betelnut and glass fibres reinforced polyester composites, Wear. 290–291 (2012) 32–40. https://doi.org/10.1016/j.wear.2012.05.022.

[5]   F.P. La Mantia, M. Morreale, Green Composites: A Breif Review, Composites Part A. 42 (2011) 579–588.

[6]   U.N.R. Council, Oil in the Sea: Inputs, Fates, and Effects. National Academy of Sciences, Washington, DC., USA. 2009, (n.d.).

[7]   L. Dongdong, L. Bin, B. Chenguang, M. Minghui, X. Yan, Y. Chunyan, Marine oil spill risk mapping for accidental pollution and its application in a coastal city, Marine Pollution Bulletin. 96 (2015) 220–225. https://doi.org/10.1016/j.marpolbul.2015.05.023.

[8]   A.P.D.E. Souza, M. Gaspar, E. Alves, D.A. Silva, E.C. Ulian, A.J. Waclawovsky, M. Yutaka, N. Jr, R. Vicentini, D.O.S. Santos, M.M. Teixeira, G.M. Souza, M.S. Buckeridge, Elevated CO2 increases photosynthesis , biomass and productivity , and modifies gene expression in sugarcane, Plant, Cell and Environment. (2008) 1116–1127. https://doi.org/10.1111/j.1365-3040.2008.01822.x.

[9]   J.R. Dufl, Y. Deng, K. Van Acker, W. Dewulf, Do fiber-reinforced polymer composites provide environmentally benign alternatives ? A life-cycle-assessment-based study, MRS Bulletin. 37 (2012) 374–382. https://doi.org/10.1557/mrs.2012.33.

[10]   OPEC. World Oil Outlook; 2013, n.d.

[11]   D. Toke, The EU Renewables Directive-What is the fuss about trading?, Energy Policy. 36 (2008) 3001–3008.

[12]   O. Faruk, A.K. Bledzki, H. Fink, M. Sain, Progress in Polymer Science Biocomposites reinforced with natural fibers : 2000 – 2010, Progress in Polymer Science. 37 (2012) 1552–1596. https://doi.org/10.1016/j.progpolymsci.2012.04.003.

[13]   H.P.S.A. Khalil, H. Ismail, H.D. Rozman, M.N. Ahmad, The effect of acetylation on interfacial shear strength between plant ® bres and various matrices, 37 (2001) 1037–1045.

[14]   R. Mahjoub, J.M. Yatim, A.R. Mohd Sam, S.H. Hashemi, Tensile properties of kenaf fiber due to various conditions of chemical fiber surface modifications, Construction and Building Materials. 55 (2014) 103–113. https://doi.org/10.1016/j.conbuildmat.2014.01.036.

Thermoset Composite   Materials Research Forum LLC
Materials Research Foundations **38** (2018)   doi: http://dx.doi.org/10.21741/9781945291876

[15] D.N. saheb, J.P. Jog, Natural Fiber Polymer Composites: A Review, Advances in Polymer Technology. 18 (1999) 351–363.

[16] B.F. Yousif. N.S.M. El-Tayeb, High-stress three body abrasive wear of treated and untreated oil palm fibre-reinforced polyester composite, Proc. Inst. Mech. Eng. J:J. Eng. Tribol. 222 (5) 637-646, (2008), (n.d.).

[17] O. Faruk, A.K. Bledzki, H. Fink, M. Sain, Biocomposites reinforced with natural fibers : 2000 – 2010, Progress in Polymer Science. 37 (2012) 1552–1596.

[18] W.M.Z.W.Y. G. Raju, C. T. Ratnam, N.A. Ibrahim, M.Z.A. Rahman, Enhancement of PVC/ENR blend properties by poly(methyl acrylate) grafted oil palm empty fruit bunch fibre, J. Appl. Polym. Sci. 110 (2008) 368-375, (n.d.).

[19] S. V Joshi, L.T. Drzal, A.K. Mohanty, S. Arora, Are natural fiber composites environmentally superior to glass fiber reinforced composites?, Composites Part A: Applied Science and Manufacturing. 35 (2004) 371–376.

[20] M. Baiardo, E. Zini, M. Scandola, Flax fibre – polyester composites, 35 (2004) 703–710. https://doi.org/10.1016/j.compositesa.2004.02.004.

[21] A. Gassan, J; Bledzki, The influence of fiber-surface treatment on the mechanical properties of jute-polypropylene composites, Composites Part A: Applied Science and Manufacturing. 28 (1997) 1001–1005.

[22] A. Singh, W. Hall, J. Summerscales, Failure strain as the key design criterion for fracture of natural fibre composites, Composites Science and Technology. 70 (2010) 995–999. https://doi.org/10.1016/j.compscitech.2010.02.018.

[23] J.F. Shackelford, Introduction to material science for engineering, (2000).

[24] W. Bolton, Engineering Materials Technology, 2nd ed., Butterworth-Heinemann Ltd., 1993.

[25] H. Unal, U. Sen, A. Mimaroglu, Dry sliding wear characteristics of some industrial polymers against steel counterface, 37 (2004) 727–732. https://doi.org/10.1016/j.triboint.2004.03.002.

[26] H. Unal, A. Mimaroglu, T. Arda, Friction and wear performance of some thermoplastic polymers and polymer composites against unsaturated polyester, 252 (2006) 8139–8146. https://doi.org/10.1016/j.apsusc.2005.10.047.

[27] F.W. Billmeyer, Textbook of Polymer Science, 3rd ed., Wiley, 1984.

[28] K.K. Chawla, Composite Materials: Science and Engineering, 3rd ed., Springer Science and Business Media, 2012.

[29] D. Gay, Composite Materials: Design and Applications, 3rd ed., CRC Press, 2014.

[30] L.H. Van Vlack, Elements of Materials Science and Engineering, 6th ed., Pearson, 1989.

[31] K. 13. Subramaniam, Asaithambi P, Friction and wear of epoxy resin containing graphite. J Reinf Plast Compos 1986; 5:200-8, (n.d.).

[32] K. Friedrich, Z. Lu, A.M. Hager, Recent advances in polymer composites' tribology, 190 (1995) 139–144.

[33] P. Jost, Tribology : How a word was coined 40 years ago, Tribology and Lubrication Technology. (2006) 24–28.

[34] Y.J. Mergler, R.P. Schaake, V.A.J. Huis, Material transfer of POM in sliding contact, Wear. 256 (2004) 294–301. https://doi.org/Https://doi.org/ 10.1016/s0043-1648(03)00410-1.

[35] J. Bijwe, J. Indumathi, A.K. Ghosh, On the abrasive wear behaviour of fabric-reinforced polyetherimide composites, Wear. 253 (2002) 768–777. https://doi.org/Https://doi.org/ 10.1016/s0043-1648(02)00169-2.

[36] J. Bijwe, J. Indumathi, J. John Rajesh, M. Fahim, Friction and wear behavior of polyetherimide composites in various wear modes, Wear. 249 (2001) 715–726. https://doi.org/Https://doi.org/ 10.1016/s0043-1648(01)00696-2.

[37] A.P. Harsha, U.S. Tewari, Tribo performance of polyaryletherketone composites, Polymer Testing. 21 (2002) 697–709. https://doi.org/Https://doi.org/ 10.1016/s0142-9418(01)00145-3.

[38] J.J. Rajesh, J. Bijwe, U.S. Tewari, Abrasive wear performance of various polyamides, Wear. 252 (2002) 769–776. https://doi.org/Https://doi.org/ 10.1016/s0043-1648(02)00039-x.

[39] U. Nirmal, J. Hashim, S.T.W. Lau, Testing methods in tribology of polymeric composites, International Journal of Mechanical and Materials Engineering (IJMME). 6 (2011) 367–373.

[40] B.F. Yousif, Design of newly fabricated tribological machine for wear and frictional experiments under dry/wet condition, Materials & Design. 48 (2013) 2–13. https://doi.org/10.1016/j.matdes.2012.06.046.

[41] S.R. Hummel, B. Partlow, Comparison of threshold galling results from two testing methods, Tribology International. 37 (2004) 291–295. https://doi.org/HTTPS://DOI.ORG/ 10.1016/j.triboint.2003.09.003.

Thermoset Composite
Materials Research Foundations **38** (2018)

Materials Research Forum LLC
doi: http://dx.doi.org/10.21741/9781945291876

[42] J. Bijwe, R. Rattan, Carbon fabric reinforced polyetherimide composites:
      Optimization of fabric content for best combination of strength and adhesive wear
      performance, Wear. 262 (2007) 749–758. https://doi.org/HTTPS://DOI.ORG/
      10.1016/j.wear.2006.08.011.

[43] C.W. Chin, B.F. Yousif, Adhesive and frictional behaviour of polymeric
      composites based on kenaf fibre, in: ICAT 274, 2nd International Conference on
      Advanced Tribology, 3-5 December 2008, Singapore, n.d.: pp. 1–3.

[44] R. Blickensderfer, G. Laird III, A pinion–drum abrasive wear test and comparison
      to other pin tests, J. Test. Eval. 16 (1988) 516–526.

[45] Y.S. Kim, J. Yang, S. Wang, A.K. Banthia, J.E. McGrath, Surface and wear
      behavior of bis-(4-hydroxyphenyl) cyclohexane (bis-Z)
      polycarbonate/polycarbonate-polydimethylsiloxane block copolymer alloys,
      Polymer. 43 (2002) 7207–7217. https://doi.org/Https://doi.org/ 10.1016/s0032-
      3861(02)00465-2.

[46] Blickensderfer R; Laird G, A pin-on-drum abrasive wear test and comparison to
      other pin test, J Test Eval. 16 (1988) 516–526.

[47] H. PihtIII, N. Tosun, Effect of load and speed on the wear behaviour of woven
      glass fabrics and aramid fibre-reinforced composites, Wear. 252 (2002) 979–984.
      https://doi.org/Https://doi.org/ 10.1016/s0043-1648(02)00062-5.

[48] R. Reinicke, F. Haupert, K. Friedrich, On the tribological behaviour of selected,
      injection moulded thermoplastic composites, Composites Part A: Applied Science
      and Manufacturing. 29 (1998) 763–771. https://doi.org/Https://doi.org/
      10.1016/s1359-835x(98)00052-9.

[49] Y.Z. Wan, Y. Huang, F. He, Q.Y. Li, J.J. Lian, Tribological properties of three-
      dimensional braided carbon / Kevlar / epoxy hybrid composites under dry and
      lubricated conditions, 453 (2007) 202–209.
      https://doi.org/10.1016/j.msea.2006.11.090.

[50] B.F. Yousif, U. Nirmal, K.J. Wong, Three-body abrasion on wear and frictional
      performance of treated betelnut fibre reinforced epoxy (T-BFRE) composite,
      Materials & Design. 31 (2010) 4514–4521.
      https://doi.org/10.1016/j.matdes.2010.04.008.

[51] A.N.J. Stevenson, I.M. Hutchings, Development of the dry sand/rubber wheel
      abrasion test, Wear. 195 (1996) 232–240. https://doi.org/Https://doi.org/
      10.1016/0043-1648(96)06965-7.

[52]   F. Guo, Z. Zhang, H. Zhang, K. Wang, W. Jiang, Tribology International
       Tribological behavior of spun Kevlar fabric composites filled with fluorinated
       compounds, Tribiology International. 43 (2010) 1466–1471.
       https://doi.org/10.1016/j.triboint.2010.02.004.

[53]   L.K. Ong, Q.K. Kien, W.K. Jye, Effect of Fibre Orientation on the Scratch
       Characteristics of E-Glass Fibre-Reinforced Polyester Composite, Recent Patents
       on Materials Science. 4 (2011) 56–62.

[54]   D.M. Nuruzzaman, M.A. Chowdhury, Friction and Wear of Polymer and
       Composites, Intech, 2012.

[55]   V. Quaglini, P. Dubini, Friction of polymers sliding on smooth surfaces, Advances
       in Tribology. 2011 (2011).

[56]   N.K. Myshkin, M.I. Petrokovets, A. V Kovalev, Tribology of polymers : Adhesion
       , friction , wear , and mass-transfer, Tribology International. 38 (2005) 910–921.

[57]   I. Sevim, Tribology in Engineering, Intech, 2013.

[58]   H.P.S.. Abdul Khalil, M. Jawaid, A. Hassan, M.T. Paridah, A. Zaidon, Oil Palm
       Biomass Fibres and Recent Advancement in Oil Palm Biomass Fibres Based
       Hybrid Biocomposites, Intech, 2012.

[59]   N.S.. El-Tayeb., B.F. Yousif., The Effect of Oil Palm Fibres As Reinforcement On
       Tribological Performance of Polyester Composite, Surface Review and Letters,
       Vol. 14, No. 6 (2007) 1095-1102, (n.d.).

[60]   B.F. Yousif, U. Nirmal, Wear and frictional performance of polymeric composites
       aged in various solutions, Wear. 272 (2011) 97–104.
       https://doi.org/10.1016/j.wear.2011.07.006.

[61]   F. Fazillah, M. Fadzli, B. Abdollah, A. Kalam, M. Hassan, H. Amiruddin,
       Tribological characteristics comparison for oil palm fi bre / epoxy and kenaf fibre /
       epoxy composites under dry sliding conditions, Tribology International. 101
       (2016) 247–254.

[62]   J. Pearsall, Concise Oxford English Dictionary. 10th ed. New York: Oxford
       University Press: 2002, (n.d.).

[63]   I.S. Ishak, M.R., Leman, Z., Sapuan, S.M., Edeerozey, A.M.M. and Othman,
       Mechanical properties of kenaf bast and core fibre reinforced unsaturated polyester
       composites', 9th National Symposium on Polymeric Materials (NSPM). (2009),
       (n.d.).

[64] T. Nishino, K. Hirao, M. Kotera, K. Nakamae, H. Inagaki, Kenaf reinforced biodegradable composite, Composites Science and Technology. 63 (2003) 1281–1286. https://doi.org/Https://doi.org/ 10.1016/s0266-3538(03)00099-x.

[65] H.G. Mohanty A.K. Misra M, Biofibres, biodegradable polymers and biocomposites; an overview. Macromol Mater Eng 2000; 276-277(1): 1-24, (n.d.).

[66] M. Zampaloni, Kenaf natural fiber reinforced polypropylene composites : A discussion on manufacturing problems and solutions, 38 (2007) 1569–1580. https://doi.org/10.1016/j.compositesa.2007.01.001.

[67] S. Narish, B.F. Yousif, D. Rilling, Proceedings of the Institution of Mechanical Engineers , Part J : Journal of Engineering Tribology Adhesive wear of thermoplastic composite based on kenaf fibres, (2011). https://doi.org/10.1177/2041305X10394053.

[68] B.F. Yousif, Investigations on wear and frictional properties of kenaf fibre polyurethane composites under dry and wet contact conditions Narish Singh * Dirk Rilling, 2 (2011) 375–387.

[69] C.W. Chin, B.F. Yousif, Potential of kenaf fibres as reinforcement for tribological applications, Wear. 267 (2009) 1550–1557. https://doi.org/10.1016/j.wear.2009.06.002.

[70] U. Nirmal, S.T.W. Lau, J. Hashim, A. Devadas, Y. My, Effect of kenaf particulate fillers in polymeric composite for tribological applications, (2015). https://doi.org/10.1177/0040517514563744.

[71] W.J. Lobovikov M, Shyam P, Piazza M, Ren H, Non Wood Forest Products 18 World Bamboo Resources. A Thematic Study Prepared In the Framework of The Global Forest Resources Assessment. Rome: Food and Agriculture Organization of the United Nations; 2007, (n.d.).

[72] H.P.S. Abdul Khalil, I.U.H. Bhat, M. Jawaid, A. Zaidon, D. Hermawan, Y.S. Hadi, Bamboo fibre reinforced biocomposites: A review, Materials & Design. 42 (2012) 353–368. https://doi.org/10.1016/j.matdes.2012.06.015.

[73] U. Nirmal, J. Hashim, K.O. Low, Adhesive wear and frictional performance of bamboo fibres reinforced epoxy composite, Tribology International. 47 (2012) 122–133. https://doi.org/10.1016/j.triboint.2011.10.012.

[74] C.R. Soccol, L.P. de S. Vandenberghe, A.B.P. Medeiros, S.G. Karp, M. Buckerridge, L.P. Ramos, A.P. Pitarelo, V.P. da S. Bon, L.M.P. de Moraes, J. de

A. Araujo, F.A.G. Torres, Bioethanol from lignocelluloses: Status and perspectives in Brazil, Bioresource Technology. 101 (2010) 4820–4825.

[75] R. Sindhu, E. Gnansounou, P. Binod, A. Pandey, Biodiversity of sugarcane crop residue for value added products-An overview, Renewable Energy. 98 (2016) 203–215.

[76] M. El, A study on the potential of sugarcane fibers / polyester composite for tribological applications, 265 (2008) 223–235. https://doi.org/10.1016/j.wear.2007.10.006.

[77] N.S.M El Tayeb, Development and characterisation of low-cost polymeric composite materials, Materials and Design. 30 (2009) 1151–1160. https://doi.org/10.1016/j.matdes.2008.06.024.

[78] J.S. Binoj, R.E. Raj, V.S. Sreenivasan, G.R. Thusnavis, Morphological, physical, mechanical, chemical and thermal characterization of sustainable Indian Areca fruit husk fibres (Areca Catechu L.) as potential alternative for hazardous synthetic fibres, Journal of Bionic Engineering. 13 (2016) 156–165.

[79] B.F. Yousif, S.T.W. Lau, S. Mcwilliam, Polyester composite based on betelnut fibre for tribological applications, Tribology International, 43 (2010) 503-511, (n.d.). https://doi.org/10.1016/j.triboint.2009.08.006.

[80] B.F. Yousif, A. Devadas, T.F. Yusaf, Adhesive wear and frictional behaviour of multilayered polyester composite based on betelnut fiber mats under wet contact conditions, 16 (2009) 407–414.

[81] J.M.L. Reis, Sisal fiber polymer mortar composites : Introductory fracture mechanics approach, Construction and Building Materials. 37 (2012) 177–180.

[82] D.O. Santos, R. Passos, D.O. Castro, A.C. Ruvolo-Filho, E. Frollini, Processing and thermal properties of composites based on recycled PET, sisal fibers, and renewable plasticizers, Journal of Applied Polymer Science. 131 (2014) 98–105.

[83] X. Xin, C.G. Xu, L.F. Qing, Friction properties of sisal fibre reinforced resin brake composites, 262 (2007) 736–741. https://doi.org/10.1016/j.wear.2006.08.010.

[84] A.K. M, R. Reddy, S. Bharathi, V. Naidu, N.P. Naidu, Friction coefficient, hardness, impact strength and chemical resistance of reinforced sisal-glass fiber epoxy hybrid composites, J Compos Mater 2010; 44 (26):3195-202 (Dec), (n.d.).

[85] N. Chand, U.K. Dwivedi, Sliding wear and friction characteristics of sisal fibre reinforced polyester composites: Effect of silane coupling agent and applied load, Polymer Composites. 29 (2008) 280–284. https://doi.org/10.1002/pc.20368.

[86]   S. Dixit, G. Dixit, V. Verma, Thermal degradation of polyethylene waste and jute fiber in oxidative environment and recovery of oil containing phytol and free fatty acids, Fuel. 179 (2016) 368–375.

[87]   T. Berhanu, P. Kumar, I. Singh, Sliding Wear Properties of Jute Fabric Reinforced Polypropylene Composites, Procedia Engineering. 97 (2014) 402–411. https://doi.org/10.1016/j.proeng.2014.12.264.

[88]   K. Sabeel Ahmed, S.S. Khalid, V. Mallinatha, S.J. Amith Kumar, Dry sliding wear behavior of SiC/Al2O3 filled jute/epoxy composites, Materials & Design. 36 (2012) 306–315. https://doi.org/10.1016/j.matdes.2011.11.010.

[89]   C.P. V.Tserki, P. Matzinos, Effect of compatibilization on the performance of biodegradable composites using cotton fibre waste as filler, J. Appl. Polym. Sci. 88 (2003) 1825., (n.d.).

[90]   S.A.R. Hashmi, U.K. Dwivedi, N. Chand, Graphite modified cotton fibre reinforced polyester composites under sliding wear conditions, Wear. 262 (2007) 1426–1432. https://doi.org/HTTPS://DOI.ORG/ 10.1016/j.wear.2007.01.014.

[91]   H. Zhang, Z. Zhang, F. Guo, K. Wang, W. Jiang, Enhanced wear properties of hybrid PTFE/cotton fabric composites filled with functionalized multi-walled carbon nanotubes, Materials Chemistry and Physics. 116 (2009) 183–190. https://doi.org/10.1016/j.matchemphys.2009.03.008.

[92]   D. Verma, P. Gope, The Use of Coir/Coconut Fibres as Reinforcements in Composites, in: Biofibre Reinforcements In Composite Materials, Woodhead Publishing Series, 2015: pp. 285–319.

[93]   R.A. Ibrahem, Friction and wear behaviour of fibreparticles reinforced polyester composites, International Journal of Advanced Materials Research. 2 (2016) 22–26.

[94]   R. Prabhu, A.K. Amin, Dhyanchandra, Development and charaterization of low cost polymer composites from coconut coir, American Journal of Material Science. 5 (2015) 62–68.

[95]   K. Bilba, M.-A. Arsene, A. Quensanga, Study of banana and coconut fibres botanical composition, thermal degradation and textural observations, Bioresource Technology. 98 (2007) 58–68.

[96]   E.C. Carlos, B. Santiago, V. Analia, G. Piedad, Wear resistance and friction behaviour of thermoset matrix reinforced with musaceae fiber bundles, Tribology International. 87 (2015) 57–64.

[97]   Food and Agriculture Organization of the United Nations, FAO Statistical
       Yearbook 2006, FAO, 2006.

[98]   S. Tamba, I. Cisse, Senegal, R. F., R. Jauberthie, France, Rice husk in lightweight
       mortars, in: S. Helland, I. Holand, S. Smeplass (Eds.), Second International
       Symposium on Structural Lightweight Aggregate Concrete, 2000: pp. 117–124.

[99]   P. Mishra, A. Chakraverty, H.D. Banerjee, Studies on physical and thermal
       properties of rice husk related to its industrial application, Journal of Material
       Science. 21 (1986) 2129–2132.

[100] T. Kapur, T.C. Kandpal, H.P. Garg, Electricity generation from rice husk in Indian
       rice mills: Pottential and Financial Viability, Biomass and Bioenergy. 10 (1996)
       393–403.

[101] N. Chand, M. Fahim, P. Sharma, M.N. Bapat, Influence of foaming agent on wear
       and mechanical properties of surface modified rice husk filled polyvinylchloride,
       Wear. 278–279 (2012) 83–86. https://doi.org/10.1016/j.wear.2012.01.002.

[102] T. Dugarjav, T. Yamaguchi, K. Shibata, K. Hokkirigawa, Friction and wear
       properties of rice husk ceramics under dry and water lubricated conditions,
       Tribology Online. 4 (2009) 78–81.

[103] K. Shibata, T. Yamaguchi, K. Hokkirigawa, Tribological behavior of polyamide
       66 / rice bran ceramics and polyamide 66 / glass bead composites, Wear. 317
       (2014) 1–7.

[104] K.D.H. Nguyen, P.H. Dang, H.X. Nguyen, M.T.T. Nguyen, S. Awale, N.T.
       Nguyen, Phytochemical and cytotoxic studies on the leaves of calotropis gigantea,
       Bioorganic and Medicinal Chemistry Letters. 27 (2017) 2902–2906.

[105] G. Dilli Babu, K. Sivaji Babu, P. Nanda Kishore, Tensile and wear behavior of
       calotropis gigentea fruit fiber reinforced polyester composites, Procedia
       Engineering. 97 (2014) 531–535.

[106] B.F. Yousif, N.S.M. El-Tayeb, The Effect of oil palm fibres as reinforcement on
       tribological performance of polyester composite, Surface Review and Letters. 14
       (2007) 1095–1102.

[107] S. Narish, B.F. Yousif, D. Rilling, Adhesive wear of thermoplastic composite
       based on kenaf fibres, Proceedings of the Institution of Mechanical Engineers Part
       J-Journal of Engineering Tribology. 225 (2011) 101–109.

[108] N.S.M. El-Tayeb, A study on the potential of sugarcane fibers/polyester composite
       for tribological applications, Wear. 265 (2008) 223–235.

[109] N.S.M. El-Tayeb, Development and characterisation of low-cost polymeric composite materials, Materials & Design. 30 (2009) 1151–1160. https://doi.org/10.1016/j.matdes.2008.06.024.

[110] B.F. Yousif, S.T.W. Lau, S. McWilliam, Polyester composite based on betelnut fibre for tribological applications, Tribology International. 43 (2010) 503–511. https://doi.org/HTTPS://DOI.ORG/ 10.1016/j.triboint.2009.08.006.

[111] S.. Narish, B.F.. Yousif, D.. Rilling, Investigations on wear and frictional properties of kenaf fibre polyurethane composites under dry and wet contact conditions, International Journal of Precision Technology. 2 (2011) 375–387.

[112] M. Ashok Kumar, G. Ramachandra Reddy, Y. Siva Bharathi, S. Venkata Naidu, V. Naga Prasad Naidu, Frictional Coefficient, Hardness, Impact Strength, and Chemical Resistance of Reinforced Sisal-Glass Fiber Epoxy Hybrid Composites, Journal of Composite Materials . 44 (2010) 3195–3202. https://doi.org/10.1177/0021998310371551.

[113] R.B. Yusoff, H. Takagi, A.N. Nakagaito, Tensile and flexural properties of polylactic acid-based hybrid green composites reinforced by kenaf, bamboo and coir fibers, Industrial Crops and Products. 94 (2016) 562–573. https://doi.org/10.1016/j.indcrop.2016.09.017.

[114] H. Fazilat, M. Ghatarband, S. Mazinani, Z.A. Asadi, M.E. Shiri, M.R. Kalaee, Predicting the mechanical properties of glass fibre reinforced polymers via artifical neural network and adaptive neuro-fuzzy inference system, Computational Material Science. 58 (2012) 31–37.

[115] R.S. Rana, A. Kumre, S. Rana, R. Purohit, ScienceDirect Characterization of Properties of epoxy sisal / Glass Fiber Reinforced hybrid composite, in: 6th International Conference of Materials Processing and Characterization (ICMPC 2016), 2017: pp. 5445–5451.

[116] M.R. Sanjay, G.R. Arpitha, B. Yogesha, Study on Mechanical Properties of Natural - Glass Fibre Reinforced Polymer Hybrid Composites: A Review, in: 4th International Conference on Materials Processing and Characterization, 2015: pp. 2959–2967.

Keyword Index

About the Editors

Dr. Anish Khan is currently working as Assistant Professor, at Chemistry Department, Centre of Excellence for Advanced Materials Research (CEAMR), Faculty of Science, King Abdulaziz University, Jeddah, Saudi Arabia. Ph.D. Completed from Aligarh Muslim University, India in 2010. He has 13 years of research experience in the field of organic-inorganic electrically conducting nano-composites and its applications in making chemical sonsor. He completed his Postdoctoral from the School of Chemical Sciences, University Sains Malaysia (USM) on electroanalytical chemistry. He has published more than 100 research articles in refereed international journals. He attended more than 10 international conferences/workshops, has published 2 books and 7 book chapters. He completed around 20 research projects. He is managerial editor of Chemical and Environmental Research (CER) Journal, a Member of the American Nano Society, his field of specialization is polymer nanocomposite/cation-exchanger/chemical sensor/micro biosensor/nanotechnology, application of nanomaterials in electroanalytical chemistry, material chemistry, ion-exchange chromatography and electro-analytical chemistry, dealing with the synthesis, characterization (using different analytical techniques) and derivatization of inorganic ion-exchanger by the incorporation of electrically conducting polymers.

Dr. Showkat Ahmed Bhawani is presently working as a Senior Lecturer at the Department of Chemistry, Faculty of Resource Science and Technology, UNIMAS Malaysia. In addition to this, he has teaching experience of two years from King Abdul Aziz University- North Jeddah and a post-doctoral experience of three years from the Universiti Sains Malaysia, Malaysia. He received his M Sc. in Analytical Chemistry and Ph D. in Applied Analytical Chemistry from Aligarh Muslim University, Aligarh, India. He is working on the synthesis of molecular Imprinting polymers for the removal/extraction of dyes, fungicides and various natural products from environmental and biological samples. In addition to this, he is also working on the development of new test methods and determining standard conditions for analysis (Separation, Isolation, and Determination) of various analytes from environmental and biological samples. He was involved in the analysis of samples like Surfactants, Amino acids, Drugs, Vitamins, Sugars and Metal ions.

Prof. Abdullah Mohammed Ahmed Asiri is Professor in Chemistry Department – Faculty of Science -King Abdulaziz University. Ph.D. (1995) From University of Walls College of Cardiff, U.K. on Tribochromic compounds and their applications. He has published more than 1000 Research articles and 20 books. He is the chairman of the

Chemistry Department, King Abdulaziz University currently and also the director of the center of Excellence for Advanced Materials Research. Director of Education Affair Unit–Deanship of Community services. He is a member of the advisory committee for advancing materials, (National Technology Plan, (King Abdul Aziz City of Science and Technology, Riyadh, Saudi Arabia). He is a member of the editorial board of the Journal of Saudi Chemical Society, Journal of King Abdul Aziz University, Pigment and Resin Technology Journal, Organic Chemistry Insights, Libertas Academica, Recent Patents on Materials Science, Bentham Science Publishers Ltd. Beside that he has professional membership of International and National Society and Professional bodies.

Dr. Imran Khan is presently working as an Assistant Professor in the Applied Science and Humanities Section, Faculty of Engineering and Technology, University Polytechnic, Aligarh Muslim University, Aligarh-202002, India. He obtained his Ph.D. in Applied Physics from Aligarh Muslim University, India in 2015. He was selected for the prestigious National Postdoctoral Fellowship (NPDF) from DST-SERB, India in 2016 and completed one year postdoctoral at the Department of Applied Sciences and Humanities, Jamia Milla Islamia, New Delhi-India. He has published more than 24 research papers, 4 book chapters, 1 books published and 1 book in progress in referred international Publishers and more than 12 international conferences/ workshop. The current research interests include magnetic nanoparticles, thin films, polymer composites, graphene composites, nanocomposites, optoelectronic materials, hybrid organic-inorganic composites materials, nanoporous nanostructured materials, etc., sol–gel, hydro–/solvo–thermal, microwave methods etc., thin film preparation through PLD,CVD, sputtering, and spin-dip coatings techniques etc.

www.ingramcontent.com/pod-product-compliance
Lightning Source LLC
Chambersburg PA
CBHW061203220326
41597CB00015BA/1305